Lecture Notes in Energy

Volume 64

Lecture Notes in Energy (LNE) is a series that reports on new developments in the study of energy: from science and engineering to the analysis of energy policy. The series' scope includes but is not limited to, renewable and green energy, nuclear, fossil fuels and carbon capture, energy systems, energy storage and harvesting, batteries and fuel cells, power systems, energy efficiency, energy in buildings, energy policy, as well as energy-related topics in economics, management and transportation. Books published in LNE are original and timely and bridge between advanced textbooks and the forefront of research. Readers of LNE include postgraduate students and non-specialist researchers wishing to gain an accessible introduction to a field of research as well as professionals and researchers with a need for an up-to-date reference book on a well-defined topic. The series publishes single and multi-authored volumes as well as advanced textbooks.

More information about this series at http://www.springer.com/series/8874

George Giannakidis · Kenneth Karlsson
Maryse Labriet · Brian Ó Gallachóir
Editors

Limiting Global Warming to Well Below 2 °C: Energy System Modelling and Policy Development

 Springer

Editors
George Giannakidis
Laboratory for Energy Systems Analysis
Centre for Renewable Energy Sources
  and Saving (CRES)
Athens
Greece

Kenneth Karlsson
DTU Management Engineering
Technical University of Denmark
Kgs. Lyngby
Denmark

Maryse Labriet
Eneris Environment Energy Consultants
Madrid
Spain

Brian Ó Gallachóir
MaREI Centre, Environmental
  Research Institute
University College Cork
Cork
Ireland

ISSN 2195-1284   ISSN 2195-1292 (electronic)
Lecture Notes in Energy
ISBN 978-3-030-08989-4   ISBN 978-3-319-74424-7 (eBook)
https://doi.org/10.1007/978-3-319-74424-7

Printed on acid-free paper

This Springer imprint is published by the registered company Springer International Publishing AG part of Springer Nature
The registered company address is: Gewerbestrasse 11, 6330 Cham, Switzerland

Foreword

Achieving an affordable, secure and sustainable energy future is one of the central challenges the world faces today. Developing robust strategies to reach these goals requires a comprehensive understanding of the energy system, taking into account the interlinkages between fuels, technologies and policy measures. To inform policymakers on these complex questions, energy modelling and scenario analysis is an invaluable means to explore possible future energy and technology pathways and to evaluate the implications of the various options that are available.

The Energy Technology Systems Analysis Programme (ETSAP) represents a unique network in this field of energy modelling and analysis. Established in 1976 as a Technology Collaboration Programme (TCP) under the auspices of the International Energy Agency (IEA), its two energy system model generators, MARKAL (MARKet ALlocation) and TIMES (The Integrated MARAL-EFOM System), are now used by modelling teams in approximately 200 institutions across 70 countries. While concerns about energy security were the main focus of many model analyses in the 1970s (and remain just as relevant today), their focus has been broadened to also include questions about social and environmental objectives. The model applications today range from the global, regional and national level, down to the local and city level. At the IEA, ETSAP's modelling tools have been used for more than a decade to help develop scenarios and assess clean energy technologies in the global context. For example, the IEA's *Tracking Clean Energy Progress* report consists of a scenario-based assessment of current technology trends against what is needed under a 2-degree pathway.

Following the first book *Informing Energy and Climate Policies Using Energy Systems Models*, this new ETSAP book provides comprehensive model-based analyses of the overarching goal of the Paris Agreement. It shows that a fundamental transformation of the global energy system is required over the coming decades, driven by rapid and clear policy action to accelerate and scale-up the deployment of clean energy technologies. The analyses in this book highlight how energy models and scenarios can provide invaluable insights into decision-makers on the interlinked policy dimensions of climate change, energy security and economic development. It also illustrates that technology and system transformations are needed not only in different parts of the energy sector but also at all levels—multilateral, national and local.

I trust readers will find within this book insights that demonstrate both the complexity and usefulness of energy systems modelling in informing technology collaboration and energy policy decision-making.

Paris, France   Dr. Fatih Birol
January 2018   Executive Director
                                              International Energy Agency

Preface

This book addresses a number of key questions arising from the transformational global political agreement reached in Paris in December 2015 to ensure human-induced global temperature increases remain well below 2 °C. How far can countries ratchet up the mitigation ambition presented in the Nationally Determined Contributions? What is the magnitude of potential and necessary carbon dioxide removal, more particularly, biomass energy with carbon capture and storage? How can stranded assets be avoided? Is carbon neutrality achievable with technology innovation alone?

These questions are at the core of the global, national and local energy systems modelling analyses in this book that explore the feasibility of roadmaps for a *well below 2 °C* future. The book is written by more than 20 teams of the IEA Energy Technology Systems Analysis Programme (IEA-ETSAP), a Technology Collaboration Programme supporting about 200 energy systems modelling teams from around 70 countries, which has operated for over 40 years. A key objective of IEA-ETSAP is to assist decision-makers in robustly developing, implementing and assessing the impact of energy and climate mitigation policies with the bottom-up techno-economic models of the MARKAL/TIMES family.

This book constitutes a natural follow-up of the first book *Informing Energy and Climate Policies Using Energy Systems Models*, edited by IEA-ETSAP in 2015, prior to the Paris Agreement on climate change. The methodologies and case studies presented here illustrate how energy systems models have been and are being used to address complex energy and climate policy questions and provide critical insights into the feasibility of enhanced ambition. This responds directly to a requirement of the Paris Agreement and its Talanoa Dialogue to take stock of the collective efforts and determine how we can, all together, move the climate policy agenda forward and turn words into action.

The editors are very grateful to the chapter authors and peer reviewers who willingly shared their expertise and contributed their valuable time, without which this book would not have been possible. In addition, we acknowledge and appreciate the English language revision support provided by Evan Boyle, Seán Collins, Paul Deane, Fiac Gaffney, James Glynn, Conor Hickey, Sean McAuliffe, Connor McGookin, Tomas Mac Uidhir, Laura Mehigan, Eamonn Mulholland and Fionn Rogan.

Madrid, Spain   Dr. Maryse Labriet
Athens, Greece   Dr. George Giannakidis
Kgs. Lyngby, Denmark   Dr. Kenneth Karlsson
Cork, Ireland   Prof. Brian Ó Gallachóir
January 2018

Contents

Introduction: Energy Systems Modelling for a Sustainable World

Maryse Labriet, George Giannakidis, Kenneth Karlsson
and Brian Ó Gallachóir

Key messages

- The energy system models are particularly well suited to provide robust information on energy pathways towards a well below 2 °C goal.
- The MARKAL/TIMES models have a long history of research and development, international collaboration and application to complex problems.
- 23 case studies assess the feasibility and characteristics of global, national and local energy transition pathways ahead for a decarbonized world.
- A well below 2 °C world is feasible but extremely challenging, and early action is essential.

1 The IEA Energy Technology Systems Analysis Program

The IEA Energy Technology Systems Analysis Program (IEA-ETSAP) is a collaboration programme of energy modelling teams in twenty one contracting party countries and an extended network of collaborating institutions from approximately

M. Labriet (✉)
Eneris Environment Energy Consultants, Madrid, Spain
e-mail: maryse.labriet@enerisconsultants.com

G. Giannakidis
Centre for Renewable Energy Sources and Saving (CRES), Athens, Greece
e-mail: ggian@cres.gr

K. Karlsson
Technical University of Denmark, Kgs. Lyngby, Denmark
e-mail: keka@dtu.dk

B. Ó Gallachóir
MaREI Centre, Environmental Research Institute, University College Cork, Cork, Ireland
e-mail: b.ogallachoir@ucc.ie

© Springer International Publishing AG, part of Springer Nature 2018
G. Giannakidis et al. (eds.), *Limiting Global Warming to Well Below 2 °C: Energy System Modelling and Policy Development*, Lecture Notes in Energy 64,
https://doi.org/10.1007/978-3-319-74424-7_1

seventy countries over the world. It has been building a bridge among the specialists of energy sectors, energy technologies, environment and economy in order to carry out joint system modelling analyses.

1.1 Build Modelling Capability for Decision-Making

IEA-ETSAP is a Technology Collaboration Programme that was initiated in 1976 under the aegis of the International Energy Agency (IEA). The focus of IEA-ETSAP was (and still is) to cooperate to establish, maintain and expand a consistent multi-country energy/economy/environment/engineering analytical capability mainly based on the MARKAL/TIMES family of models.

The objective is to build modelling capability in order to assist and support government officials and decision-makers in increasing the robustness of the evidence base underpinning energy and environmental policy issues by applying these tools for energy technology assessment and analysis. IEA-ETSAP developed, through co-operation, the MARKAL (MARKet ALlocation) and—subsequently—the TIMES (The Integrated MARKAL-EFOM System) energy systems model generators, both based on a multi-regional, multi-period, bottom-up, linear programming, optimization paradigm. These bottom-up techno-economic models have been used to build long term energy scenarios and to provide in-depth local, national, multi-country, and global energy and environmental analyses.

1.2 From Information Exchanging to Understanding the Energy Transition to Achieve the Well Below 2 °C Goal

IEA-ETSAP has promoted since its very beginning a continuous exchange of information related to energy modelling, energy systems analysis and energy technology assessment in order to share experiences, stimulate co-operative studies and benefit from common analysis. Those principles have remained essential to IEA-ETSAP.

Rapidly, a broad consensus has developed that environmental aspects of energy systems were susceptible to analytical study. During the 1980s, pollutants contributing to acid rain were the main concern. Greenhouse gases and climate change turned to be a priority from about 1990, when the International Energy Agency itself called sharp attention to the linkages between energy supply and consumption and climate change. The focus on climate change has been reinforced to analyze the conditions of implementation of the Kyoto Protocol (UN 1998), more particularly the possible common actions and emission trading, thanks to the global TIMES Integrated Assessment Model (Loulou 2008; Loulou and Labriet 2008). The overall goal was clearly to serve national decision-makers and support the work of international organisations on the topics. In that sense, IEA-ETSAP members have developed tight collaborations with other major energy modelling groups, such as the Stanford Energy Modelling Forum. In addition, the global TIMES Integrated Assessment Model was used, jointly with other integrated assessment models, to the emission scenario assessment presented in the Fifth Assessment Report of the Intergovernmental Panel on Climate Change to understand how actions and policies affect emissions and energy pathways (IPCC 2014).

More recently, always driven by the global environmental and energy challenges, the tool development moved towards the combination of the energy flow optimisation approach with macroeconomic modelling, technology learning, uncertainty modelling as well as the better representation of behaviour, variable renewable sources, and linkages with water, land and ultimately food.

The current work programme of ETSAP focuses on the energy transition to achieve the well below 2 °C target, more particularly by exploring the interplay between long term goals, such as the well below 2 °C target, and short term policies and dynamics, such as the Nationally Determined Contributions (NDCs) or the international fuel prices. Special attention is given to the collaboration with non-OECD countries for the assessment of their development of NDCs.

2 MARKAL/TIMES Models

2.1 Exploring Technology Pathways

Two model generators have been developed and made available via the IEA-ETSAP collaboration, namely MARKAL and subsequently TIMES.

MARKAL and TIMES models compute an inter-temporal dynamic partial equilibrium on energy and emission markets based on the maximization of total surplus, defined as the sum of suppliers' and consumers' surpluses while respecting environmental and many technical constraints. They underpin scenario analyses used to explore future possible evolutions of the energy system to meet energy service demands at least cost. This least-cost and perfect foresight perspective is that of a perfect rationality and full cooperation amongst all agents. Additional constraints can be imposed to the models to help represent the non-rational behaviours and ensure the microeconomic robustness of the analyses.

A key and distinct feature of MARKAL and TIMES models is the extent of technology detail. This allows precise tracking of optimal capital turnover and provides a precise description of technology and fuel competition for estimating energy dynamics over a long-term, multi-period time horizon. The technological explicitness is also particularly relevant to assess the role of technology focused instruments (technology standards, subsidies) in energy pathways.

The model's variables include the investments, capacities, and activity levels of all technologies at each period of time, as well as the amounts of energy, material, and emission flows in and out of each technology, and the quantities of traded energy.

Linking MARKAL and TIMES models with complementary models can provide useful additional insights into the results from standalone models. Examples of coupling with macroeconomic, climate, power system and land-use models were provided in Giannakidis et al. (2015). Some further coupling applications are presented in this book.

Both MARKAL and TIMES are written in GAMS (General Algebraic Modelling System) code and CPLEX and XPRESS are typically the solvers used. The code is transparent and well documented (Loulou et al. 2016).

2.2 From Global to Local Decision-Making Support

The Global Models built using MARKAL or TIMES are particularly useful for exploring global technology roadmaps, energy trades and prices, carbon markets and feedbacks between the energy and climate systems. The IEA Energy Technology Perspectives (ETP) analyses, published each year, use the global ETP-TIMES Supply model, which covers 28 regions (IEA 2017). The original TIMES Integrated Assessment Model (TIAM) and several different derived TIAM models (developed by various modelling teams) are being frequently used in European research projects, international institutions and modelling groups; several applications are presented in this book.

Regional Models have also been developed. Amongst them, the JRC-EU-TIMES model (EC 2018) represents the energy system of the EU28 member States, Iceland, Norway and Switzerland. It is also set up to include Balkan countries. The model is used to assess the role of technologies in decarbonizing Europe's energy system, and to identify key parameters that can influence the early competitiveness and the large scale deployment of key low-carbon technologies.

Most of the applications of IEA-ETSAP modelling tools are based on National Models, which cover many individual countries in all continents, and on Sub-National and Local Models. Several applications at national and local levels are presented in this book.

3 This Book: Towards More Ambitious Climate Mitigation Targets

Through the Agenda for Sustainable Development (United Nations 2015) and the Paris Agreement on climate change (UNFCCC 2015), almost all countries now acknowledge the need for implementing ambitious and integrated sustainable development pathways in industrialized and developing countries. These pathways are multifaceted, seeking to deliver poverty eradication, food security, health, water and energy access, employment and the other dimensions covered by the sustainable development goals. Climate change is considered to have an influence on all of them, and the connections, be these synergies or trade-offs, between these dimensions are complex.

The integrated national energy and climate plans to be developed by the European Union's Member States for the period 2021–2030 also reflect the multidimensional nature of energy pathways by covering the five dimensions of the EU's Energy Union (EC 2016): security, solidarity and trust; a fully-integrated internal energy market (resulting to lower prices and therefore reduced energy poverty, amongst other factors); energy efficiency; climate action—decarbonising the economy; and finally research, innovation and competitiveness.

In addition, the current climate plans, as reflected in the Nationally Determined Contributions (NDCs) of the Paris Agreement, are not sufficient to achieve greenhouse gas emissions reductions by 2030 that are compatible with a long term well below 2 °C ambition. Emissions reductions of the NDCs may represent only one third of the emissions reductions needed to be on a least cost pathway for the goal of staying well below 2 °C with 50–66% chance (UNEP 2017). The Facilitative Dialogue of the Paris Agreement, now called the Talanoa Dialogue, is the first step of the ratchet mechanism to increase mitigation ambition.

How much faster does the economy need to be decarbonized, compared to less stringent goals? What is the required or potential contribution of negative emissions, more particularly, biomass energy with carbon capture and storage? How can stranded assets be avoided? Is carbon neutrality achievable with technology innovation alone? What are the consequences of a temporary overshoot over the longer term mitigation actions? How to define a national emission goal which is fair and compatible with a well below 2 °C future? How can a *just transition* be achieved? Decision-makers need evidence-based and robust information to elaborate more ambitious climate and energy plans.

This book collates insights and possible roadmaps from different perspectives and scales. These case studies complement the case study applications described in Giannakidis et al. (2015). The results here reinforce the insights from energy systems modelling for energy and climate policies in the context of the Paris Agreement. The book is structured in four parts, which focus on the magnitude of the transformation of the global energy system required to maintain the global temperature increase well below 2 °C (Part I), the diversity of the national energy transitions and, in some cases, the difficulties in reaching a deep decarbonisation of the energy systems in Europe (Part II) and outside Europe (Part III), and finally, how cities and local communities can also contribute to the global decarbonisation (Part IV).

3.1 Part I—The Radical Transformation of the Global Energy System

Building on the framework of the World Energy Scenarios 2016, Chapter by **Kober et al.** concludes that a 1.5 °C compatible pathway is feasible, assuming a carbon budget of 480 $GtCO_2$ for the energy sector between 2020 and 2060. The global combustion-related CO_2 emissions are negative by 2060. Net CO_2 removal through bioenergy conversion with CO_2 capture and storage (CCS) relieves the pressure to reduce emissions in sectors with high mitigation costs, such as aviation and industry. But, the decarbonisation of energy demand requires immediate action with strong enabling measures for energy conservation and promotion of electricity to substitute oil and natural gas, both at the core of the additional efforts to achieve a stringent climate target. More than half of the power generation investments need to be dedicated to wind and solar technologies in the period 2011–2060. By contrast, coal, with or without CCS, is not an option, it must be phased-out from the power sector.

Chapter by **Lehtilä and Koljonen** raises the question whether the objective of limiting global warming to well below 2 °C (represented by an appropriate radiative forcing trajectory) is achievable without considerably overshooting the target of the estimated carbon budget within the current century. Immediate mitigation action is absolutely required for avoiding excessive overshooting. Negative emissions have a crucial role in making the well below 2 °C target achievable, especially in the case of an initial overshooting of the carbon budget. Capturing (including direct air capture) and utilizing carbon dioxide as a feedstock also appears to offer a key element for a viable technology pathway to the post-fossil economy during the current century.

Chapter by **Winning et al.** assesses the impacts of delaying ratcheting until 2030 on global emissions trajectories compatible with the 2 and 1.5 °C targets, represented by carbon budgets of 915 and 400 $GtCO_2$ between 2015 and 2100. The analysis suggests that delaying action makes pursuing the 1.5 °C (and the 2 °C) goals unachievable without extremely high levels of negative emissions technologies (NETs), such as carbon capture and storage combined with bioenergy. Depending on the availability of biomass, other NETs beyond bioenergy will be even required. Of course, the risks of lock-into a high fossil future is high if NETs cannot be scaled. Finally, a higher focus on emission reductions in the demand sectors is indispensable.

Chapter by **Mousavi and Blesl** explores the relative roles of supply-side and demand-side measures in 1.5 °C consistent scenario, assuming a carbon budget of 600 $GtCO_2$ for the energy sector between 2015 and 2100. Carbon neutrality is reached in 2040. While technological measures are essential to meet the decarbonisation target, reducing energy-service demands, including modifying consumer behaviour, is also essential to facilitate a cost-effective transition and offset the macroeconomic impacts of the climate policy. The overshoot of the energy sector carbon budget as soon as in 2040 must be compensated by a significant deployment of negative emissions technologies.

Chapter by **Karlsson et al**. shows that focusing only on technological development is likely not to be sufficient to limit the increase in global mean temperature to well below 2 °C. Increasing educational attainment, family planning and less meat intensive diets will reduce environmental impacts and increase the probability of stabilising temperature increase well below 2 °C, while attaining an equitable distribution of economic growth between the regions of the world can have positive and negative impacts on how challenging it will be to reach a well below 2 °C world.

The diversity of carbon budgets used by the global analyses shows the importance of timeframe, assumptions on non-CO_2 emissions and probability of keeping the long term temperature increase well below 2 °C.

3.2 Part II—The Diversity of the National Energy Transitions in Europe

Chapter by **Seljom and Rosenberg** demonstrates the feasibility of a cost-optimal transition towards a carbon neutral energy system in Scandinavia (Denmark, Norway and Sweden) in 2050 without imported biofuels and no use of CO_2 storage. Since the Scandinavian electricity sector is already highly renewable, carbon neutrality requires extensive changes in the building, transport and industry sectors. Hydrogen is the dominant fuel used in the transport sector, and hydrogen production without access to CO_2 storage requires extensive investments in wind power and photovoltaics. The results emphasise the importance of considering the entire energy system when designing policy to reach carbon neutrality: the required investments in electricity capacity depend on the degree of electrification, and the future electricity consumption depends on the availability and competiveness of biofuels.

Biomass is today an important part of the Swedish energy supply and has the potential to increase even further. Sweden's biomass feedstock is mainly linked to the harvest of logs used for non-energy purposes and, thus, almost exclusively consists of forestry residues. Chapter by **Krook-Riekkola and Sandberg** explores different net zero emission pathways with an emphasis on where the extensive (but still finite) domestic biomass resources and restricted carbon capture could be used most cost-efficiently. Electricity, district heating and space heating can be close-to-zero emissions by as early as 2025; achieving carbon neutrality in transport and industry is more challenging however. Using existing biomass more efficiently, integrating biofuel production with the pulp and paper industry to use residues, and using waste heat for district heating are promising measures. Competitiveness of biofuels in transport would require a carbon price that is flexible with respect to fluctuations in oil price.

Based on a model coupling approach to incorporate a detailed representation of the transportation sector, Chapter by **Tattini et al**. proposes pathways to decarbonise the Danish inland passenger transport sector. The results indicate that a ban on the sale of the internal combustion engines enforced in 2025 would enable the largest cut in cumulative greenhouse gas emissions of all the policies considered. However, none of the policies analysed is compliant with an equitable Danish carbon budget capable of limiting the increase of global temperature to well below 2 °C.

Chapter by **Panos and Kannan** compares the current commitments of the Swiss National Determined Contribution to a low carbon scenario consistent with a below 2 °C goal, under the framework of the Swiss energy strategy objectives: gradual phase-out of nuclear power, energy efficiency gains and deployment of renewables. Electrification and efficiency appear to be the key pillars in achieving decarbonisation, and new business models involving microgrids, smart grids, virtual power plants, storage and power-to-gas technologies need to emerge that enable active participation of consumers in the energy supply. Early action and continuous policy response are necessary to avoid lock-in of emission-intensive infrastructure and stranded assets.

Chapter by **Millot et al**. explores the impact of two contrasting lifestyles for France: the digital lifestyle represents an individualistic and technological society, and the collective one depicts a society with strong social ties and cooperation between citizens. The digital society involves significant growth of both GDP and the unemployment rate, and does not result in carbon neutrality, whereas the collective society leads to smaller growth of GDP and a decrease in the unemployment rate, but makes it possible to reach a nil carbon target. In other words, policymakers must ensure consistency between the evolving lifestyles and the need to decarbonize the economy.

Chapter by **Yue et al**. examines the feasible ratcheting of decarbonisation ambition of Ireland, currently based on a 2 °C goal. The results indicate that a national carbon budget compatible with a 1.5 °C target would need to be almost three times smaller than the carbon budget resulting from the current national climate policy. This budget is technically feasible, but extremely challenging with the current technology assumption. A carbon budget midway between 1.5 and 2 °C appears much more realistic. Finally, ambitious carbon budget targets can only be achieved through much stronger near-term mitigation efforts than suggested by the current Nationally Determined Contribution.

Chapter by **Seixas et al**. assesses the impacts on energy system of deep decarbonisation in Portugal up to 2050. The electrification of final energy consumption of the Portuguese economy contributes significantly towards decarbonisation, especially through electric vehicles, heat pumps (both in residential and commercial buildings) and dryers and kilns in some industrial sectors. These technologies are not addressed in the current energy and climate national policies, nor the major investments that will be required. Such aspects need to be specifically addressed in the Portugal Carbon Neutrality Roadmap studies that have recently started, and the focus of the Portuguese energy and climate decision-makers should not be limited anymore to broad national emission targets and incentives to the decarbonisation of the power sector.

3.3 Part III—The Decarbonisation Pathways Outside Europe

Chapter by **Vaillancourt et al**. identifies different decarbonisation pathways for Canada corresponding to increasing levels of mitigation efforts for 2050. Beyond the significant energy conservation and efficiency improvements, greater penetration of electricity and bioenergy, and the rapid decarbonisation of electricity production, the chapter shows how Canada would benefit from greater cooperation between Canadian jurisdictions because of the large diversity in the composition of regional energy systems. The difficulty to reach carbon neutrality by 2050 without additional mitigation options in industry and negative emissions is finally discussed.

Chapter by **Zakerinia et al**. describes the pathways towards decarbonisation in California and other Western States of United States to 2030 and 2050, as well as the impact of California's policies on the Western Electricity Coordinating Council grid. A climate target on California only and not on the other states could contribute to the greening of power plants in the Western States, driven by the possibility to export electricity to California. When a carbon target is extended to all regions, the grid of all Western States, as well as the entire energy system of California, zero emissions are not feasible without adopting carbon capture and storage. In order to take advantage of the spatial and temporal differences of the Western US, it is therefore very important to completely deregulate the Western Electricity Coordinating Council grid.

Australia's high greenhouse gas (GHG) emissions per capita reflects the relatively high proportion of fossil fuels in energy consumed, high usage of less efficient private transport and high production of non-ferrous metals per capita. However, the dominance of coal-fired electricity generation masks Australia's rich diversity of renewable energy resources. Chapter by **Reedman et al**. shows that the electricity and transport sectors of Australia can achieve the greatest emissions reductions of 70–80% by 2050. High carbon prices and the successful management of high shares of variable renewable electricity generation appear important in ensuring Australia's electricity sector achieves deep emission reduction. The direct combustion sector has a harder abatement task owing to fewer directly substitutable low emission energy sources. Hydrogen and solar thermal heat warrant further research to ensure emissions can be completely eliminated without the need to purchase potentially higher cost emission credits from other domestic sectors or the international market.

Chapter by **Chepeliev et al**. provides an assessment of low-emission development scenarios for the Ukrainian economy, including the Ukrainian low-carbon development strategy initiative (consistent with 2 °C target), and a more ambitious transition towards 92% share of renewables in gross final energy consumption by 2050 (consistent with 1.5 °C). Results show that further maintenance of the existing highly inefficient energy system is more expensive than transition towards a high renewables share. Moreover, key differences between 2 and 1.5 °C scenarios arise after 2035–2040, which enables the possibility of smooth transitions from less to more stringent pathways during this period. Measures towards efficient pricing of fossil fuels, in particular price signals for industrial users, more transparent and market-oriented approach to residential consumers, elimination of cross subsidization in the electricity sector, move to competitive energy markets must be considered by policymakers.

Kazakhstan's energy system is highly carbon intensive and insufficient progress is being made towards achieving the Nationally Determined Contribution of the country. Chapter by **Kerimray et al**. assesses how to achieve the NDC goals, and how a ban on coal across all sectors could help. It shows that NDC goals requires an almost full phase-out of coal consumption in power generation by 2050, and that the overall target as set by Kazakhstan's Strategy 2050 and Green Economy Concept to reach 50% of renewable and alternative energy sources by 2050 is very close to the least-cost 25% emissions reduction pathway, which corresponds to the higher mitigation target of the NDC. However, a coal ban alone is not sufficient to reduce GHGs, additional actions are needed to promote renewables.

Mexico has positioned itself as a leader among emerging countries for its efforts to mitigate climate change. However, the Energy Reform bill approved in 2014 promotes the production of hydrocarbons and the use of natural gas for electricity generation in order to reduce electricity prices in the short term. In this context, Chapter by **Solano-Rodríguez et al**. shows that a deep decarbonisation of the power system appears techno-economically feasible and cost-optimal through renewables (mainly solar PV and wind), and that decarbonisation paths post-2030 are largely dependent on the investment decisions made in the 2020s. It is therefore essential that Mexico's energy planning decision-makers avoid a natural gas "lock-in" that would either cause carbon targets to be missed or risk leaving some natural gas infrastructure stranded.

The biggest developing country, China, will face a particularly severe challenge to balance the increasing energy demand and the CO_2 mitigation required by a below 2 °C target. Chapter by **Chen et al**. analyzes the key challenges for China's end-use sectors. Emission reduction beyond the Nationally Determined Contribution of China is feasible but challenging, especially in the end-use sectors. Electrification drives the CO_2 mitigation reduction in end-use sectors, resulting in deep changes in power source structure and supply capacity. The penetration of high-efficient technologies requires the implementation of sufficient incentives like targeted policies or market mechanisms. For example, energy efficiency standards are also essential to promote the transition of end-use sectors to low-carbon sectors. They need to be supported by market tools such as carbon trading.

Inspired from the Thirsty Energy initiative of the World Bank, Chapter by **Goldstein et al**. offers a wider perspective focused on the importance of the water-energy nexus in South Africa and China when moving towards below 2 °C. The most fundamental conclusion is that policies being pursued to mitigate climate change impacts reduce both CO_2 emissions and water needs by the energy sector— with only modest increase in energy system cost. Moreover, including the supply and cost of water has a dramatic effect on the upstream technology choices. For example, government mandated policies forcing dry cooling for new coal-fired power plants was reaffirmed as wise and appropriate, though at odds with achieving Nationally Determined Contributions (NDC) which quickly disincentives the use of coal while promoting renewables and nuclear, as a major step towards achieving below 2 °C emission reductions. The influence of climate change on energy and water planning differs across countries, as water availability in China's northern Energy Bases may actually increase slightly, while in South Africa the water system is stressed forcing more dramatic changes in the energy sector.

3.4 Part IV—The Role of Cities and Local Communities

Based on the InSMART project (Integrative Smart City Planning), Chapter by **Giannakidis et al.** explores the concerted action needed in four EU cities to manage energy consumption and reduce greenhouse gas emissions: Évora (Portugal), Cesena (Italy), Nottingham (UK) and Trikala (Greece). It concludes that it is important not to overestimate the contribution and the area of influence of city-agents to the global greenhouse gas target; but it is undoubted that municipalities are extremely well positioned for actions related to households, and their consumption in buildings and transport, for bridging locally the gap between what is perceived/known and what would be economically and technically feasible and for urban planning with a focus on significant benefits for GHG emissions reduction.

Sustainability has been integrated in the planning of many Nordic cities, and the Nordic capitals have the potential to lead the low-carbon transition by example. Oslo is a small city in a global context, but it wants to show how cities can take the responsibility for the development of sustainable energy systems with innovative ideas and solutions for the future. Chapter by **Lind and Espegren** analyzes how various energy and climate policies and measures can transform the city of Oslo into a low-carbon city. One of the key findings is that the majority of the emissions from the stationary sector can be removed at a low abatement cost, and most of these actions are relatively easy to implement. The phase-out of fossil-fuels in buildings occurs in all climate mitigation scenarios explored. The transport sector completes its full transition to non-fossil fuels only when fossil fuels are banned by specific policies. In all cases, support to fuel and technology innovation appears essential to the low-carbon transition of Oslo.

Chapter by **Yazdanie** evaluates decarbonisation pathways for Switzerland through local-scale energy systems planning under a national energy policy which reflects Swiss CO_2 emission reduction targets within the Paris Agreement. It finds that locally generated CO_2 emissions are reduced by 85% in 2050 relative to 2015, on average, across the evaluated archetypes and sectors. The uptake of efficiency measures (including renovations and efficient end-use devices) is clear, and these measures should be encouraged by local governments as part of local climate strategies. Decision-makers should also encourage the local-scale deployment of heat pump and solar PV technologies, which are found to generate significant shares of heat and electricity by 2050, cost optimally, across the archetypes. The utilization of local energy resources, including biomass, also plays an important role in achieving significant local-scale emission reductions in the long-term.

4 Conclusion

This book collates a range of concrete analyses of energy systems at different scales from around the globe, undertaken by ETSAP teams. It proposes feasible roadmaps of countries and local communities to make a well below 2 °C world a reality. The analyses are the result of a unique, collaborative, international research effort over the past 40 years. It illustrates how energy systems models have been and are being used to answer complex policy questions such as climate change mitigation and the need for more ambitious climate commitments by all countries within the Paris Agreement.

References

EC (2016) Proposal for a regulation of the European parliament and of the Council on the Governance of the Energy Union. COM(2016) 759 final/2, European Commission. Available at http://eur-lex.europa.eu/legal-content/EN/TXT/?uri=COM:2016:759:REV1

EC (2018) The JRC-EU-TIMES model. European Commission, EU Science Hub. Available at https://ec.europa.eu/jrc/en/scientific-tool/jrc-eu-times-model-assessing-long-term-role-energy-technologies

Giannakidis G, Labriet M, Ó Gallachóir B, Tosato GC (eds) (2015) Informing energy and climate policies using energy systems models: insights from scenario analysis increasing the evidence base. "Energy System" series. Springer.

IEA (2017) ETP-TIMES supply model. International Energy Agency. Available at http://www.iea.org/etp/etpmodel/energyconversion/

IPCC (2014) Climate Change 2014: Mitigation of Climate Change. Contribution of working group III to the fifth assessment report of the Intergovernmental Panel on Climate Change [Edenhofer O, Pichs-Madruga R, Sokona Y, Farahani E, Kadner S, Seyboth K, Adler A, Baum I, Brunner S, Eickemeier P, Kriemann B, Savolainen J, Schlömer S, von Stechow C, Zwickel T and Minx JC (eds)]. Cambridge University Press, Cambridge, United Kingdom and New York, NY, USA

Loulou R (2008) ETSAP-TIAM: the TIMES Integrated Assessment Model. Part II: mathematical formulation. CMS 5(1–2):41–66

Loulou R, Labriet M (2008) ETSAP-TIAM: the TIMES Integrated Assessment Model. Part I: model structure. CMS 5(1–2):7–40

Loulou R, Goldstein G, Kanudia A, Lehtila A, Remme U (2016) Documentation for the TIMES model. Part I. Part II. Part III. Energy Technol Syst Anal Program (IEA-ETSAP). Available at https://iea-etsap.org/index.php/documentation

UN (1998) Kyoto Protocol to the United Nations Framework Convention on Climate Change. United Nations, available at http://unfccc.int/resource/docs/convkp/kpeng.pdf

UN General Assembly (2015) Transforming our world: the 2030 Agenda for Sustainable Development. United Nations, A/RES/70/1, available at http://www.refworld.org/docid/57b6e3e44.html

UNEP (2017) The Emissions Gap Report 2017 United Nations Environment Programme (UNEP), Nairobi. Available at https://www.unenvironment.org/resources/emissions-gap-report

UNFCCC (2015) The Paris Agreement. United Nations Framework Convention on Climate Change, available at http://unfccc.int/paris_agreement/items/9485.php

Part I
The Radical Transformation of the Global Energy System

Energy System Challenges of Deep Global CO₂ Emissions Reduction Under the World Energy Council's Scenario Framework

Tom Kober, Evangelos Panos and Kathrin Volkart

Key messages

- Moving from *Unfinished Symphony* (the most stringent scenario among the World Energy Scenarios 2016) to a 1.5 °C climate pathway is feasible.
- Decarbonisation of energy demand requires immediate action with enabling measures for energy conservation and promotion of electricity to substitute oil and natural gas.
- In addition to nuclear and variable renewable energy technologies, CCS is a critical technology for natural gas and bioenergy based electricity production. Bioenergy with CCS becomes cost-competitive under a stringent climate stabilisation pathway.
- Systemic approaches (based on TIMES/MARKAL) are crucial to identify least-cost mitigation pathways and technology trade-offs in systems with complex interdependencies.

1 Introduction

Over the past four decades the global energy system has undergone a substantial growth and large infrastructure systems have been built, which are mainly based on centralised large-scale technologies. Looking forward until the year 2060, the World Energy Scenarios 2016, as performed by the World Energy Council (WEC) in collaboration with the Paul Scherrer Institute and Accenture, represent a

T. Kober (✉) · E. Panos · K. Volkart
Laboratory for Energy Systems Analysis, Energy Economics Group,
Paul Scherrer Institute (PSI), Villigen PSI, Villigen, Switzerland
e-mail: tom.kober@psi.ch

E. Panos
e-mail: evangelos.panos@psi.ch

© Springer International Publishing AG, part of Springer Nature 2018
G. Giannakidis et al. (eds.), *Limiting Global Warming to Well Below 2 °C: Energy System Modelling and Policy Development*, Lecture Notes in Energy 64,
https://doi.org/10.1007/978-3-319-74424-7_2

scenario framework of possible future developments under different boundary conditions. The future energy pathways investigated under the World Energy Scenarios 2016 are illustrated through three scenarios named *Modern Jazz, Unfinished Symphony* and *Hard Rock*, which are described in detail in WEC (2016). All three scenarios are explorative scenarios for which the emissions of carbon dioxide (CO_2) are a scenario result determined by the model's optimization algorithm and the scenarios' boundary conditions. One of the findings to be concluded from the World Energy Scenarios 2016 is that none of the emissions trajectories of the scenarios is in line with the ambitious long-term climate goal to limit the increase of the global mean temperature at no more than 1.5 °C compared to pre-industrial level, which has been agreed by the parties of the United Nations Framework Convention on Climate Change (UNFCCC) at the Conference Of the Parties (COP) 21 in Paris. Unfinished Symphony, the most stringent scenario, translates roughly to a temperature rise that is slightly above 2 °C in 2100 (WEC 2016). In this chapter we build on the scenario framework of the World Energy Scenarios 2016 and investigate the additional energy sector transformation needed to achieve this ambitions climate goal as well as we provide insights in regional and technological changes involved.

2   Methodology and Scenario Framework

For the analysis we employ the Global Multi-regional MARKAL (GMM) model, a technology-rich bottom-up model of the MARKAL (MARKet ALlocation) family of models. The GMM model represents the global energy system disaggregated into 15 world regions including region-specific characteristics of energy supply and demand as well as the corresponding CO_2 emissions. It is a linear optimization model, with minimisation of total system costs and perfect foresight until 2100. The GMM model is based on a partial equilibrium approach with exogenous demands for energy services. It is operated with a comprehensive technology database that contains many possible fuel transformation and energy supply pathways, and encompasses technologies based on fossil, nuclear, and renewable energy resources. Both currently applied technologies and future applicable advanced technologies, such as advanced fossil-fuelled power plants, hydrogen technologies, a broad variety of renewable energy options, and CO_2 capture and storage (CCS) techniques in power plants and industrial applications are available in the model's technology portfolio.

For our analysis we develop a climate target scenario compatible with limiting the increase of the global mean temperature at no more than 1.5 °C by the end of this century compared to pre-industrial level. For this new scenario, which complements the existing World Energy Scenarios 2016, we build on the general scenario framework as defined for WEC's *Unfinished Symphony* scenario. *Unfinished Symphony* represents a world in which more 'intelligent' and sustainable economic growth models emerge as the world drives to a low carbon future.

Unfinished Symphony assumes the existence of governmental promoted support mechanisms to mitigate climate change mitigation and to deploy low carbon technologies. As such, *Unfinished Symphony* distinguishes from the other two scenarios: *Modern Jazz* which stands for a market-driven future where digitalization unlocks new potential for innovative decentralized energy technologies, as well as *Hard Rock* in which a future development is assumed in which economies are rather fragmented and less cooperative and national policy agenda prioritize security of supply over environmental goals. Further information on the scenario definitions of the World Energy Scenarios 2016 is provided in WEC (2016).

The climate target scenario we focus in this chapter is named *Symphony 1.5C*. Based on Millar et al. (2017) we allow in this scenario a cumulative carbon budget for all energy-related CO_2 emissions of no more than 480 $GtCO_2$ for the period 2020–2100. Even though we focus our results discussion on the time horizon until 2060, we have run the model for the modelling horizon until 2100 in order to consider the carbon budget until end of the century as provided by Millar et al. (2017). The scenario assumes a development of the socio-economic indicators in line with the scenario *Unfinished Symphony*, in which the global GDP grows at 2.9% per year on average from 2010 to 2060 while the global population increases to 10.2 billion (UN 2015). Similar to *Unfinished Symphony*, in scenario *Symphony 1.5C* a broad technology portfolio is available to achieve the stringent mitigation target. Besides the consumer's willingness to apply various measures to shift away from fossil fuels and to improve the energy efficiency in the energy demand sectors, the scenario *Symphony 1.5C* also assumes public acceptance of both decentralized as well as centralized low-carbon energy conversion technologies, including nuclear technologies and technologies with CCS. Negative net emissions can be realized through the conversion of bioenergy with CCS.

In this study, we use an aggregated maximum global potential for various types of biomass of about 160 EJ, which reflects our judgement that limited biomass may be available when sustainability criteria are accounted for (GEA 2012; de Hoogwijk et al. 2009; Thrän et al. 2010). According to Hendriks et al. (2004) we limit the total cumulative storage potential for CO_2 in geological formations to 1660 $GtCO_2$. Related to the implementation of the emissions reduction obligations compatible with the 1.5 °C climate target, we presume the existence of a perfect carbon market which allows mitigation to be deployed at least costs independent of the economic condition of the region. We acknowledge that appropriate effort allocation schemes are necessary to achieve a fair distribution of climate change mitigation costs across the world regions. This has been investigated, for instance, by Tavoni et al. (2013) and Kypreos and Lehtila (2015), but it was not scope of our study.

3 Results

In this section we compare the *Symphony 1.5C* scenario results with the results of the World Energy Scenarios 2016 with a focus on the time horizon until 2060.

3.1 CO₂ Emissions and Primary Energy

3.1.1 Going Beyond NDC Pledges

Comparing the combustion related CO_2 emissions of the *Symphony 1.5C* scenario with the emission trajectories of the three World Energy Scenarios 2016 indicates that additional mitigation efforts are needed to achieve the global long-term climate target agreed at COP 21 in Paris. The total cumulative carbon budget in the period 2010–2060 of the *Symphony 1.5C* scenario (880 $GtCO_2$) is roughly about half of the budget of *Hard Rock* and *Modern Jazz*, and about 30% less than the budget of *Unfinished Symphony* (Fig. 1). In the scenario *Symphony 1.5C*, CO_2 emissions reduce after a stabilisation from today until 2020 at 33–30 $GtCO_2$ in 2030 and further to −1 $GtCO_2$ in 2060. For 2030, this implies an emission reduction beyond the pledged Nationally Determined Contributions (NDCs) of COP 21 in Paris by about 5 $GtCO_2$ or by 15% on global average.

3.1.2 Future Rates of Emissions Decline Exceed Emissions Growth Rates of the Past

In order to decarbonize the energy sector globally as depicted in *Symphony 1.5C* emissions need to reduce by 1%/year. until 2030, 4%/year. between 2030 and 2040, and even 9%/year. in the period 2040–2050. Comparing these emissions decline rates required in the period past 2030 with the emissions growth rates observed over the past 45 years (maximum 3%/year on a decadal average) reveals the challenge involved in realizing the speed of decarbonisation for achieving the Paris climate goal. While the global average CO_2 emissions per capita reduce by one third

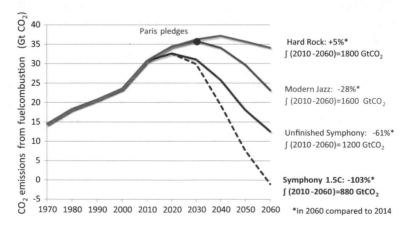

Fig. 1 Global CO_2 emissions across all scenarios (based on data from IEA 2016a; WEC 2016, own calculations)

between 2014 and 2050 in *Modern Jazz* and by one fifth in *Hard Rock*, they would need to decline by four fifths in the scenario *Symphony 1.5C* in the same timeframe. The second half of this century would be the period to compensate for CO_2 emitted in the first half of the century. In the scenario *Symphony 1.5C*, global emissions are below zero from 2060 onwards which means net carbon removals from the atmosphere. In this scenario, the CO_2 emissions accumulate to 880 $GtCO_2$ for the period 2010–2060, which implies cumulative negative emissions of 400 $GtCO_2$ (on average 10 $GtCO_2$/year) for the period 2060–2100 in order to stay within the total (2010–2100) carbon budget of 480 $GtCO_2$ to reach the climate target.

3.1.3 Some Fossil Fuel Reserves Stay Unexploited

Meeting stringent climate goals has substantial implications for the use of fossil fuels and hence for resource extraction, not only creating stranded assets but also stranded resources. Compared to the resource production in the *Hard Rock* scenario between 2020 and 2060 33% of crude oil, 15% of natural gas and 50% of coal would remain in the underground in the scenario *Symphony 1.5C*. For coal, the results of *Symphony 1.5C* are very similar to those of *Unfinished Symphony*, while for oil and gas resources production reduces much stronger in *Symphony 1.5C* than in *Unfinished Symphony*. With about 50% less extraction until 2060 in *Symphony 1.5C* (as well as in *Unfinished Symphony*) compared to *Hard Rock*, unconventional oil and gas production is in relative terms stronger affected by the shrinking demand than production from conventional reserves. In absolute quantities, conventional oil production reduces in *Symphony 1.5C* by 260 EJ (cumulatively 2020–2060) compared to *Hard Rock*, while unconventional oil production declines by about 80 EJ. Almost half of these total reductions in crude oil extraction of 340 EJ are attributable to declines of the production in North America (United States, Canada and Mexico) and Europe (OECD and non-OECD Europe, incl. Russia). Conventional natural gas production in the period until 2060, accumulate in these two scenarios to the same volume of about 600 EJ. However, natural gas produced from unconventional resources reduce from 220 EJ in *Hard Rock* to 100 EJ in *Symphony 1.5C*. The reductions of natural gas extraction observed for North America represent four quarters of the global production decline between the two scenarios and are mainly driven by the abandonment of unconventional gas production in the region.

3.1.4 Primary Energy to Peak by 2030

In the long run (2060), the primary energy consumption exceeds the consumption level of 572 EJ in 2014 in all World Energy Scenarios as well as in scenario *Symphony 1.5C*. However, all scenarios show a peaking of primary energy consumption per capita before the year 2030. In the *Symphony 1.5C* scenario total primary energy also reaches its absolute maximum in 2030. By using less energy in a more efficient way, per capita primary energy consumption in 2060 in the *Symphony 1.5C* scenario is declined roughly at the level of 1970 (60 GJ per capita and year). The most notable change for this indicator is to be observed for North America where specific consumption is reduced by half in the period 2010–2060 reaching a consumption of 120 GJ of primary energy per capita in 2060 in the scenario *Symphony 1.5C*. This indicates the significant energy efficiency improvements to be realized in this region by adopting consumption patterns similar to other industrialised regions. Despite the convergence in technology adaptation and energy usage patterns across regions, some regional differences prevail in future. In North America, the specific per capita primary consumption is still twice as high as the global average in 2060 in the same scenario and about 25% higher than the specific consumption in Europe (European Union, Norway, Iceland and Switzerland).

3.2   Final Energy Consumption

3.2.1   Decelerated Energy Consumption Growth

The disaggregation of the results for final energy consumption in sectors and fuels highlights two important developments: (1) increased overall energy savings at more ambitious climate targets, and (2) the role of electricity for decarbonizing the demand sectors (Fig. 2). Compared to the scenarios *Modern Jazz* and *Hard Rock* where the final energy consumption grows between 2010 and 2060 by about 50% with major increases in the transport sector and in industry, meeting the 1.5 °C climate target requires accelerated decoupling of energy consumption and economic growth. In the scenario *Symphony 1.5C*, the final energy consumption in these two sectors increases from 2014 until 2030 on average by 0.9–1.1%/year. and reduces afterwards towards 2060 on average at 0.8–1.0%/year. A major contributor to the improvements of energy intensity is the switch from fossil fuel based technologies towards electric appliances, such as electric heat pumps, electric mobility and increased electric processes and appliances in industry. Compared to *Modern Jazz* and *Unfinished Symphony*, where the use of natural gas increases over time in order to substitute more carbon intensive fuels, the results for *Symphony 1.5C* show the opposite trend for natural gas in the long run. This is driven by an accelerated substitution of all kinds of fossil fuels by electricity.

3.2.2   Decarbonising Energy Demand Through Increased Electrification

While the total final energy consumption in 2060 reduces from *Unfinished Symphony* towards *Symphony 1.5C* by about 20%, the demand for electricity in *Symphony 1.5C* exceeds the demand observed under *Unfinished Symphony* by

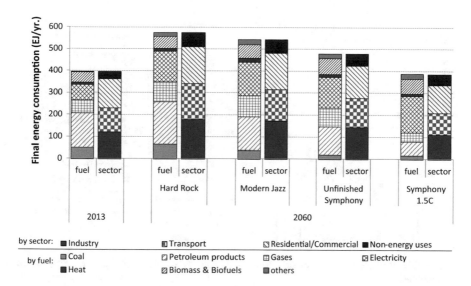

Fig. 2 Global final energy consumption by fuel and sector (based on data from WEC 2016; IEA 2016b, own calculations)

7.5 PWh (+19%). In the scenario *Symphony 1.5C*, the share of electricity of total final consumption grows from 18% in 2014 to a maximum of 43% in 2060, translating into an absolute increase of electricity demand by about 30 PWh. Almost 30% of this increase relates to the increased electrification of the transport sector. The share of electricity of total final consumption of the transport sector grows from around 1% in 2014 to 30% by 2060 in this scenario. Electrifying road transport and substituting gasoline and diesel engines reduces the energy intensity for road transport services by factor 2–3 which contributes significantly to the overall energy savings in the transport sector. In *Symphony 1.5C*, more than half of the global car fleet in 2060 is based on pure battery electric vehicles and plug-in hybrid vehicles. This development represents an acceleration of the transformation observed for the scenario *Unfinished Symphony*, in which the share of electricity of the transport sector's final energy consumption in 2060 is three times less than in *Symphony 1.5C*, and the share of hydrogen is about two thirds less. The substitution of fossil transport fuels through electricity and hydrogen, as well as further fuel saving measures reduce the emissions of the transport sector by about 3 $GtCO_2$ in 2060 between the two scenarios.

3.3 Power Sector Developments

Over the past 45 years, the global electricity production has seen a five-fold increase with about 18 PWh more produced in 2014 compared to 1970. Over the

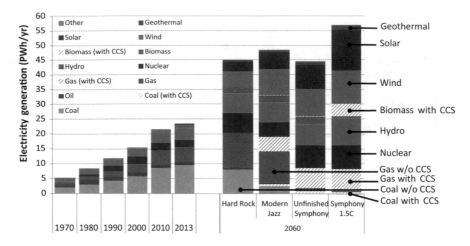

Fig. 3 Global electricity generation (based on data from WEC 2016; IEA 2016b, own calculations)

45 years to come, the scenario results indicate an increase in the total electricity generation by factor 1.9–2.4, which equals worldwide to additional electricity amounts of 22–34 PWh in 2060 compared to 2013 (Fig. 3). Resulting from the electricity based decarbonisation of energy demand sectors in scenario *Symphony 1.5C*, electricity generation is highest across all scenarios in *Symphony 1.5C* reaching 57 PWh in 2060. Compared to the scenario *Unfinished Symphony*, electricity generation in *Symphony 1.5C* needs to increase up to 30% (2060) in order to fulfil the more ambitious climate change mitigation obligations through enforced electrification of energy demand. Our results indicate that, essentially the additional electricity production could be provided by renewable energy technologies while an increase of fossil fuels to largely cover the growth electricity demand is not cost-efficient.

3.3.1 Gas and Biomass Based CCS and Nuclear to Complement Solar, Wind and Hydro

The electricity generation structure changes in all four scenarios from two thirds fossil fuel based generation in 2013 towards higher shares of renewable energy and declining shares of fossil fuels. The share of renewable energy of the total electricity production in scenario *Symphony 1.5C* amounts to 70% in 2060, which equals to 170% of the total electricity generation from all fuels in 2013. Worldwide, solar and wind become the main pillars of the electricity system. Even though these power technologies are proven, enforced efforts are needed for the technologies' system integration. The additional efforts for the system transformation from *Unfinished Symphony* to *Symphony 1.5C* are primarily realized by a 20% increase of wind power production and a 50% solar based power production in the period until 2060. A major difference when comparing *Unfinished Symphony* and *Symphony 1.5C* is the enhanced deployment of technologies leading to net negative CO_2 emissions. While in the scenario *Unfinished Symphony* power plants using bioenergy in combination with CCS are hardly deployed until 2060, this technology contributes with 8% (4 PWh) to total electricity generation in 2060 in *Symphony 1.5C*, which allows removing almost 10 Gt CO_2 from the atmosphere and helping to offset CO_2 emissions from other sectors.

Besides renewable energy technology, nuclear power plants also represent a cost-efficient decarbonisation option and triple their production from 2013 to 2060 generating 8 PWh (14% of total generation) in 2060 in scenario *Symphony 1.5C*. In comparison to *Unfinished Symphony*, the production in nuclear power plants is relatively constant in *Symphony 1.5C* as a result of the limitations on the future deployment of this technology as assumed in this study.

In the near term (2020/2030), early emissions reductions can be realized in *Symphony 1.5C* by substituting coal-based electricity generation with electricity generated from natural gas, mainly enabled by a higher utilisation of natural gas power plant capacities. In the long-run, coal has hardly any significant contribution to the total electricity production, even if equipped with CCS, while natural gas could stay in the electricity mix if equipped with CCS. When comparing electricity generation in natural gas power plants between the two scenarios *Symphony 1.5C* and *Unfinished Symphony*, our results show roughly the same production level in 2060, however, the developments in earlier periods show that under *Symphony 1.5C* scenario, natural gas power plants need to deploy two to five times faster shifting investments from natural gas power plants without CO_2 capture to CCS technology. Independent of being equipped with CCS or not, at high shares of variable renewable energy, natural gas power plants need to be designed for flexible operation (ca. 3500 full load hours in 2060). This development can be observed for *Symphony 1.5C* and for the less stringent scenario *Unfinished Symphony*.

3.3.2 In Contrast to Coal, Gas Power Plant Capacity Additions in Future to Exceed Historic Levels

The transformation of the power sector translates into changes in the investment patterns of new power plants on global and regional level (Figs. 4 and 5). Across all four scenarios, the results for coal power plants show average annual future capacity additions below the historic capacity additions observed between 2000 and 2010, which was a boom time for new coal-based installations. In particular, on the way to achieve the 1.5 °C climate goal, coal power plants are not an investment option, even with CCS. This explains the reduced coal extraction in the scenarios *Unfinished Symphony* and *Symphony 1.5C* compared to *Hard Rock* and the resulting stranded coal resources. The investments in coal power plants over the period 2011–2060 depicted in *Symphony 1.5C* are one third of the corresponding investments in *Hard Rock*.

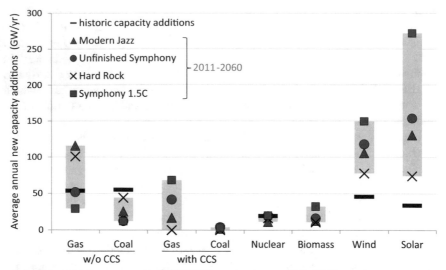

N.B.: Historical data correspond to 2000-2010 for coal and gas, to 1980-1990 for nuclear energy, and to 2010-2015 for wind and solar. The data is assembled from: EPIA (2014, 2016), GWEC (2016), IEA-PVPS (2016), IEA-CCS (2012) and Platt's (2013).

Fig. 4 Global new capacity additions for selected electricity generation technologies

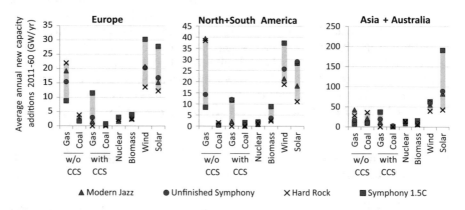

Fig. 5 Regional new capacity additions for selected electricity generation technologies

Conversely to the trend of coal power plants, future annual capacity additions of new gas power plants (across all scenarios 90–130 GW/year. over the period 2011–2060) exceed historic capacity additions of the period 2000–2010. The results for the scenarios *Symphony 1.5C* reveal that natural gas power plants with CCS need to start to deploy in 2030 and increase their penetration until 2060 with annual new capacity additions higher than what has been observed for natural gas power plants without CCS between 2000 and 2010. Also in the scenario *Unfinished Symphony*, natural gas power plants would start to deploy in 2030, however with about half of the installed capacity than in scenario *Symphony 1.5C*. In *Symphony 1.5C*, the installed capacity of natural gas power plants with CCS in 2040 and beyond is higher than the installed capacity of natural gas power plants without CCS. Such a development cannot be observed in *Unfinished Symphony* before 2060. The massive deployment of CCS not only requires power plant manufacturing industry to be able to produce, install and maintain these new technologies but also to build up the corresponding CO_2 transport and storage infrastructure which implies a need to overcome substantial regulatory and social barriers in many countries.

3.3.3 Gas CCS Equally Important in Europe and North and South America

Depending on the scenario, between 900 and 1080 GW of natural gas capacity are installed in Europe (incl. Russia) over the time horizon until 2060, which equals to average annual new capacities of 18–22 GW/year. (Fig. 5). If deep CO_2 emissions reductions are to be achieved at least cost, and if nuclear power deployment remains limited, a discussion about the attitude towards CCS would be needed in the European context accompanied with building up know-how on commissioning and operation of CCS technology. In North and South America, the market for natural gas power plants in the *Symphony 1.5C* scenario has a similar size (in terms of new installed capacity) to Europe with CCS becoming increasingly important past 2030. In North America, the share of natural gas from unconventional resources of total domestic gas consumption is higher than in other world regions with up to 63% (*Hard Rock*) for the cumulative quantities over the period 2020–2060. Gas produced from unconventional resources is a driver of new gas power installations if climate change mitigation has lower priority. In the scenario *Symphony 1.5C*, however, environmental concerns related to unconventional resource extraction and stringent climate policy translate into one third less natural gas being produced from unconventional resources in North America compared to *Modern Jazz* and *Hard Rock*. The capacities of new natural gas power plants installed from 2011 to 2060 reduce by about 50% between the latter scenarios and *Symphony 1.5C* in favour of increased wind and solar PV capacities.

3.3.4 Future Growth of Nuclear on Historic Levels, Mainly Taking Place in Asia

Future global annual capacity additions of nuclear power plants could be of similar magnitude as observed during 1980–1990 with on average about 20 GW/year. with 6% of the total global power generation investments in the period 2011–2060 being invested in nuclear technologies in the scenario *Symphony 1.5C*. Contrary to the 1980s and 1990s, when nuclear power plants were primarily built in Europe (including Russia) and North America, future nuclear capacity investments concern to more than 60% Asian regions. For Europe and North America the scenarios assume the possibility of retrofit and replacement of existing nuclear power plants, and omit the large-scale deployment of new nuclear power plants well beyond currently installed capacities.

3.3.5 Unprecedented Growth of Solar and Wind Technologies

The worldwide average annual new capacity additions for wind and solar technologies for the period 2011–2060 not only exceed average historic capacity additions of wind and solar technologies of the period 2010–2015 in all four scenarios, they also exceed the level of annual new coal power plant capacities of the timeframe 2000–2010. The results for *Hard Rock* show for the period 2011–2060 average annual installations of 80 GW/year. for wind and 75 GW/year. for solar technologies, which is 45% higher than the 55 GW/year. of new coal power plant capacities between 2000 and 2010. Requirements to install more renewable power technologies increase with tightening the climate target and reach annual averages of 150 GW/year. for wind and more than 250 GW/year. for solar technologies in scenario *Symphony 1.5C*. These annual capacity additions represent approximately four times the current solar PV module production capacities of about 80 GW peak in 2016 (Fraunhofer Institute for Solar Energy Systems 2017) which would need to be increased substantially in the coming decades to satisfy the demand for new solar modules. In Europe, annual average new capacities over the 2011–2060 period reach about 30 GW/year. for each solar and wind technologies in the scenario *Symphony 1.5C*. In North and South America together, annual average wind capacity additions in *Symphony 1.5C* could be as high as natural gas capacity additions in the scenarios *Modern Jazz* and *Hard Rock*. Our results indicate that more than half of the global solar capacity additions and more than 40% of the global wind capacity additions are to be realized in the Asian market. Driven by a high demand for low-carbon electricity, such significant capacity additions for solar and wind technologies also result from the fact that solar and wind technologies usually have a lower annual availability and shorter lifetime compared to conventional thermal power plants. This necessitates on the one hand to install a comparably higher capacity to produce the same amount of electricity and, on the other hand, more frequent capacity replacements to retain a certain amount of installed capacity.

3.3.6 Half of the Global Power Generation Investments in Wind and Solar Power

Cumulatively over the period 2011–2060, the global power sector requires investments in generation equipment worth USD_{2010} 59 trillion in scenario *Symphony 1.5C* (investments refer to undiscounted investment costs over the given period). Roughly half of the global investments concern power plants in Asia, with India accounting for 12% and China for 24% of the global investments. Relative to the regions' gross domestic products, the investments represent a share of 1.1% for India and 0.9% for China. On a global level, slightly more than half of the total cumulative power generation investments of the period 2011–2060 would be needed to install wind and solar technologies, while fossil fuelled power plants represent an investment share of 18% only if the 1.5 °C climate target is to be achieved at least cost. Across all four scenarios, the investments in hydropower and nuclear technologies are each in a range of 6 to 9% of global total power generation investments.

4 Conclusion

In this chapter we investigate the long-term energy system transformation in line with the long-term climate goal to limit the increase of the global mean temperature at no more than 1.5 °C compared to pre-industrial level. For this analysis we build on the framework of the World Energy Scenarios 2016 (of which none of the scenarios is compatible with the 1.5 °C climate goal) and complement the existing set of scenarios with an additional, more stringent, climate target scenario (*Symphony 1.5C*). Our first main finding is that on a global least-cost mitigation pathway, energy-related CO_2 emissions must become negative by 2060 if the 1.5 °C climate target is to be reached cost-efficiently. Decarbonisation of the energy sector under this climate target requires global CO_2 emissions reductions to decline past 2030 at higher speed than emissions grew over the past 45 years. The energy sector transformation will result in a significant share of the fossil fuel reserves unexploited, i.e. in *Symphony 1.5C* scenario, resource extraction reduces by 33% for crude oil and by 50% for coal in the period 2020–2060 compared to a scenario in which climate change mitigation has a low priority. Given the capital intensity of fuel production the risk of stranded assets and stranded resources should be seriously considered if stringent climate policy is anticipated.

Reducing fossil fuel consumption in the energy end-use sectors is primarily driven by two effects: (1) increasing the sectors' energy efficiencies, and (2) enforced use of electricity-based technologies for decarbonising the demand sectors. The transport sector and industry need to achieve the highest energy demand reductions. While the total final energy consumption in 2060 reduces from *Unfinished Symphony* (the most stringent scenario among the World Energy Scenarios 2016) towards *Symphony 1.5C* by about 20%, the demand for electricity in *Symphony 1.5C* exceeds the demand observed under *Unfinished Symphony* by 19%, which underlines the role of electricity when tightening the climate target. On a global level, the share of electricity of total final consumption grows from 18% in 2014 up to 43% in 2060 in *Symphony 1.5C*. For policies focussing on energy demand side measures this means to further promote highly efficient electricity-based end-use technologies to substitute fossil fuels besides other energy efficiency improvement measures, such as building insulation and appliances performance standards. General (technology unspecific) policy targets on electricity savings in energy end-use sectors, however, are seen as critical, because they might oppose increased electrification and hinder the shift from fossil fuels to electricity. The fact that in the scenario *Symphony 1.5C* the share of electricity of the global transport sector's final energy consumption in 2060 is three times higher than in *Unfinished Symphony*, and the share of hydrogen is about 50% higher, reveals the importance of policy measures to unlock mitigation options in the mobility sector, and in particular related to passenger transport.

In 2060, 85% of the total global electricity generation is generated from renewable and nuclear resources in the scenario *Symphony 1.5C*. Comparing *Unfinished Symphony* and *Symphony 1.5C* shows that the additional efforts needed for the system transformation are primarily realized by a 20% increase of wind power production and a 50% solar based power production in the period until 2060. In the long-run, coal has hardly any significant contribution to the total electricity production, even not with CCS, while natural gas remains in the electricity mix if equipped with CCS. In contrast to *Unfinished Symphony* bioenergy with CCS becomes cost-competitive under the condition of a 1.5 °C climate stabilisation pathway, allowing for net CO_2 removals from the atmosphere (almost 10 $GtCO_2$ in 2060) which offsets emissions from other sectors where emissions mitigation is more difficult and costly, such as aviation and industry.

The transformation of the electricity system goes along with fundamental changes in the investment patterns in the power sector. For the time until 2060 average annual new electricity capacity installations for wind and solar are required which exceed the capacity addition rates observed for fossil fuelled power plants during the boom time in the early 2000's by factor 4. Worldwide, more than half of the total power generation investments in the period 2011–2060, which amount to USD 59 trillion in *Symphony 1.5C*, need to be dedicated to wind and solar technologies. In this time frame, the electricity markets in Asia have the highest growth and require about half of the total global generation investments. In this regard, it is important to emphasise that financing mechanisms and an appropriate investment climate need to be fostered especially for low-carbon technologies in emerging economies.

References

de Hoogwijk M, Faaij A, de Vries B, Turkenburg W (2009) Exploration of regional and global cost—supply curves of biomass energy from short-rotation crops at abandoned crop land and rest land under four IPCC SRES land-use scenarios. Biomass Bioenerg 33:26–43

Fraunhofer Institute for Solar Energy Systems (2017) Photovoltaics report. www.ise.fraunhofer.de. Accessed 12 July 2017

GEA (2012) Global energy assessment—towards a sustainable future. Cambridge University Press, Cambridge

Hendriks C, Graus W, Bergen FV (2004) Global carbon dioxide storage potential and costs. Ecofys in cooperation with TNO, EEP-02001, Utrecht

Kypreos S, Lehtila A (2015) Assessment of carbon emissions quotas with the integrated TIMES and MERGE Model. In: Giannakidis G, Labriet M, Ó Gallachóir B, Tosato G (eds) Informing energy and climate policies using energy systems models—insights from scenario analysis increasing the evidence base, vol 30, pp 111–124

IEA (2016a) CO$_2$ Emissions from fuel combustion 2016. International Energy Agency (IEA), Paris

IEA (2016b) Energy balances statistics 2016. International Energy Agency (IEA), Paris

Millar RJ, Fuglestvedt JS, Friedlingstein P et al (2017) Emission budgets and pathways consistent with limiting warming to 1.5C. NGEO. 10. https://doi.org/10.1038/ngeo3031

Thrän D, Seidenberger T, Zeddies J, Offermann R (2010) Global biomass potentials—resources, drivers and scenario results. Energy Sustain Dev 14:200–205

United Nations (UN) Department of Economic and Social Affairs, Population Division (2015) World Population prospects: the 2015 revision, DVD Edition

Tavoni M, Kriegler E, Aboumahboub T, Calvin K, De Maere G, Jewell J, Kober T, Lucas P, Luderer G, McCollum D, Marangoni G, Riahi K, vanVuuren D (2013) The distribution of the major economies' effort in the Durban platform scenarios. Clim Change Econ 4(4):1340009

WEC (2016) World energy scenarios 2016—the grand transition. World Energy Council (WEC), London

Pathways to Post-fossil Economy in a Well Below 2 °C World

Antti Lehtilä and Tiina Koljonen

Key messages

- Negative emissions have a crucial role in making the well-below 2 °C target achievable, despite the practically inevitable temporary overshooting.
- Utilizing carbon dioxide as a feedstock appears to offer a key element for a viable technology pathway to the post-fossil economy during the current century.
- The technology-rich approach of the ETSAP TIMES modelling framework is especially well-suited for analysing the complex inter-actions between emerging technologies within the overall energy system in a long-term context.

1 Introduction

The IPCC Fifth Assessment Report (2013) collected about 900 quantitative green-house gas (GHG) mitigation scenarios, representing different mitigation pathways with different tech-nology portfolios and socio-economic assumptions. Most of the scenarios meeting the global 2 °C temperature target require over-shooting the carbon budget and later removal of the excess carbon with large negative emissions, typically 400–800 $GtCO_2$ by 2100. The remaining carbon budget for the 2 °C target has been estimated at 900–1500 $GtCO_2$ from 2015. That would be exhausted in a couple of decades with the current rates (Friedlingstein et al. 2014), but for a well-below 2 °C target even far lower carbon budgets have been reported, illustrating the high urgency of the challenge.

A. Lehtilä (✉) · T. Koljonen
VTT Technical Research Centre of Finland, VTT, Espoo, Finland
e-mail: antti.lehtila@vtt.fi

T. Koljonen
e-mail: tiina.koljonen@vtt.fi

© Springer International Publishing AG, part of Springer Nature 2018
G. Giannakidis et al. (eds.), *Limiting Global Warming to Well Below 2 °C: Energy System Modelling and Policy Development*, Lecture Notes in Energy 64,
https://doi.org/10.1007/978-3-319-74424-7_3

For complying with the well-below 2 °C target, alternative technologies or measures are thus evidently needed, for removing CO_2 from the atmosphere. These include (1) increasing carbon sinks by forestation, (2) direct CO_2 capture from air (DAC), (3) capture and utilization of biogenic carbon from combustion processes (BECCU), and (4) capture and storage of biogenic carbon from combustion processes (BECCS). In all these cases, mitigation occurs only if carbon is stored permanently in underground storage, forests, soils or materials, or if fossil carbon is replaced by CCU or DAC in the production of energy, fuels, chemicals, or other materials. In this chapter we specially focus on options with net negative CO_2 emissions, enabling sustainable pathways into a carbon neutral or carbon negative world, where fossil fuels are totally replaced with renewable energy and synthetic materials. The transition leads to what may be called a "post-fossil" economy, where the source of carbon is from air or biomass instead of fossils.

Our chapter presents two well-below 2 °C global mitigation scenarios, which are compared to a reference scenario with known policies and measures until 2030. In the first scenario setting, we explore the mitigation targets attainable with a rapid techno-logical change towards low carbon and carbon neutral technologies across all sectors. We also illustrate the range of forcing trajectories reached as a function of marginal GHG abatement costs, taking into account estimates about other forcing agents, such as the aerosol masking effect.

The primary focus of our chapter, however, is the emerging post-fossil era of 2050–2100 studied with a second scenario, where renewable natural resources are not only used for producing energy but also synthetic fuels, chemicals, and materials. There is increasing literature on alternative pathways to 100% renewable systems, but the studies generally do not consider replacement of fossils in the entire economy, as in our chapter.

2   Gateways to a Post-fossil Economy

CO_2 can be considered as a versatile building block for a broad range of applications to produce CO_2-based fuels for transport and aviation, proteins, chemicals and polymers. First pilot and commercial production plants have been started, where CO_2 is, for example, used as a polymer building block or to produce synthetic fuels.

In a post-fossil economy, abundant renewables, especially solar and wind, are combined with direct air capture and carbon utilization to store energy and to produce renewable fuels and other materials (Hannula 2015; Köning et al. 2015; Schemme et al. 2017; Brynolf et al. 2017). We call this C1 chemistry to highlight the production of new sustainable fuels and other products with a single carbon molecule as a building block instead of fossil fuels.

More commonly, the terms Power-to-Gas, Power-to-Liquid, and Power-to-X are used to describe producing synthetic hydrocarbons from renewable electricity. The concepts are based on producing hydrogen (H_2), which reacts with CO_2 via the reverse water-gas shift reaction to syngas that may be synthesized further to liquid hydrocarbons by Fisher-Tropsh (FT) synthesis. The downstream product separation and upgrading allows the production of different fractions of synthetic hydrocarbons for specific applications. These processes also generate valuable by-products, e.g. high-purity oxygen and heat.

FT synthesis can be defined as a process to convert syngas to synthetic crude oil, which can be composed of alkanes, alkenes, alcohols, carbonyls, and carboxylic acids. Depending on the reaction conditions and catalysts, different products can be produced, like methane (CH_4), DME (CH_3OCH_3), methanol (CH_3OH), higher hydrocarbons (e.g. gasoline) and higher alcohols (Köning et al. 2015; Brynolf et al. 2017). Based on the existing knowledge on routes to produce hydrocarbons from renewable electricity and CO_2, we have assumed that by 2100 all fossil fuels may be replaced and thus leading us to a post-fossil economy.

3 Modeling Methodology and Scenario Assumptions

3.1 Global TIMES-VTT Energy System Model

In the scenario analysis we employed the TIMES-VTT energy system model. It is a global multi-region model originally developed from the global ETSAP TIAM model (Loulou 2008; Loulou and Labriet 2008). It is based on the IEA TIMES modeling framework (Loulou et al. 2016), and is characterized as a technology-rich, bottom-up type partial equilibrium model. The model consists of 17 regions, namely the Nordic countries (Denmark, Finland, Norway, Sweden), Western Europe, Eastern Europe, CIS (Former Soviet Union excluding the Baltics), Africa, the Middle East, India, China, Japan & South Korea, Other Developing Asia, Canada, the USA, Latin America and Australia & New Zealand (Koljonen et al. 2009, Koljonen and Lehtilä 2015).

The TIMES modeling system incorporates also an integrated climate module, with a three-reservoir carbon cycle for CO_2 concentrations and single-box decay models for the atmospheric CH_4 and N_2O concentrations, and the corresponding functions for radiative forcing. The forcing functions for CO_2, CH_4 and N_2O follow the non-linear formulations presented in the IPCC Fifth Assessment Report (Myhre et al. 2013), but are linearized around user-defined points. If necessary, by using an iterative approach the accuracy of the linearization can be improved to an arbitrary level. Additional forcing induced by other natural and anthropogenic causes is taken into account by means of exogenous projections. Finally, the changes in mean temperature are simulated for two layers, surface, and deep ocean (Loulou et al. 2016). When modeled, the emissions of F-gases (HFCs, PFCs and SF_6) can also be taken into account in the climate model by converting them into equivalent CO_2 emissions. Although both the carbon cycle and the concentrations of CH_4 and N_2O are represented by quite simple models, the radiative forcing from anthropogenic GHG emissions is reasonably well approximated by the TIMES climate module, and is calibrated to reproduce the 2010 level 2.1 W/m^2.

3.2   Scenarios Modeled

3.2.1   Overview

We illustrate alternative pathways to a low carbon economy (LCE) and a post-fossil economy (PFE) with three main scenarios: the Baseline, 1.5C-LCE and 1.5C-PFE scenarios. The Baseline serves as a reference case and assumes only current poli-cies, the EU 2030 energy and climate policies (40% reduction in GHG emissions by 2030 compared to 1990), and the effect of the intended nationally determined contri-butions (INDCs). The 1.5C-LCE and the 1.5C-PFE scenarios represent our two main variants for the pathways into a well-below 2 °C world, of which the 1.5C-PFE depicts a rapid transition into a post-fossil economy, whereas the 1.5C-LCE may be characterized as only reaching a low-carbon economy, despite also featuring rapid technological change.

The global GHG emission levels resulting from the INDCs have been estimated to be 53–55 $GtCO_2e$) in 2030 (UNFCCC 2015), which is in agreement with the emissions in the Baseline, 54.5 Gt. However, in the absence of additional global climate policies, in the longer term the Baseline departs far from the pathways required for a well below 2 °C world. In terms of CO_2 concentrations, it is close to the RCP 6.0 scenario and thereby to a 4 °C world, in similarity to the IEA ETP 4DS scenarios (e.g. IEA 2016).

The 1.5C-LCE scenario assumes that the technical and economic performance of a wide scale of commercially available and demonstrated clean energy technologies continues to develop favourably, as commonly estimated in the literature, especially concerning renewable energy technologies and CCS. In the 1.5C-PFE scenario we further assume a rapid development in technologies utilizing so-called C1 chem-istry, hydrogen and fuel cell technologies, and related energy storage technologies (Table 1).

Table 1   Scenarios considered in the analysis

Scenario	Description	Climate target	GHG price variants
Baseline	Existing policies + INDCs, EU 2030 policies	None	€200, €250, €300, €350, €400, €450 per tonne CO_2 eq
1.5C-LCE	Representative well-below 2 °C with overshooting, low-carbon economy	Forcing trajectory corresponding to a climate sensitivity of 2.75 °C (Table 2)	None
1.5C-PFE	Representative well-below 2 °C with overshooting, rapid transition to a post-fossil economy	Forcing trajectory corresponding to a climate sensitivity of 2.75 °C (Table 2)	None

3.2.2 Modeling of Well-Below 2 °C Scenarios

For the representative well below 2 °C scenarios (1.5C-LCE and 1.5C-PFE), we chose to use radiative forcing from anthropogenic sources as the target climate indicator, mainly to avoid speculation about climate sensitivities (equilibrium and transient climate sensitivity), and also because the temperature model of TIMES is quite simple.

So-called Representative Concentration Pathways (RCPs) have been constructed for the emission, concentration and land-use trajectories that will reach selected radiative forcing values in the year 2100, from 2.6 to 8.5 W/m^2. Specifically, the RCP2.6 scenario limits forcing to 2.6 W/m^2 in 2100, but in such a way that it leads to a higher forcing of around 3 W/m^2 mid-century and to a decline afterwards, thereby being also called RCP3-PD, for peak-and-decline (van Vuuren et al. 2011). This is the lowest RCP, and has been estimated leading to a global mean temperature change of around 1.7 °C in 2100, assuming a mid-range equilibrium climate sensitivity of 3 °C (Weyant et al. 2009).

Given also that the representative value of 2.75 °C has recently been considered appropriate for mid-range climate sensitivity (e.g. Millar et al. 2017), one may estimate that a forcing pathway similar to the RCP2.6 scenario would have a high probability of keeping the temperature change at most 1.5 °C by 2110, and below 1.7 °C between 2050 and 2100. We have therefore chosen such a peak-and-decline forcing pathway to represent a well-below 2 °C scenario, with a peak value of 3.1 W/m^2 in 2050–2060. The target trajectory for the total forcing and its components are presented in Table 2.

Another critical exogenous assumption that has to be made is the projection for CO_2 emissions from land-use change, which are mostly caused by deforestation.

Table 2 Assumed total target for anthropogenic radiative forcing and the exogenous components derived for the scenario modeling, based on the RCP scenarios (IPCC 2013)

Year	Total forcing target	Components of exogenous forcing				Total exogenous forcing
		Montreal gases	Aerosols	Ozone	LUC	
2000	1.73	0.31	−0.92	0.33	−0.15	−0.43
2010	2.10	0.33	−0.90	0.35	−0.15	−0.37
2020	2.50	0.33	−0.73	0.39	−0.17	−0.18
2030	2.80	0.29	−0.60	0.42	−0.19	−0.08
2040	3.00	0.24	−0.44	0.36	−0.21	−0.05
2050	3.10	0.20	−0.32	0.31	−0.23	−0.04
2060	3.10	0.17	−0.26	0.26	−0.25	−0.08
2070	3.05	0.15	−0.19	0.22	−0.27	−0.09
2080	2.90	0.13	−0.15	0.19	−0.29	−0.12
2090	2.75	0.11	−0.12	0.16	−0.31	−0.16
2100	2.60	0.10	−0.12	0.14	−0.33	−0.21

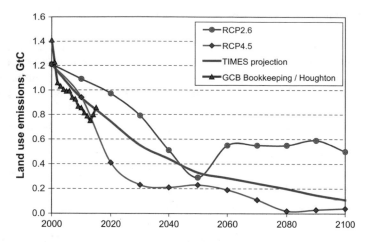

Fig. 1 Historical land use emissions estimated by the global carbon project (2000–2010, Houghton 2012), extrapolation to 2015 based on Houghton and Nassikas (2017), and decadal AFOLU emission projections of the RPC2.6 and RPC4.5 scenarios of the AR5 (IPCC 2013)

We assume that it is necessary to achieve deep reductions in the deforestation rates in order to achieve the strict climate change mitigation targets. The projection used in our scenarios is shown in Fig. 1, and is based on the RPC2.6 and RPC4.5 scenarios (IPCC 2013).

3.2.3 Demand Driver Projections

The model uses exogenous projections for factors such as GDP, population, household size and sectoral outputs to estimate the development of future energy service demands for the residential, commercial, agricultural, industrial, and transport sectors in each region, in the same way as in the ETSAP TIAM model (Loulou and Labriet 2008). The regional population and GDP projections are in good agreement with the UN and OECD projections (OECD 2014), showing the global PPP adjusted GDP growing to about 3.7 times higher in 2010–2050 and to about 10.5 times higher by 2100, the global GDP growth rates gradually declining from the present level of 4 to 2%/a by 2080.

We can expect that the decoupling between GDP growth and energy use will become stronger in the future, but it is uncertain how much stronger. Our assumptions on the elasticities between the economic drivers and the demands for useful energy services may be relatively conservative in that respect, but appear quite comparable to those in the IEA ETP studies. In the Baseline scenario global primary energy supply approaches the level of 800 EJ and final energy use about 600 EJ in 2050 (including non-energy use), while in the 2016 ETP 4DS scenario the projections were 830 EJ for primary energy and 590 EJ for final energy in 2050 (IEA 2016).

As the TIMES model is a partial equilibrium model, we also make use elasticities of the demands to their own prices. The price elasticities used in our scenarios are mostly the same as those in the ETSAP TIAM model database (Loulou and Labriet 2008).

3.2.4 Resource and Technology Projections

Assumptions concerning the future potential of bioenergy supply are very important for the well-below 2 °C scenarios. Our data on regional bioenergy potentials are based on various international sources and our earlier research on global biomass resources (e.g. Kallio et al. 2015). The assumed total bioenergy potential is in agreement with the IPCC SRREN base estimates for 2050 (IPCC 2012), being about 200 EJ in 2050 and 260 EJ in 2100. Among different bioenergy sources, energy crops have the largest future potential, about 100 EJ in 2050 and increasing to about 140 EJ in 2100 along with productivity growth. Sustainable supply of energy crops is strongly tied to the available surplus land and food and feed crop production, which in turn are affected by productivity changes, changing climatic conditions and human diets. Relatively low potential estimates may therefore be justified by the uncertainties involved. Sustainable global forest biomass potential is estimated at about 50 EJ in 2050, without notable further potential by 2100.

Carbon capture and storage (CCS) technologies are among the key options enabling low-emission energy production and industrial processes. The data for CCS technology in power production and industry are mainly based on the ZEP and IEAGHG studies (ZEP 2011; IEAGHG 2011) and the IEA technology roadmaps (IEA 2011, 2013, 2015). The estimates for the regional potential of CO_2 storage capacity is based on the ESTAP TIAM model database and our earlier work (Koljonen et al. 2009).

For achieving the well-below 2 °C trajectory for radiative forcing, negative emission technologies (NETs) seem necessary, due to the exhausting remaining carbon budgets. There are four main NET options: bioenergy with carbon capture and storage (BECCS), afforestation, direct air capture of CO_2 from ambient air (DAC), and enhanced weathering of minerals (Smith et al. 2016). For the BECCS technologies, we mainly use the IEA GHG projections for power plants and studies by Hannula for various biofuel conversion processes (Hannula 2015, 2016). For afforesta-tion, we continue using the potentials estimated in the EMF-21 study (de la Chesnaye and Weyant 2006), as in the ETSAP TIAM. In the 1.5C-PFE scenario we include the DAC technology, which has the advantage of having a small direct land footprint. Our base data for DAC are from Schmidt et al. (2016), but the future development is roughly based on Breyer et al. (2017), with cost reductions of 2%/a during 2015–2060. Enhanced weathering to remove CO_2 from the atmosphere was not considered due to high material and land-use requirements.

Methane and methanol are the key feedstocks for the C1 chemistry in the 1.5C-PFE scenario. The data on methanation and methanol synthesis from H_2 and CO_2, as well as the further upgrading processes, are based on Breyer et al. (2017)

and Hannula (2015, 2016). Both methanol and methane can also be utilized in SOFC fuel cells for power and heat production, the high fuel exhaust CO_2 concentrations facilitating energy efficient CO_2 capture at relatively low additional cost. We assumed methanol SOFCs with capture having 15% higher costs and 10% lower efficiencies than systems without capture.

4 Basic Results from the Scenario Variants

4.1 Emission and Forcing Trajectories

Many integrated assessment models have been reported to produce extremely high carbon price trajectories when imposing tight mitigation policies such as the RCP2.6 scenario. In the IPCC AR5 WG III Scenario Database, carbon prices in 430–480 ppm scenarios rise to 100–6000 USD/tCO$_2$e by 2100, with a median of 1500 USD/tCO$_2$e. In some models, even a theoretical backstop technology is needed for achieving the targets. For example, the TIAM-Grantham and MESSAGE-GLOBIOM models have recently been reported to produce CO_2 prices rising to 7000–10000 USD(2005)/tCO$_2$ by 2100, in a 2 °C scenario with global mitigation action delayed until 2030 (Gambhir et al. 2017).

The results from our scenarios show more moderate price trajectories, as illustrated in Fig. 2. In the 1.5C-LCE scenario the CO_2 price rises to about 110 €/t in 2050 and to 470 €/t in 2100. Correspondingly, in the 1.5C-PFE scenario the price rises to about 100 €/t in 2050 and to 220 €/t by 2100. No backstop technologies were needed in our scenarios, and one may also note that on the basis of our analysis, the DAC technology would set an upper limit for the CO_2 price to at most about 500 €/tonne even when using much more conservative estimates about the technology learning rates.

As a sensitivity analysis, we also analysed the forcing trajectories that would be achieved with a range of linearly increasing price trajectories in 2020–2100, shown in Fig. 2. The sensitivity analysis was carried out using otherwise the same assumptions as in the 1.5C-LCE scenario. The results from these scenarios indicate that the overshoot-and-decline trajectory assumed sufficient for achieving the well below 2 °C targets is roughly realized also by the €350 CO_2 price scenario (Fig. 3). In addition, we can conclude that the €450 price trajectory, peaking at about 2.9 W/m^2 in 2050 and reaching the level of 2.35 W/m^2 in 2100, would have a very high confidence of remaining well-below 2 °C, but also a much higher early cost burden.

According to the Baseline results, total GHG emissions would reach about in 60 GtCO$_2$e in 2050 and about 70 GtCO$_2$e in 2100, leading to a forcing of almost 6 W/m^2 in 2100 (Fig. 4). The Baseline scenario thus corresponds reasonably well to a scenario with a temperature increase of 4 °C. In our representative well below 2 °C scenarios, strong mitigation actions must start immediately from the year 2020, such that the net emissions decline almost linearly between 2020 and 2080,

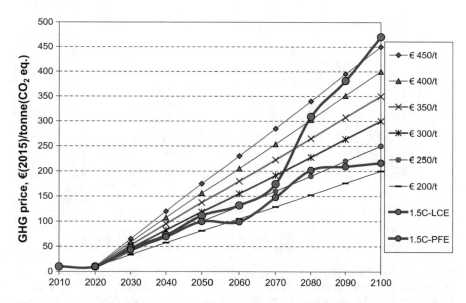

Fig. 2 Emission prices in the climate change mitigation scenarios considered. The prices in the 1.5C_LCE and 1.5C-PFE scenarios are model results, and the other cases are sensitivity analysis cases with exogenous price trajectories

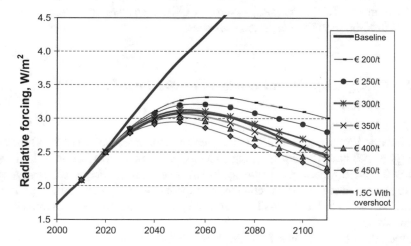

Fig. 3 Radiative forcing in the Baseline and the climate change mitigation scenarios. The forcing in the 1.5 °C overshoot scenario is constrained to peak at 3.1 W/m^2 around 2050–2060 and is then required to decrease to 2.6 W/m^2 by 2100

reaching negative CO_2 emissions of 11–13 Gt between 2080 and 2100, the zero-emission level being reached around 2070. In terms of all GHG emissions of the Kyoto gases, total emissions would have to reach zero by 2080 and negative by

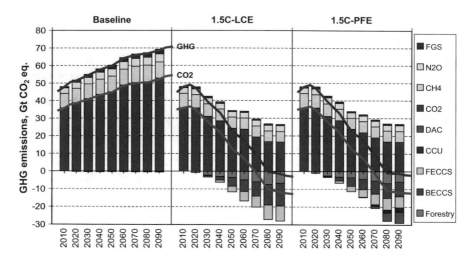

Fig. 4 Development of greenhouse gas emissions (Kyoto gases) in the main scenario variants. The red and blue lines represent the total net emissions of GHGs and CO_2, respectively, and the vertical bars show the gross emissions (positive) and removals (negative) either from flue gases or the atmosphere

2–3 Gt in 2100. The results clearly indicate the crucial importance of FECCS and BECCS in the 1.5C-LCE case, where BECCS alone would account for about 40–50% of the total CO_2 emission removals.

4.2 Energy Supply and Demand in the Low-Carbon Economy

For moving into a well below 2 °C low carbon economy, the use of fossil fuels for energy would have to start declining strongly, as soon as possible, in order to stay below the target trajectory for forcing. In terms of primary energy, bioenergy, solar energy, and energy system efficiency improvements would have the largest contri-butions to reducing fossil fuels on the global scale, followed by nuclear and wind power (Fig. 5).

According to the results from the 1.5C-LCE scenario, energy efficiency measures and bioenergy would bring the major contributions to lower GHG emissions from energy supply and use until 2050. The share of solar energy in total primary energy supply would increase from the 0.4% share in 2015 to 13% by 2050, and to almost 50% in 2100, while in the Baseline the shares are 9 and 27%. The impact of the well-below 2 °C target on solar energy expansion would thus be greatly pronounced after 2050.

However, in electricity generation the contribution of solar power is increasing much more prominently, from the modest 1% share in 2015, to about 42% in 2050

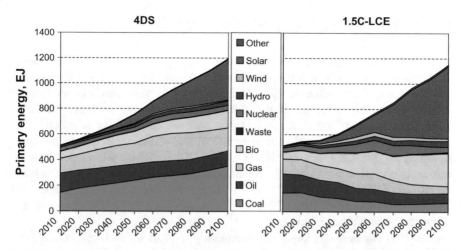

Fig. 5 Development of global primary energy consumption in the Baseline (4DS) and the 1.5C-LCE scenario. "Other" includes geothermal energy

and up to about 78% in 2100 in the 1.5C-LCE scenario, while in the Baseline only to 34% in 2050 and 67% in 2100. Consequently, through the decreasing costs of PV systems, solar energy becomes the dominant source of electricity already in the Baseline, which makes it more evident why solar does not gain a much higher share in primary energy. One should also note that we assume very low solar power capacity credits at the time of peak demand, from 1% in Finland up to 7% in close to equatorial regions. Therefore, the demand for peak capacity must almost fully be satisfied by other generation options or storage. Minimum stable load levels are also required for thermal power plants, thereby ensuring that the partial load levels of dispatchable generation plants remain realistic.

The contribution of wind and nuclear power to the global electricity supply both peak in 2060–2070 and then turn into decline. This can be attributed to the development of energy storage and power-to-fuel systems, which together with the low costs of PV systems (fixed-tilt systems around €300/kWp by 2080) further improve the competitive position of solar power.

Major uncertainties are, however, related to bioenergy supply in the longer term. The impact of global warming on biomass yields per hectare varies strongly by region, and can be positive in some regions while quite negative in many others. On the global scale, the impacts of climate change on yields are likely to be negative, unless the CO_2 fertilization effect is not taken into account, which might counterbalance the net impacts. In addition, one can expect an increasing demand of biomass for material use and for various chemicals, and introduction of stricter sustainability criteria, all having adverse impacts on biomass energy use in the long term. Therefore, the 1.5C-LCE scenario, where the reliance on bioenergy is quite high, includes substantial risks of failing to achieve the negative emissions by the presumed large-scale utilization of BECCS.

5 Outlook for the Post-fossil Economy

In the 1.5C-LCE scenario we have made our first attempt to model the transition to a post-fossil economy, where the use of fossil fuels both for energy and chemicals is much more radically reduced. We anticipate that this transition will most likely require large scale utilizing of direct air capture technology. There are many technology concepts for DAC, but in our analysis, we assume a generic CO_2 scrubbing technology with the future technical and economic performance developing along the estimates by Breyer et al. (2017). While the thermodynamic minimum energy require-ments for capture and compression have been estimated at about 0.66 GJ/tCO_2 (APS 2011), our generic scrubbing technology reaches the specific energy consumption of 5 GJ/tCO_2 in 2080–2100, i.e. remaining over 7 times the thermodynamic minimum.

The CO_2 captured via DAC is subsequently utilized for methane or methanol synthesis, which can be directly used as fuels or processed further e.g. into other hydro-carbon fuels (gasoline, diesel oil) or used as feedstock in the chemical industry, eventually replacing all mineral oil by synthetic hydrocarbons. In addition, carbon captured from flue gases can also be used for methanation or methanol synthesis, representing then carbon capture and use option (CCU).

The overall development of the global GHG emissions 1.5C-PFE scenario does not differ much from the 1.5C-LCE scenario, as was already illustrated in Fig. 4. However, the contributions to negative emissions are, indeed, significantly different (Fig. 6). While the gross emissions remain at a similar level in both scenarios, 26–27 GtCO_2 eq.) in 2080–2100, the 1.5C-PFE scenario features a much wider range of technology options for CO_2 emission removal (CCS, CCU, DAC and afforestation), resulting in the total net GHG emissions falling below zero around 2080, and CO_2 emissions already by 2070.

The contribution of DAC to negative emissions becomes quite significant at the end of the horizon, and the role of CCU is also notable. Together, these additional options would thus contribute to negative emissions of almost 10 GtCO_2, and thereby the need for BECCS and FECCS remains at a considerably lower level. While in the 1.5C-LCE scenario the total amount of CCS was about 23 GtCO_2 in 2100, in the post-fossil economy case it is reduced to 14 GtCO_2, which makes a significant reduction both in the need for CO_2 storage capacity and the uncertainties related to long-term CO_2 storage.

At first sight, the overall primary energy consumption does not look very different between the 1.5C-LCE and 1.5C-PFE cases (Fig. 7). In both cases solar energy becomes the dominant primary energy source after 2050. However, important differences appear in the use of mineral oil for energy, which in the post-fossil case ends around 2090, and in the total use of fossil fuels, which decreases from the present levels by over 80% by 2100. Furthermore, an additional significant difference is in the growth in the demand for biomass for energy, which is substantially smaller in the 1.5C-PFE case, in particular concerning dedicated energy crops. Bearing in mind the large uncertainties related to sustainable potential

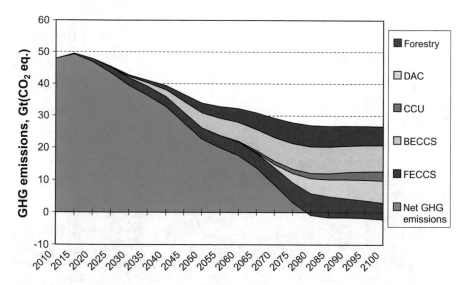

Fig. 6 Development of global greenhouse gas emissions in the 1.5-PFE scenario

of global biomass supply, and the increasing needs for substituting renewable and recycled materials for non-renewable minerals, the reduced bioenergy requirements are in line with the visions for a post-fossil economy.

The production of the synthetic fuels is based on producing hydrogen, mainly with solar energy. The post-fossil economy depicted by our 1.5C-PFE scenario is thus also largely a hydrogen economy, where we have assumed favourable technology develop-ment for both one-way and reversible fuel cell technologies and hydrogen storage, as well as methanol fuel cells and electricity storage. With the seasonal storage systems for hydrogen and synthetic fuels, the seasonal variation in solar energy can in fact be better counter-balanced in the 1.5C-PFE scenario, as compared to the 1.5C-LCE scenario, while other technologies were can be assumed to provide sufficient short-term electric storage.

As mentioned earlier, the further decoupling between economic activity and demand for energy services may be relatively conservative in our scenarios. Nonetheless, in the well-below 2 °C scenario, total final energy demand in 2100 is nearly 20% less than in the Baseline, and shifts remarkably to electricity throughout the century, within all sectors, as illustrated in Fig. 8. Traditional biomass use is gradually replaced by modern solid, liquid and gaseous biofuels. Oil still remains in the post-fossil scenario, but is synthetic oil, and it is worth noting that also non-energy uses of fossil fuels (~ 100 EJ in 2100) are completely replaced by synfuels in this scenario.

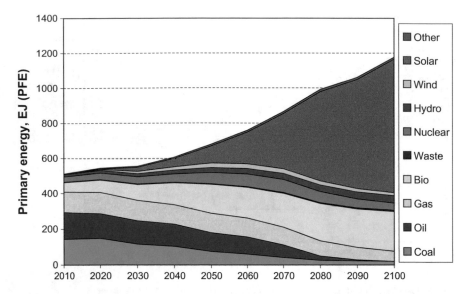

Fig. 7 Development of global primary energy consumption in the 1.5 °C-PFE scenario for the post-fossil economy. "Other" includes geothermal energy

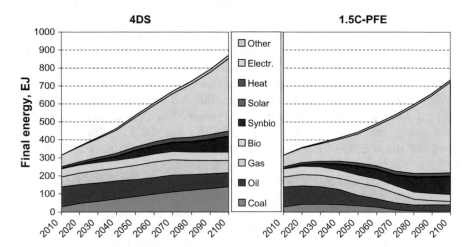

Fig. 8 Global final energy consumption in the Baseline and the 1.5 °C-PFE scenario for the post-fossil economy (excluding non-energy use). "Other" includes e.g. district cooling

6 Conclusion

Our scenario assessments show that a global well-below 2 °C climate trajectory is achievable, if strong mitigation actions are implemented immediately, with a deep decline in fossil fuel use and almost linear decline of net GHG emissions reaching negative CO_2 emissions by 2070. Negative emission technologies thus seem ever more necessary, due to the continuously shrinking remaining carbon budget. There are several technical alternatives for removing carbon from the atmosphere, such as afforestation and bio-CCS. However, mitigation can be remarkably enhanced by employing so-called C1 chemistry, where synthetic hydrocarbons are produced from water and CO_2 captured from air. Hydrogen production would be effectively integrated with variable renewable energy, and the synthetic fuels may offer large-scale seasonal energy storage.

The scenario depicted for a post-fossil economy highlights several advantages over the more conventional low-carbon alternative: it substantially reduces the uncertainties related to large-scale and long-term CO_2 storage systems that are required for CCS, it also considerably reduces the need for expanding the use of biomass for energy, it improves the feasibility of integrating very large amounts solar power in energy systems through the flexibility provided by power-to-fuel systems, and it paves the way to a truly post-fossil economy by phasing out fossil fuels almost completely.

Limited biomass resources are to be produced sustainably and used efficiently, primarily for producing industrial bio-material products and to a lesser extent energy and transport fuels. Future challenges with land use emissions indicate that higher levels of using biomass for energy may increase the uncertainties in mitigation. In some regions, CO_2 storage capacity may also become a limiting factor, which, in fact, occurred in the 1.5C-LCE scenario. However, the global storage potentials were assumed relatively low in our scenarios (about 2500 Gt).

Finally, if technologies related to direct CO_2 capture from air develop sufficiently favourably, the marginal and overall costs of climate change mitigation may be significantly reduced. With direct CO_2 capture from air, CO_2 would also be readily available at the point of manufacture. However, one should recognize that CO_2 is a difficult molecule to work with for the same reasons that it accumulates in the atmosphere: it is not reactive. In the chemicals industry, finding suitable catalysts for utilizing carbon dioxide has been a challenge. Hence, considerable further work is required for in-depth analysis of the various technologies needed in the transition to a post-fossil economy.

References

APS (2011) Direct air capture of CO_2 with chemicals. A Technology Assessment for the APS Panel on Public Affairs, American Physical Society

Breyer, C et al (2017) On the role of solar photovoltaics in global energy transition scenarios. Prog Photovolt: Res Appl 25:727–745. Model data available at https://www.researchgate.net/publication/316883121_100_Renewables_Scenarios_Model_Data_and_Results

Brynolf S et al (2017) Electrofuels for the transport sector: a review of production costs. Renew Sustain Energy Rev (in press)

de la Chesnaye FC, Weyant JP (2006) Multigas mitigation and climate policy. Energy J Spec Issue on Multi-Greenhouse Gas Mitig Clim Policy

Friedlingstein P et al (2014) Persistent growth of CO_2 emissions and implications for reaching climate targets. Nat Geosci 7:709–715

Gambhir et al (2017) Assessing the feasibility of global long-term mitigation scenarios. Energies 10(1):89

Hannula I (2015) Co-production of synthetic fuels and district heat from biomass residues, carbon dioxide and electricity: Performance and cost analysis. Biomass Bioenerg 74:26–46

Hannula I (2016) Hydrogen enhancement potential of synthetic biofuels manufacture in the European context: a techno-economic assessment. Energy 104:199–212

Houghton RA et al (2012) Chapter G2 carbon emissions from land use and land-cover change. Biogeosciences 9:5125–5142

Houghton RA, Nassikas AA (2017) Global and regional fluxes of carbon from land use and land cover change 1850–2015. Global Biogeochem Cycles 31:456–472

IEA (2011) Technology roadmap: carbon capture and storage in industrial applications. International Energy Agency, Paris

IEA (2013) Technology roadmap: carbon capture and storage. International Energy Agency, Paris

IEA (2015) Technology roadmap: hydrogen and fuel cells. International Energy Agency, Paris

IEA (2016) Energy technology perspectives 2016. International Energy Agency, Paris

IEAGHG (2011) Potential for biomass and carbon dioxide capture and storage. IEAGHG. Report 2011/6. http://ieaghg.org/docs/General_Docs/Reports/2011-06.pdf

IPCC (2012) Renewable energy sources and climate change mitigation: special report of the Intergovernmental Panel on Climate Change (ISBN 978-1-107-02340-6)

IPCC (2013) Annex II: climate system scenario tables. In: Climate Change 2013: the physical science basis. Contribution of working group i to the fifth assessment report of the Inter-governmental Panel on Climate Change. Cambridge University Press, Cambridge and New York

Kallio M, Lehtilä A, Koljonen T, Solberg B (2015) Best scenarios for the forest end energy sectors—implications for the biomass market. Cleen Oy. Research report no D 1.2.1. https://www.researchgate.net/publication/284414046_Best_scenarios_for_forest_and_energy_sectors_-_implications_for_the_biomass_market

Koljonen T et al (2009) The role of CCS and renewables in tackling climate change. Energy Procedia 1:4323–4330

Koljonen T, Lehtilä A (2015) Modelling pathways to a low carbon economy for Finland. In: Giannakidis G et al (eds) Informing energy and climate policies using energy systems models, vol 30. Lecture Notes in Energy. Springer, Cham

Köning DH et al (2015) Simulation and evaluation of a process concept for the generation of synthetic fuel from CO_2 and H_2. Energy 91:833–841

Loulou R (2008) ETSAP-TIAM: the TIMES integrated assessment model. Part II: mathematical formulation. CMS 5(1–2):41–66

Loulou R, Labriet M (2008) ETSAP-TIAM: the TIMES integrated assessment model. Part I: model structure. CMS 5(1–2):7–40

Loulou R, Remme U, Kanudia A, Lehtilä A, Goldstein G (2016) Documentation for the TIMES model, energy technology systems analysis programme (ETSAP). http://iea-etsap.org/docs/Documentation_for_the_TIMES_Model-Part-I_July-2016.pdf

Millar RJ, Nicholls ZR, Friedlingstein P, Allen MR (2017) A modified impulse-response representation of the global near-surface air temperature and atmospheric concentration response to carbon dioxide emissions. Atmos Chem Phys 17:7213–7228

Myhre G et al (2013) Anthropogenic and natural radiative forcing—supplementary material. In: Climate change 2013: the physical science basis. Contribution of working group i to the fifth assessment report of the intergovernmental panel on climate change

OECD (2014) Economic outlook No 95—May 2014—Long-term baseline projections. http://stats.oecd.org/Index.aspx?DataSetCode=EO95_LTB

Schemme S et al (2017) Power-to-fuel a key to sustainable transport systems—an analysis of diesel fuels produced from CO_2 and renewable electricity. Fuel 205:198–221

Schmidt, PR et al (2016) Renewables in transport 2050. Empowering a sustainable mobility future with zero emission fuels from renewable electricity. FVV, Frankfurt am Main, Report 1086

Smith P et al (2016) Biophysical and economic limits to negative CO_2 emissions. Nature Climate Change 6(1):42–50

UNFCCC (2015) Synthesis report on the aggregate effect of the intended nationally determined contributions. Technical annex—synthesis report on the aggregate effect of the intended nationally determined contributions. In: United nations framework convention on climate change. http://unfccc.int/focus/indc_portal/items/9240.php

Van Vuuren DP et al (2011) RCP2.6: exploring the possibility to keep global mean temperature change below 2 °C. Clim Change 109:95–116

Weyant J et al (2009) Report of 2.6 versus 2.9 Watts/m2 RCPP evaluation panel. Inter-govern-mental panel on climate change. http://www.ipcc.ch/meetings/session30/inf6.pdf

ZEP (2011) The costs of CO_2 capture. Post-demonstration CCS in the EU. European technology platform for zero emission fossil fuel power plants. http://www.zeroemissionsplatform.eu/library/publication/166-zep-cost-report-capture.html

How Low Can We Go? The Implications of Delayed Ratcheting and Negative Emissions Technologies on Achieving Well Below 2 °C

Matthew Winning, Steve Pye, James Glynn, Daniel Scamman
and Daniel Welsby

Key messages

- Delaying action will require substantially increased emissions reductions efforts between 2030 and 2050. In particular, achieving targets moving towards 1.5 °C require immediate action regardless of technology or resource assumptions.
- Development of significant levels of biomass are required in all instances as BECCS deployment is essential even for a 2 °C target.
- If policy makers want to achieve targets towards 1.5 °C (or even 2 °C with delayed ratcheting) then consideration of the development of NETs is necessary. However, there are risks of lock-in to a high fossil future if NETs fail to scale, the implication of which must be fully understood.
- TIMES modelling applied in this chapter allows for the exploration of levels of negative emissions technologies not yet known.

M. Winning (✉) · D. Welsby
UCL Institute for Sustainable Resources, London, UK
e-mail: m.winning@ucl.ac.uk

D. Welsby
e-mail: daniel.welsby.14@ucl.ac.uk

S. Pye · D. Scamman
UCL Energy Institute, London, UK
e-mail: s.pye@ucl.ac.uk

D. Scamman
e-mail: d.scamman@ucl.ac.uk

J. Glynn
MaREI Centre, Environmental Research Institute, University College
Cork, Cork, Ireland
e-mail: james.glynn@ucc.ie

© Springer International Publishing AG, part of Springer Nature 2018
G. Giannakidis et al. (eds.), *Limiting Global Warming to Well Below 2 °C: Energy
System Modelling and Policy Development*, Lecture Notes in Energy 64,
https://doi.org/10.1007/978-3-319-74424-7_4

1 Introduction

The current short-term pledges proposed by countries in the Paris Agreement on Climate Change are an important first step to hold the increase in the global average temperature to well below 2 °C above pre-industrial levels and pursuing efforts to limit the temperature increase to 1.5 °C. They are, however, insufficient to achieve this long-term goal (Hof et al. 2017). Increased ambition (or ratcheting as it is referred to in the Paris Agreement), occurring sooner rather than later, will be crucial in determining whether the ambitious temperature targets, to stay well below 2 °C and move towards 1.5 °C by 2100, are even possible.

In this chapter, we explore the consequences of a delay until 2030 in the implementation of more ambitious pledges towards the long-term goals of a well-below 2 °C target and a stricter target towards 1.5 °C. We also assess how much additional effort would be required depending upon when, and if, international effort is increased. A special focus is on global biomass potential given the key role of bioenergy in the feasibility of emission reductions. To undertake the analysis, we utilise the TIAM-UCL model.

2 Methods

2.1 TIAM-UCL Model

TIAM-UCL (TIMES Integrated Assessment Model) is a global energy systems model developed at University College London based on the TIMES framework and is developed from an early version of the ETSAP-TIAM model (Loulou and Labriet 2007). TIAM-UCL is a partial equilibrium model which uses linear programming and cost-minimisation to calculate the optimal energy system for a chosen set of energy service demands and technology constraints and is therefore often used for counterfactual policy analysis. The model is technology rich and as such represents all primary energy sources from extraction, through to refinement and conversion, and finally to end-use demand.

The model covers 16 global regions (UK is a separate region) and is calibrated to a base year of 2005, with consideration of the 2010 and 2015 energy balances, using International Energy Agency and UK statistical data with a time-frame which allows model runs out to 2100 in five year time steps. End use energy demands are determined exogenously using regional and sectoral drivers such as GDP, population, households etc. In this analysis, we use the Shared Socioeconomic Pathway (SSP2) projections for GDP and population which drive energy service demands (Riahi et al. 2017). Regions in TIAM-UCL are linked by trade in a number of commodities such as hard coal, crude oil, gas via pipelines, LNG, and a variety of petroleum products and energy crops.

A climate module is run in conjunction with TIAM-UCL; it is calibrated to a recent version of MAGICC (Meinshausen et al. 2011). In this analysis, we allow net negative emissions of 50 Gt CO_2 a year from 2050 onwards. While there is no known limit on negative emissions, in practice there will be some limitations of scaling. Therefore we assume high levels of net negative emissions are possible by allowing for around the same level as current positive emissions are at present. It is unlikely this constraint will ever become binding except in the most extreme of scenarios.

Recent uses of TIAM-UCL include the geographical consideration of unburnable fossil fuels (McGlade and Ekins 2015), fossil trade under climate constraints (Pye et al. 2016), and exploring uncertainty in energy modelling (Price and Keppo 2017). Further details of the model are available in the model documentation (Anandarajah et al. 2011).

2.2 Scenarios

Our scenarios are developed to allow for consideration of the impacts of delayed ratcheting of the NDC commitments, to ascertain how important timing is in achieving the long-term goals. In other words, how low can anthropogenic warming be limited if we delay more ambitious emission reductions and what are the additional changes required in the energy system when ratcheting is delayed?

Conducting further analysis with a "no climate policy" assumption is rather hypothetical since countries have started implementing their plans as proposed in their NDCs. Therefore, our baseline includes the greenhouse gas (GHG) commitments of the NDCs, considering the midpoint between conditional and unconditional commitments. These are for CO_2, N_2O and CH_4 across all sectors except LULUCF. Other specific commitments, such as renewable energy targets, are not considered. Where emissions reductions are against a hypothetical BAU and the BAU level is not provided, we calculate our own BAU using a regional average from TIAM-UCL. These assumptions allow us to have specific GHG targets for all TIAM-UCL regions during the NDC period to 2030.

Beyond 2030 the assumption in the NDC scenario is to maintain a constant ratio of emissions per GDP/capita. In other words, we consider that emissions do not simply revert to a no-policy case but nor do they keep reducing at the same rate as during the NDC period as in Vandyck et al. (2016). This can be seen as a vaguely pessimistic assumption in that no further climate policy is implemented beyond the NDCs but that the level of ambition does not decrease below that of the NDC commitments.

Alternative scenarios are defined as follows (Table 1). The "Below 2 °C" (B2D) and "Towards 1.5 °C" (T15) scenarios consider the emissions budget over 2015–2100 as proposed by IPCC (2014) and Rogelj (2016) for policy analysis. The B2D scenario corresponds to a 66% chance of reaching 2 °C (OECD/IEA 2017); in other words, this may simply reflect below 2 °C rather than "well below". Non-CO_2

Table 1 Scenario description

Scenario	Description
NDC	NDC commitments for each region until 2030 followed by no further increase in ambition
B2D	915 Gt CO_2e carbon budget (2015–2100) + RCP 2.6 + 2 °C temp limit. Fixed to NDC run until 2020. More ambitious reductions can start from 2020
B2D-2030	915 Gt CO_2e carbon budget (2015–2100) + RCP 2.6 + 2 °C temp limit. Fixed to NDC run until 2030. More ambitious reductions can start from 2030 only
B2D-2030-HB	B2D-2030 with increased biomass (From 150 to 300 EJ)
T15	400 Gt CO_2e carbon budget (2015–2100) + RCP 2.6 + 1.7 °C temp limit. Fixed to NDC run until 2020. More ambitious reductions can start from 2020
T15-2030	400 Gt CO_2e carbon budget (2015–2100) + RCP 2.6 + 1.7 °C temp limit. Fixed to NDC run until 2030. More ambitious reductions can start from 2030 only
T15-2030-HB	T15-2030 with increased biomass

GHGs are limited to the Representative Concentration Pathway corresponding to RCP2.6 (van Vuuren et al. 2011). We also constrain the climate module to upper-temperature limits of 2 and 1.7 °C in 2100 for the B2D and T15 scenarios, respectively. The CO_2 and non-CO_2 emissions budgets are set separately in the model because TIMES, being an energy systems model, is stronger at analyzing CO_2 specifically, than it is at other GHGs. Therefore the addition of a temperature limit ensures the Paris target is met.

Sensitivity analyses are proposed with higher global biomass availability applied to T15 in order to show the impact that resource availability and resultant technology assumptions can have upon model results and feasibility. The biomass, all of which is considered sustainable, can be used in the system directly or combined with CCS to allow for negative emissions. There are a significant range of estimations for biomass availability in the literature (Resch et al. 2008; Cho 2010; Tomabechi 2010). Dornburg et al. (2010) estimate that a range of 200–500 EJ is likely once considerations of food security, biodiversity, water availability etc. are internalised. However, Smith et al. (2014) show that there is low agreement on availability being over 100 EJ p.a. Therefore, in line with other more conservative assumptions of TIAM-UCL, the central biomass availability assumption in our analysis is 150 EJ per year from 2050 onwards; the amount reaches 300 EJ per year in the high availability scenario.

2.3   Negative Emission Technologies (NETs)

The development of CCS appears to be crucial in achieving deep long-term emissions reductions. Without CCS, achieving the Paris goals is extremely costly at best and at worst infeasible (Hughes et al. 2017). In particular a considerable majority of 2 °C scenarios runs (104 of 116) in the IPCC AR5 database utilise carbon capture and storage in conjunction with bioenergy (BECCS) as a means of carbon dioxide removal (CDR) (Clarke et al. 2014). BECCS results in net negative emissions since residual carbon dioxide emissions is considerably less than the carbon dioxide captured in the biomass growth phase (Smith et al. 2016). Many Integrated Assessment Model (IAM) scenarios rely on NETs to cumulatively capture levels of carbon dioxide that are in the same order of magnitude of the remaining 2 °C carbon budget (Anderson and Peters 2016). However, negative emissions technologies (NETs) are a contentious topic of research (Jackson et al. 2017). As a result of a lack of demonstration of these NETs at scale to date, a call for an open discussion for the policy implications of reliance on NETs to achieve temperature stabilisation in IAMs is ongoing (van Vuuren et al. 2017). While BECCS is one such NET, other NETs include direct air capture (DAC), enhanced weathering (EW), deep ocean direct injection, subsurface mineralisation, alongside peatland rewetting, biochar, afforestation and reforestation (Tavoni and Socolow 2013).

TIAM-UCL has an upper limit of primary supply of bioenergy of 150–300EJ per year in line with other IAMs (Bauer et al. 2017). TIAM-UCL has detailed technology specifications for a range of BECCS across power generation, industry CHP and process heat, as well as upstream biofuels transformation.

Since these do not capture the full spectrum of potential future NETs, and in the absence of a certain set of nascent future NETs, a CO_2 backstop technology, which captures CO_2 at a high mitigation cost of \$5000/tCO_2 and sequesters CO_2 into the available geological storage space in TIAM-UCL, is included in the model to explore marginal mitigation requirements to meet the stringent climate goals. The utilisation of a backstop technology highlights the difficulty in meeting the Paris agreement goals with available mitigation options and suggests that these goals are near the limits of technological mitigation within our current understanding of mitigation technologies, expectations of future technological learning, and assumptions of future energy service demands responses to energy prices. The potential further NET options listed, such as DAC and EW, are estimated to be available at marginal abatement costs considerably lower than \$5000/tCO_2 and at mitigation volumes between 3 and 12GtCO_2/year (Fuss et al. 2016) i.e. considerably less than the CO_2 backstop technology used in the final period of scenarios in TIAM-UCL. However, research, development and demonstration at scale is required to narrow the range of uncertainty and systemic impacts on electricity and heat requirements, investment and operation costs, land use changes and food competition to appropriately specify more developmental NETs in TIAM-UCL.

3   Results

3.1   The Need for Negative Emissions

The aggregate global NDC CO_2 emissions in 2030 are 40.1 Gt CO_2 which represents an increase of 11% from 2010 levels. For those scenarios that delay ratcheting
until 2030, this is the emissions level in 2030 where increased ambition beyond
NDCs can begin. Significant emissions reductions are required in all scenarios in
order to achieve the long-term Paris goals compared to a NDC baseline beyond
2030 (Fig. 1). Overall net CO_2 emissions are around 42% (39%) lower in the T15
(B2D) scenario when compared against the NDC run in 2030. During the 2020 to
2030 period, the NDC net emissions increase by 0.2% a year (as do all the delayed
scenarios) whereas B2D and T15 reduce at 4.6 and 5.2% a year, respectively. In
particular, there is always a sharp drop in emissions in the period immediately after
the policy is tightened i.e. between 2020 and 2025 for the B2D and T15 scenarios
and 2030 to 2035 for the other scenarios. It is debatable how realistic these quick
drops in global emissions are in practice given lock-in of existing capacity but it
provides an important insight into how critical near term emission reduction are.

In 2030 the majority of emissions savings occur in the electricity sector—
electricity CO_2 emissions in the T15 (B2D) scenario are 65% (64%) lower than they
are in the NDC scenario while transport, industry and buildings sectors are 43% (35%),
33% (31%) and 22% (21%) lower, respectively. Furthermore the biggest cut in
emissions in moving from B2D to the T15 target occurs in the transport sector showing
that, in the near-term, higher and earlier reductions would be needed in the transport
sector when attempting to achieve a more stringent long-term emissions goal i.e. going
from well below 2 °C to towards 1.5 °C, and are a cheaper means of achieving global
reductions compared to the extra effort in the electricity or other sectors.

Fig. 1 Global CO_2
emissions, 2010–2080
(Gt CO_2)

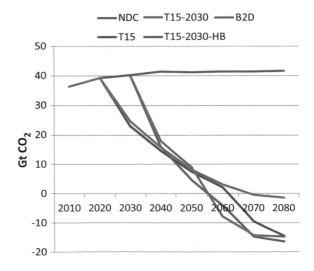

In terms of the effort required after 2030, the T15 (B2D) scenario requires yearly net CO_2 emissions reductions of around 5.5% (5.4%) between the years of 2030 and 2050, whereas an extra 10 years of delayed action mean the yearly reductions in the same period rise to 7.3% p.a. (6.8% p.a.). The high biomass scenario T15-2030-HBio (B2D-2030-HBio) shows that deeper emissions reductions are possible between 2030 and 2050 as CO_2 reductions in the region of 10.2% (6.8%) per year are possible due to greater the ability to employ larger amounts of biomass and BECCS throughout the model run.

All B2D and T15 emissions reduction scenarios have net CO_2 emissions in 2050 lower than 10 Gt CO_2. Indeed, net emissions become negative between 2060 and 2070 in all scenarios, which is consistent with the Paris Agreement's goal of "achieving net zero emissions by the second half of this century," unlike the NDC baseline. Net negative emissions are then achieved by 2060 for T15-2030 and T15-2030-HB, where the most stringent emissions reductions are required in the shortest time-frame. Cumulatively, these net negative emissions over the time horizon amount to 437 Gt for T15 and around 580 Gt in both the delayed 1.5 °C scenarios. In all the well-below 2 °C scenarios there are low levels of net negative emissions which occur from 2070 and beyond with only 17 cumulative Gt CO_2 below zero in the B2D instance and around 230 Gt in the other two 2 °C delayed scenarios. The technologies which allow for net negative emissions are discussed further in Sect. 3.2.

3.2   CCS, BECCS and Other NETs

By the middle of the century, both fossil CCS and BECCS will play a crucial role in decarbonisation efforts regardless of stringency of emissions reductions (Fig. 2). In 2050, CO_2 capture in electricity and upstream sectors become important in all reduction scenarios, capturing around 50% of these sectors' emissions, while around a fifth of industry CO_2 emissions are captured. Importantly, as a negative emissions technology, BECCS not only sequesters carbon emissions from bio-energy when it is combusted but gains negative credit based on the assumption that the combusted emissions would have been sequestered in the biosphere, in trees/plants etc. (Smith et al. 2014). All of the non-NDC scenarios utilize BECCS in 2050 to the order of 6–8 Gt CO_2. However, there are significant uncertainties in the availability and scaling of these technologies which suggests that performing a sensitivity analysis around them is crucial to model results (Anderson and Peters 2016).

An important finding is that NETs, not yet specified, at a high marginal abatement cost ($5000 per tCO_2), are being utilised in all scenarios except NDC and B2D-2030-HB (Table 2). These NETs are deployed as a last resort given its extremely high cost, and reflects the absence of any known alternative mitigation opportunity in the model or possibly the fact that known technologies cannot be deployed at a sufficiently high rate due to constraints imposed.

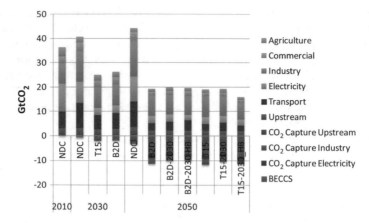

Fig. 2 CO_2 emissions by sector in 2030 and 2050 (Mt CO_2)

Table 2 Cumulative emissions from BECCS and other NETs compared against carbon budget

Scenario	Carbon Budget (Gt CO_2) 2015–2100	BECCS (Gt CO_2) 2015–2100	% BECCS versus carbon budget	Other NETs (Gt CO_2)	% Other NETs versus carbon budget	Total NETs (Gt CO_2)	% Total versus carbon budget
T15	400	543	136	415	104	958	213
T15-2030-HB	400	792	198	279	70	1072	238
T15-2030	400	496	124	674	168	1170	260
B2D	915	531	58	11	1	541	60
B2D-2030-HB	915	784	86	0	0	784	87
B2D-2030	915	509	56	257	28	766	85
NDC	N/A	134	N/A	0	–	134	N/A

For the 2 °C scenarios, other NETs are not needed when biomass availability assumptions are higher than our standard assumption. In fact it appears the 150 EJ p.a. biomass availability assumption is for the point at which BECCS is the only negative emissions technology required i.e. if assumption is any lower than around 150 EJ then Other NETs are required in the system. This finding is apparent as there is only around 10 Gt of cumulative CO_2 emissions captured by Other NETs in B2D, which for context is about 1% of the total B2D carbon budget, and therefore a small increase in the biomass assumption would remove the need for any Other NETs. In fact, there are no Other NETs required in the B2D-2030-HB scenario at all, against 257 Gt CO_2 required in the lower biomass scenario to achieve the target. These results show the importance of sensitivity analysis around biomass availability in modelling. For policymakers it reflects the criticality of communicating such results so these issues are not hidden, and the necessary efforts required on NETs research and future planning for adequate bioenergy resources.

The results indicate that moving to well below 2 °C is achievable with either no or low levels of Other NETs as long as there is no delay in ratcheting. When biomass availability is higher, some delay is even possible. However, substantial amounts of Other NETs are required in all the scenarios which are aiming towards 1.5 °C. In 2070 they are required for the T15 scenario and even earlier—2055 and 2060—for the T15-2030 and T15-2030-HB scenarios, respectively.

In the T15 scenario, the 'Other NETs' technology is used to mitigate some 415 Gt CO_2 towards the end of the time horizon. This means that the whole of the budget under the T15 scenario relies on a yet-to-be specified mitigation technology. T15-2030-HB has a lower yet still significant amount of Other NETs emissions at 279 GtCO_2. However, the T15-2030 has considerably larger deployment as 674 Gt CO_2 between 2055 and 2100. These can be seen as the amount of emissions that would have to be reduced elsewhere in the energy system in order to meet the stipulated emissions target. It must be stressed that considerable amounts of investment in R&D will be required to ensure these NETs become available in the short-term and in order to reduce costs, or will need to see much more rapid action in the near term on fossil fuel phase out. Therefore if policymakers are serious about moving towards 1.5 °C then action must occur without delay in order to make sure technologies are available and for costs to be reasonable.

3.3 Some Fossil Fuels Survive Due to CCS and NETs

Given the emissions outlook presented above, and the role of NETs, it is insightful to explore the use of fossil fuels in future years. What is clear is that their continued use is permitted due to the availability of CCS, and the offsets provided by negative emissions generated by BECCS. This is an important issue to remember when interpreting the results below. For example, B2D has a carbon budget of 915 GtCO_2, which is increased by 60% through NETs; without NETs, emissions from fossil fuels would need to be 60% lower, in other words, not consumed. Table 3 shows the reductions in consumption for coal, oil and natural gas, relative to the NDC baseline.

3.3.1 Coal

Due to its high carbon intensity, coal sees the largest reductions of all fuels across all scenarios, relative to the NDC baseline. In the NDC baseline global coal consumption remains relatively stable around 150 EJ until 2035, before increasing linearly to around 250 EJ by 2080; regionally this is due to a phase out of coal in the developed economies of Western Europe, the UK, Canada, and the US, whilst allowing India and China more time to transition (i.e. consumption in India and China offsets the reductions made by the developed economies).

Table 3 Percentage change in global fossil fuel consumption for each scenario relative to the NDC baseline

Scenario	2030			2050			2080		
	Coal (%)	Oil (%)	Gas (%)	Coal (%)	Oil (%)	Gas (%)	Coal (%)	Oil (%)	Gas (%)
B2D	−69	−17	−12	−83	−44	−33	−78	−56	−33
B2D-2030	–	–	–	−82	−41	−42	−73	−59	−42
B2D-2030-HB	–	–	–	−79	−36	−30	−75	−53	−30
T15	−73	−29	−19	−84	−47	−41	−81	−59	−52
T15-2030	–	–	–	−81	−46	−44	−74	−56	−44
T15-2030-HB	–	–	–	−85	−55	−55	−87	−60	−68

Reminder: In B2D-2030 and T15-2030 scenarios, the results are fixed to the NDC baseline until 2030 so there is no difference in consumption up to 2030

When the NDC commitments move to below 2 °C or towards 1.5 °C, huge reductions in coal consumption are required, particularly in India and China, where unprecedented phasing out of coal is required to meet carbon reduction targets. In other words, coal does not survive even with CCS and NETs available in the scenarios.

Of particular interest in India is the implication of delaying the ratcheting up of their NDC commitments to 2030, with coal consumption increasing between 2020 and 2030, and thus requiring an even greater reduction from 2030 (in both of the B2D-2030 and T15-2030 scenarios); in India as in other developing regions, the flexibility of the energy system, and in particular power generation, to adapt to this is absolutely fundamental, and the policy implications of this study encourages action now in order to facilitate consumption 'smoothing', as coal is phased out (i.e. to avoid a shock reduction in demand in 2030).

The importance of a global effort to substantially reduce coal as soon as possible can be seen in the B2D and T15 scenarios where the energy system turns rapidly towards electrification; in the first ten years after more stringent carbon reduction targets are introduced above the NDC baseline, delaying action to 2030 requires the installation of around 30% more generation capacity in all B2D and T15 scenarios, when compared to electricity capacity investments when carbon budgets are ratcheted up in 2020. India and China account for between 56 and 67% of total remaining global coal consumption by 2050 in the B2D and T15 scenarios, predominantly as an input to the industrial and electricity generation sectors. Additionally, by 2050, at least 70% of electricity generation from coal relies on CCS capabilities (for reference, across the B2D and T15 scenarios electricity generation from coal accounts for less than 2% of total electricity generation).

3.3.2 Gas

The highest level of gas consumption, when aggregated globally, is in the NDC baseline scenario, reaching just over 200 EJ by 2050. In short, the more stringent carbon budgets imposed in the B2D and T15 scenarios means that global gas consumption is consistently below the NDC baseline across the entire modelling time horizon. However, the relative reduction in global gas consumption below the NDC baseline is far less than the reductions required for coal consumption. Additionally, delaying action to 2030 requires more rapid reductions in natural gas consumption across all of the B2D and T15 scenarios.

Globally, natural gas consumption in the electricity generation sector remains relatively stable across all the B2D and T15 scenarios with medium biomass availability until 2045, relying on CCS for around 40% of this generation, before declining sharply from 2045 to 50. Of particular interest for natural gas consumption in the power generation sector is the impact of higher biomass availability; in both the B2D-2030 and T15-2030 scenarios with high biomass availability, electricity generation from natural gas falls rapidly from 2030, before stabilizing from 2060. Additionally gas consumption in the residential sector reduces as energy service demand (e.g. residential heating, cooling, and cooking) is increasingly satisfied by electricity.

The industrial sector (chemicals, iron and steel, paper and pulp, non-ferrous metal production, etc.) maintains a relatively steady level of gas consumption in all the B2D and T15 scenarios (between 50-65 EJ out to 2080). Focusing on the industrial sector, three key issues are as follows:

Firstly, the relatively stable level of gross natural gas consumption in the industrial sector is predominantly due to consistently large, and relatively stable, consumption by the chemical and petrochemical sub-sectors.

Secondly, global consumption of natural gas in heavy industries (i.e. iron and steel production) is higher in the B2D and T15 scenarios than in the NDC baseline for all years post-2030, and experiences growth in all three T15 scenarios until 2045, as coking coal/other coal (generally lower energy content) inputs into the production process are replaced in favor of natural gas.

Thirdly, the regional variation in industrial natural gas consumption demand depends to a large extent on future economic restructuring and exogenous drivers, which will have significant implications for global natural gas trading patterns and individual regions/countries emissions profiles (i.e. where will heavy industry and chemical/petrochemical production be situated).

3.3.3 Oil and the Transport Sector

In 2015, the transportation sector accounted for around 64% of global oil consumption (IEA 2017). As decarbonisation policies are ratcheted up, oil is more rapidly substituted in the transportation sector by natural gas, hydrogen, and electricity. However, significant uncertainties surround this transition, particularly for hydrogen and electric vehicles, which require large infrastructural investments including for production, distribution and refueling (Morton et al. 2014; Scamman 2017). In all of the B2D and T15 scenarios, oil consumption in the transportation sector is significantly reduced by 2050 to between 30 and 45 EJ against 120 EJ in the NDC scenario. This requires a reduction of between 75 and 80% from 2015 levels. Oil consumption is highly sensitive to how quickly carbon budgets are ratcheted in ambition, as well as the availability of biomass in the T15 case (oil consumption is 10 EJ higher in the T15-2030 scenario in 2050, compared to the T15-2030 scenario with high biomass availability).

Transportation energy service demand in all of the B2D and T15 scenarios is below the NDC baseline, with an increased shift to pooling/public transportation modes, as well as efficiency improvements across the vehicle fleet. From an energy input vector perspective, natural gas, and in particularly liquefied natural gas combustion engines are utilized far less in the T15 scenarios, when compared to the B2D cases. Part of this natural gas is replaced by hydrogen, including fuel cell technologies, which become increasingly prevalent in the T15 scenarios, representing between 28 and 36% of transport sector energy consumption by 2050. Across all scenarios, electricity maintains a relatively stable level of gross input to satisfy transport energy demand by 2050, with between 15 and 20 EJ. It should be noted that much of the above analysis depends on the direction of policy support, which can assign 'winners' and 'losers' before the diffusion of a technology.

The insights above are therefore heavily dependent on the level of policy support, the level at which the externalities (i.e. emissions) of existing vehicles are taxed, and the availability and cost of different technologies across regions, including the required feedstock fuels.

4 Conclusion

This chapter uses the TIAM-UCL model to consider the effects on the global energy system of delaying the ratcheting of the NDCs when attempting to achieve the long-term goals of below 2 °C and towards 1.5 °C.

Ratcheting-up commitments without delay becomes critical to go beyond 2 °C towards 1.5 °C. The rate of emissions reductions required when effort is delayed to 2030 is significantly higher as well as the reliance on large-scale NETs deployment, despite its high cost. Delaying also requires larger levels of bioenergy resource to have any chance of meeting the target, due to its use in combination with CCS (BECCS) to generate negative emissions.

An availability of slightly above 150 EJ per year allows for the 2 °C target without delay to be met using BECCS but without any other NETs required. The delayed scenarios are highly sensitive to the availability of biomass. The 2 °C target with a high biomass assumption also does not require any other NETs beyond BECCS. However, when biomass is at its medium availability with delay then around 257 Gt CO_2 of NETs are required cumulatively, which is employed in combination with BECCS deployment meaning BECCS and NETs combined amount to 84% of the total carbon budget. This reliance on biomass availability implies significant policy implications, not least due to uncertainties over land-use and land-use changes.

Delay increases the reliance not only on BECCS but on other NETs that are not yet well known. Without those other NETs, the 1.5 °C target appears to be unachievable. It is critical that the levels of negative emissions are revealed to the policy community so that they are clear as to what their current strategies are premised on. It is also critical, as we have shown in this analysis, to reveal that for some levels of ambition our models do not solve for the most ambitious targets, unless we introduce proxy NETs options.

If large scale amounts of negative emissions are to be avoided in moving towards 1.5 °C, efforts must focus on those sectors where residual emissions occur in the later periods, such as industry and electricity, whose positive emissions require offsetting via negative emissions. Clearly technological innovation or demand reduction in these sectors should be a priority for those who do not believe that BECCS and NETs can be deployed at such a scale.

However, it is inevitable that some level of CCS/NETs will be required in the future. This is the case for two reasons; (i) both levels of ambition see emissions going net-negative, due to a large part of the budget being used prior to 2050, and (ii) some sectors will be extremely challenging to decarbonize. Therefore, policy makers need to start orientating the agenda and action towards such technologies whose commercial development is still uncertain.

It is also important that decision makers understand that the prospects for fossil fuels are very much tied to the outlook for NETs. If NETs cannot be scaled, the levels of fossil fuels suggested in this analysis are not compatible with the Paris Agreement goals. In other words, there are risks of lock-in to a high fossil future if NETs fail to scale. Policy makers must therefore understand well the risks of different strategies. To reduce the risks of lock-in to a high fossil future, policy makers may be interested in exploring pathways with lower levels of CCS/NETs, to help develop strategies that are more robust to the risks of failure.

To help inform the debate, it is also critical that greater attention is paid to the demand side of the model to balance the focus away from 'supply-side only' solutions. This means a focus on the underlying drivers, non-technical measures focused on societal behavior, and the changing delivery of energy services e.g. mobility as a service (MaaS) in the transport sector, and measures to reduce industrial emission intensity e.g. changes to materials used in the economy (circular economy etc.). Recognising the uncertainty, particularly in the case of economic drivers, which tend to project current economic forecasts and structures forward in time, is crucial given the tendency for significant changes in economic regimes, both nationally and globally. In TIAM-UCL, the projections suggest continuous economic growth driving income levels and demand for energy services e.g. air travel. However, other futures are very much possible and should be explored.

References

Anandarajah G, Pye S, Usher W, Kesicki F, Mcglade C (2011) TIAM-UCL global model documentation. UKERC working paper UKERC/WP/ESY/2011/001; 2011

Anderson K, Peters G (2016) The trouble with negative emissions. Science 354(6309):182–183. https://doi.org/10.1126/science.aah4567

Bauer N et al (2017) Shared socio-economic pathways of the energy sector—quantifying the narratives. Glob Environ Change 42:316–330

Cho A (2010) Energy's tricky tradeoffs. Science 786–7. https://doi.org/10.1126/science.329.5993. 786

Clarke L et al (2014) Mitigation of climate change. In: Edenhofer et al (eds) Climate change, Chap. 6. Cambridge University Press, Cambridge

Dornburg V, van Vuuren DP et al (2010) Bioenergy revisited: key factors in global potentials of bioenergy. Energy Environ Sci. https://doi.org/10.1039/b922422j

Fuss S et al (2016) Research priorities for negative emissions. Environ Res Lett 11(11). https://doi.org/10.1088/1748-9326/11/11/115007

Global CCS Institute (2010) Global status of BECCS projects 2010. http://hub.globalccsinstitute.com/sites/default/files/publications/13516/gccsi-biorecro-global-status-beccs-110302-report.pdf. Accessed 12 Jan 2018

Hof AF et al (2017) Global and regional abatement costs of nationally determined contributions (NDCs) and of enhanced action to levels well below 2 °C and 1.5 °C. Environ Sci Policy 71 (1):30–40. https://doi.org/10.1016/j.envsci.2017.02.008

Hughes et al (2017) The role of CCS in meeting climate policy targets: understanding the potential contribution of CCS to a low carbon world, and the policies that may support that contribution. http://www.globalccsinstitute.com/publications/report-university-college-london-role-ccs-meeting-climate-policy-targets. Accessed 12 Jan 2018

OECD/IEA (2017) Key world energy statistics. https://www.iea.org/publications/freepublications/publication/KeyWorld2017.pdf. Accessed 12 Jan 2018

IPCC (2014) Climate Change 2014: synthesis report. In: Pachauri RK, Meyer LA (eds) Contribution of working groups I, II and III to the fifth assessment report of the Intergovernmental Panel on Climate Change. IPCC, Geneva, Switzerland, 151

Jackson RB et al (2017) Focus on negative emissions. Environ Res Lett 12. https://doi.org/10.1088/1748-9326/aa94ff

Loulou R, Labriet M (2007) ETSAP-TIAM—the TIMES integrated assessment model Part 1: model structure. CMS 5:7–40. https://doi.org/10.1007/s10287-007-0046-z

McGlade C, Ekins P (2015) The geographical distribution of fossil fuels unused when limiting global warming to 2 °C. Nature 517:187–190. https://doi.org/10.1038/nature14016

Meinshausen M, Raper SCB, Wigley TML (2011) Emulating coupled atmosphere-ocean and carbon cycle models with a simpler model, MAGICC6—Part 1: model description and calibration. Atmos Chem Phys 11(4):1417–1456. https://doi.org/10.5194/acp-11-1417-2011

Morton C, Anable J, Brand C (2014) UKERC energy strategy uncertainties. perceived uncertainty in the demand for electric vehicles: a qualitative assessment. http://www.ukerc.ac.uk/publications/ukerc-energy-strategy-under-uncertainties-perceived-uncertainty-in-the-demand-for-electric-vehicles-a-qualitative-assessment.html. Accessed 12 Jan 2018

Price J, Keppo I (2017) Modelling to generate alternatives: a technique to explore uncertainty in energy-environment-economy models. Appl Energy 195:356–369

Pye S et al (2016) Exploring national decarbonization pathways and global energy trade flows: a multi-scale analysis. Clim Policy 16(1):92–109. https://doi.org/10.1080/14693062.2016.1179619

Resch G, Held A, Faber T, Panzer C, Toro F, Haas R (2008) Potentials and prospects for renewable energies at global scale. Energy Policy 36:4048–4056. https://doi.org/10.1016/j.enpol.2008.06.029

Riahi K et al (2017) The shared socioeconomic pathways and their energy, land use, and greenhouse gas emissions implications: an overview. Global Environ Change 42:153–168. https://doi.org/10.1016/j.gloenvcha.2016.05.009

Rogelj MS (2016) Differences between carbon budget estimates unravelled. Nat Climate Change 6:245–252. https://doi.org/10.1038/nclimate2868

Scamman D (2017) The Transport Sector. In: Staffell I, Dodds PE (eds) The role of hydrogen and fuel cells in future energy systems. H2FC SUPERGEN, London, p 53–70

Smith P et al. (2014) Agriculture, Forestry and Other Land Use (AFOLU). In: Edenhofer OR et al (eds) Climate change 2014: mitigation of climate change. Contribution of Working Group III to the fifth assessment report of the intergovernmental panel on climate change. Cambridge University Press, Cambridge, United Kingdom and New York

Smith P et al (2016) Biophysical and economic limits to negative CO_2 emissions. Nat Climate Change 6:42–50. https://doi.org/10.1038/nclimate2870

Tavoni M, Socolow R (2013) Modeling meets science and technology: an introduction to a special issue on negative emissions. Climatic Change 118:1–14. https://doi.org/10.1007/s10584-013-0757-9

Tomabechi K (2010) Energy resources in the future. Energies 3:686–695. https://doi.org/10.3390/en3040686

van Vuuren DP et al (2011) The representative concentration pathways: an overview. Clim Change 109(1):5–31. https://doi.org/10.1007/s10584-011-0148-z

Vandyck T, Keramidas K, Saveyn B, Kitous A, Vrontisi Z (2016) A global stocktake of the Paris pledges: implications for energy systems and economy. Glob Environ Change 41:46–63

van Vuuren DP et al (2017) Open discussion of negative emissions is urgently needed. Nat Energy 2:902–904. https://doi.org/10.1038/s41560-017-0055-2

Analysis of the Relative Roles of Supply-Side and Demand-Side Measures in Tackling the Global 1.5 °C Target

Babak Mousavi and Markus Blesl

Key messages

- Since biomass with CCS is a necessary option to move towards a 1.5 °C target, strong policies are needed to support efficient supply and consumption of sustainable biomass for energy purposes and to prioritise the use of biomass in sectors with limited mitigation options (e.g. industry).
- A cost-effective mitigation strategy would benefit from policy instruments that aim at modifying consumer behaviour to further support the transition beyond technological mitigation measures.
- Reducing energy-service demands plays a critical role in offsetting the additional pressure on the power sector resulting from the rapid and strong electrification of the energy sector.
- TIAM-MACRO proves to be a suitable tool for analysis of the required energy system transformations and the associated macroeconomic implications in order to meet decarbonisation targets, at global and regional scales.

1 Introduction

Climate change is one the most critical global concerns that must be addressed by all nations in the world. In 2015, the Paris Agreement under the United Nations Framework Convention on Climate Change (UNFCCC) codified two ambitious

B. Mousavi (✉) · M. Blesl
Institute of Energy Economics and Rational Energy Use (IER),
Stuttgart University, Stuttgart, Germany
e-mail: babak.mousavi@ier.uni-stuttgart.de

M. Blesl
e-mail: markus.blesl@ier.uni-stuttgart.de

© Springer International Publishing AG, part of Springer Nature 2018
G. Giannakidis et al. (eds.), *Limiting Global Warming to Well Below 2 °C: Energy System Modelling and Policy Development*, Lecture Notes in Energy 64,
https://doi.org/10.1007/978-3-319-74424-7_5

long-term global targets: holding the increase in the global average temperature well below 2 °C above pre-industrial levels and pursuing efforts to limit the temperature increase to 1.5 °C. It is recognized that the latter goal is a notably safe guardrail and will significantly reduce negative impacts of climate change. As a result, policy instruments for achieving the 1.5 °C target have become of central interest in the current climate change debate.

To meet such an ambitious goal, there is no single mitigation option; instead, the energy sector offers a wide range of technological options, namely, energy efficiency improvement, shifting from high carbon-intensive fossil fuels to less carbon-intensive alternatives (e.g., switching from coal to natural gas), and the enhanced use of renewables, nuclear, and Carbon Capture and Storage (CCS). On the other hand, additional investments in cleaner technologies will, ceteris paribus, result in higher price of energy services and consequently reduce demand for energy-services which is considered as a mitigation measure (Fujimori et al. 2014). Hence, for a more systematic assessment of decarbonisation strategies, it is of crucial importance to take into account not only the contribution of technological options, but also the role that price-induced energy-service demands reductions will play.

This chapter aims at exploring the relative roles of energy-service demands reductions and technological measures in meeting the 1.5 °C target by 2100. To provide a deeper evaluation of future decarbonisation strategies, the interconnections between the energy system and the rest of the economy is taken into account to quantify the macroeconomic implications of the strategies. The analysis relies on a model-based energy scenario analysis and is centred at the global level which is critical in the context of climate change policies. To this end, the hybrid TIAM-MACRO model is applied, which combines the technological explicitness of the global TIAM model with the macroeconomic representation of the MACRO model.

One of the key parameters of the MACRO model is elasticity of substitution reflecting the degree to which energy-service demands can be replaced by the aggregate of capital and labour, as their relative prices change. In fact, this parameter determines how strong the economy reacts to energy-service price changes, caused by climate policy. In the original MACRO model, it is assumed that this parameter is constant over the planning time and across all regions. However, there is no evidence supporting this simplifying assumption. Therefore, for a better representation of the price-induced energy-service demands reduction, in one of the scenarios presented in this chapter, elasticity of substitution varies across regions and over the planning horizon as a function of regional incomes. The basic idea behind this function is that as a country/region becomes richer, it has higher flexibility to substitute its production's inputs in order to prevent negative impacts of the climate policy on the economy.

2 Literature Review

Many studies have addressed the required energy system transformations in order to meet future climate targets. According to Dessens et al. (2016), who conducted a comprehensive review of model based studies, most of the reported scenarios present a significant role for electrification of end-use sectors coupled with a fast decarbonisation of the electricity sector. Moreover, they shed light on the necessity of deploying Biomass with CCS (BECCS) in combination with a full portfolio of other mitigation technologies in the power sector, especially in the second half of the century.

In contrast to studies that focused on technological mitigation measures, few attempts addressed the possible role of price-induced energy-service demand reduction in achieving the stringent 2 °C target. Examples include Kesicki and Anandarajah (2011) who applied the demand-elastic version of the TIAM model to analyse the possible role of energy-service demand reduction to tackle global climate change, and Pye et al. (2014) who used the same methodology to address the uncertainty associated with such service demand responses. The findings of these studies show that reducing energy-service demands plays a critical role in meeting such an ambitious climate target, ensuring a more cost-effective transition to a low carbon energy system.

Despite the need for more ambitious climate targets, only few studies have so far discussed the required energy system transformations to hold warming well below 2 °C, or even further below 1.5 °C by 2100. In fact, the only existing model-based analysis that discussed the energy system characteristics of mitigation pathways consistent with the 1.5 °C limit, was given by Rogelj et al. (2016). According to this study, achieving the target will require immediate attention to push mitigation in every individual sector of the economy. However, to be best knowledge of the authors no study has so far analysed the contribution of technological mitigation options with regard to price-induced energy-service demand reduction in meeting a 1.5 °C consistent decarbonisation goal. The present chapter aims at filling this gap in the literature.

On the other hand, there exist a large number of empirical studies that employed the VES production function with capital and labour inputs. Examples include Lovell (1973), Roskamp (1977), Bairam (1991), and Zellner and Ryu (1998). Most of the studies found the VES a more realistic approach compared to other approaches. However, no chapter so far has employed a VES production function in the context of energy system modelling. In this sense, this chapter opens avenues for future research by introducing a macroeconomic model with the VES specification.

3   Methodology: Focus on the Elasticity of Substitution

TIMES Integrated Assessment Model (TIAM) is a global multi-regional, technology-rich, bottom-up energy system model, maintained by Energy Technology Systems Analysis Program (ETSAP) (Loulou and Labriet 2008). The model encompasses current and future energy technologies in a detailed manner and aims to find the least-cost mix of technologies to fulfil a given set of energy-service demands under different energy and climate related policies. Therefore, it is considered as an appropriate tool to chapter the role of technological measures in combating climate change. However, as the model ignores feedback between energy-service demands and their prices, it is not able to fully address the role of energy-service demands. Furthermore, being restricted to the energy sector limits the ability of the model to account the repercussions on the rest of the economy. To bridge these gaps, TIAM is linked to MACRO which is a top-down macroeconomic model with an aggregated view of long-term economic growth. Since Kypreos and Lehtila (2015) extensively discussed the original TIAM-MACRO, we only discuss how the MACRO's production function is generalized so as to support the VES specification.

The production function of the standard MACRO model is a nested, constant elasticity of substitution (CES) function between the aggregate capital and labour inputs and the energy-service demands, which allows substitution between the pair capital-labour and the energy-service demands.

$$Y_{r,t} = \left[a_r \left(K_{r,t}^{kpvs_r} \cdot L_{r,t}^{(1-kpvs_r)} \right)^{\rho_r} + \sum_i b_{r,i} \cdot DM_{r,t,i}^{\rho_r} \right]^{1/\rho_r} \tag{1}$$

$$\rho_r = 1 - 1/\sigma_r \tag{2}$$

where,

$Y_{r,t}$	Annual production of region r in period t
$K_{r,t}$	annual capital of region r in period t
$L_{r,t}$	annual labour growth index of region r in period t
$DM_{r,t,i}$	annual energy-service demand in MACRO for commodity i of region r in period t
a_r	capital-labour constant for region r (determined in a base-year benchmarking procedure)
ρ_r	energy-service demand constant for commodity i in region r (determined in a base-year benchmarking procedure)
$kpvs_r$	share of capital in the value-added aggregate of region r
ρ_r	substitution constant for region r
σ_r	elasticity of substitution for region r

In the standard MACRO model, this parameter is assumed to be a constant for all model regions and over the long-term planning horizon (2015–2100). Under the constancy assumption, all regions in the model react similarly to a unit change in the relative prices of the production inputs. However, considering the structural differences between the world regions, this cannot reflect the reality. Moreover, the assumption represents that the behaviour of an economy does not change over time. Notwithstanding, considering the long-term horizon of the model, it can be argued that as time passes, the economy may react differently to a unit change in the relative prices. Therefore, for a deeper analysis of mitigation pathways, we generalize the production function of the MACRO model in a way that it allows the elasticity of substitution to vary not only across the model regions, but also over the time horizon.

Basically, to implement the VES approach, first step is to generalize the production function of the standard MACRO model (Eq. (1) in order to support time dependent elasticity of substitution:

$$Y_{r,t} = \left[a_r \cdot K_{r,t}^{kpvs_r \cdot \rho_{r,t}} \cdot L_{r,t}^{(1-kpvs_r) \cdot \rho_{r,t}} + \sum_i b_{r,i} \cdot DM_{r,t,i}^{\rho_{r,t}} \right]^{1/\rho_{r,t}} \qquad (3)$$

$$\rho_{r,t} = 1 - 1/\sigma_{r,t} \qquad (4)$$

where,

$\rho_{r,t}$ Substitution constant for region r in time period t

$\sigma_{r,t}$ Elasticity of substitution for region r in time period t

Once the assumption of constancy is dropped and the variability of the substitution elasticity is admitted, we face a variety of alternatives for defining the functional form of the elasticity of substitution. The resultant production function, therefore, depends on the assumptions involved in this function. To develop the function, it is assumed that as an economy becomes richer, it will have higher flexibility to replace energy-service demands with the other production inputs. To specify the regional elasticity of substitutions in the first planning period (2015–2020), adapted from World Bank (2017) and the investigation of Remme and Blesl (2006), the model regions are divided into high-income, middle-income and low-income regions with elasticities of 0.25, 0.2 and 0.15, respectively. To determine the elasticity values for the next years (i.e. 2021–2100) it is assumed that the elasticity of substitution of each region is a function of GDP (Gross Domestic Production) per capita growth of that region. However, in order to avoid overestimations and stay within the range proposed by Remme and Blesl (2006), the maximum elasticity of substitution (for India in 2100) is set to 0.5. Accordingly, the elasticities for all the other regions/time periods are adopted. Figure 1 depicts the given values for substitution elasticities.

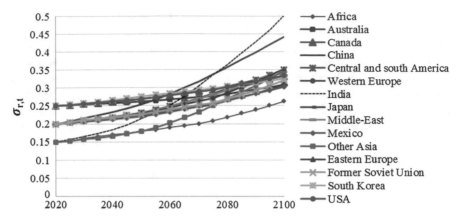

Fig. 1 Applied elasticity of substitutions in the VES case

4 Scenario Definition

Four different scenarios are constructed for this chapter. While one of the scenarios does not involve any CO_2 reduction policy, all others are subject to a carbon budget consistent with at least a 33% chance of limiting the average global temperature increase to 1.5 °C by 2100. In this context, the Base scenario is employed as a benchmark for the decarbonisation scenarios. For the purposes of the decarbonisation scenarios, the energy sector carbon budget (including emissions from industrial processes as well as fuel combustion) is calculated from an estimate of the total CO_2 emissions budget for the given temperature limit (see IPCC 2014 and IEA 2017). Non-energy sector CO_2 emissions that mainly arise through land-use, land-use change and forestry are given according to IEA (2017). The assumed carbon budget for the energy sector over the period 2015–2100 is presented in Table 1.

Due to uncertainties in the contribution of price-induced energy service demands reductions, the decarbonisation scenarios vary depending on their assumption concerning this mitigation measure. One of the decarbonisation scenarios (1.5D-DEM) ignores the interconnection between energy-service demands and their prices in order to rely only on technological mitigation measures. In contrast, the other two decarbonisation scenarios consider both types of mitigation options. While in one of them (1.5D) the elasticity of substitution is a constant value (0.25)

Table 1 CO_2 budget assumptions in this chapter

Net anthropogenic warming with a probability of more than 33%	Total CO_2 budget (2015–2100)	Non-energy CO_2 emissions (2015–2100)	Energy sector CO_2 budget (2015–2100)
<1.5 °C	570 GtCO$_2$	−30 GtCO$_2$	600 GtCO$_2$

for all regions/periods, to provide more insights into the role of price-induced energy-service demand reduction, in the other scenario (1.5D-VES), the elasticity varies as in Fig. 1. The start year of mitigation action in all decarbonisation scenarios is set to 2020. Table 2 reviews the scenarios of this chapter.

5 Results

5.1 Primary Energy Consumption

Reaching the ambitious 1.5 °C target requires a significant and rapid transition in the global energy system. In the Base, primary energy consumption more than doubles the 2014 levels to reach 1282 EJ in 2100 (Fig. 2). While fossil fuels continue to dominate primary energy consumption, their total share decreases from 82% in 2014 to 72% in 2100. The remaining primary energy mix in 2100 consists of 15% biomass and waste, 7% other renewables and 5% nuclear.

In the 1.5D, growth in primary energy consumption in 2100 compared to 2014 levels is limited to 80% and is around 256 EJ lower than in the Base. This features

Table 2 Observed scenarios

Scenario name	Description
Base	No carbon budget limitation
1.5D	The carbon budget is consistent with the 1.5 °C target (Table 1) Elasticity of substitution is set to 0.25 for all regions/time periods
1.5D-VES	The carbon budget is consistent with the 1.5 °C target (Table 1) Elasticity of substitution varies as in Fig. 1
1.5D-DEM	The carbon budget is consistent with the 1.5 °C target (Table 1) Service demands do not react to changes in their prices

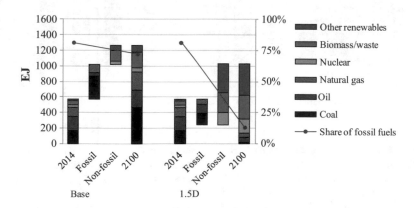

Fig. 2 Global primary energy consumption in the Base and 1.5D scenarios, 2014–2100

the need to accelerate the decoupling of primary energy consumption from economic growth. In this scenario, reliance on fossil fuels falls dramatically over the century. Most of the remaining fossil fuels in 2100 are used either in combination with CCS technologies or for feedstocks and non-energy purposes. Renewables overtake fossil fuels to dominate the primary energy mix by the share of 69% (565 EJ) in 2100. This is around 25% higher than today's total primary energy consumption. The remainder of the primary energy mix in 2100 is nuclear with a share of 18%, highlighting that despite its challenges, nuclear is considered to be an essential part of the transition.

5.2   The Rapid Decarbonisation of the Power Sector

An initial step in an effective transformation of the energy system is decarbonisation of the power sector. Figure 3 shows that moving from the Base to the 1.5D entails deep and rapid changes in this sector.

In the 1.5D, electricity generation from renewables (in this chapter, renewables refer to all renewables excluding BECCS) grows dramatically over the century. This highlights the importance of policy instruments focusing on the challenges associated with renewables deployment. According to Mousavi et al. (2017), to develop renewable energies, especially in less-developed countries, governments should address not only economic barriers (e.g. fossil-fuel subsidies and high capital costs of renewables), but also non-economic challenges (e.g. lack of sufficient awareness and confidence) that they may face.

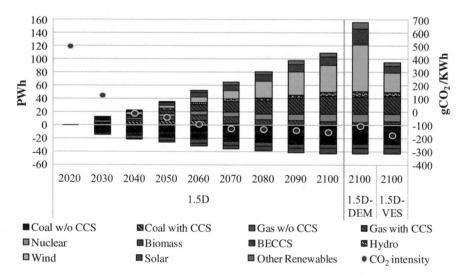

Fig. 3 Changes in the global electricity generation mix in the 1.5D over 2020–2100 and in the 1.5D-DEM and 1.5D-VES in 2100, relative to the Base

On the other hand, variable renewable sources which are non-dispatchable due to their fluctuating nature, account for around 75% of the increase in renewables-based generation in 2100. Considerable system flexibility is needed to integrate such a high level of variable renewables. In the 1.5D, this issue is reflected in the increased deployment of geothermal, hydro, biogas and Solar Thermal (STE) plants with storage, as well as fossil-fired plants with CCS.

Despite the long technical lifetime of conventional coal power plants, in the 1.5D, they are almost completely phased out by 2035. This raises concerns of stranded assets and affects current and near future investment decisions. However, CCS retrofits allow the continued use of fossil-fuels in the power sector for some decades.

One of the key measures in the shift from the Base to the 1.5D is BECCS. In fact, 20% share of BECCS turns the power sector into a source of net-negative emissions over the period 2040–2100. The negative emissions are crucially important to offset residual emissions in other sectors where direct mitigation is either technically too difficult or more expensive. Despite the advances made in CCS technology in recent years, no large BECCS power plant operates at a commercial scale. Therefore, achieving the ambitious target of the 1.5D requires continued investment in Research, development and deployment (RD&D) in this area. However, ahead of any of the concerns related to large-scale deployment of BECCS, the availability of sustainable and sufficiently large biomass supply over the world regions is critical. This elevates the importance of an effective system to support efficient production of biomass for energy purposes.

Figure 3 also illustrates that changes in energy-service demands have a notable impact on the electricity generation from renewables, in a long-term perspective. In the 1.5D-DEM, the majority of the additional generation in 2100 is supplied by wind and solar, which is primarily due to their relatively high technical potential. For instance, the global technical potential of solar energy is estimated to be 1600 EJ (Resch et al. 2008), which is around 280% of the global primary energy consumption in 2014.

5.2.1 End-Use Electrification and Decarbonising the Power Sector

In the 1.5D, the power sector is almost completely decarbonised by 2050 and share of electricity in final energy consumption increases from around 20% in 2020 to 62% in 2100 (Fig. 4). This conveys a clear message that a cost-effective strategy to reach the ambitious 1.5 °C target is decarbonising the power sector and substituting fossil fuels in end-use sectors with the decarbonised electricity. More than 83% of the increase in global final electricity consumption by 2100 is derived by non-OECD countries, especially India. In relative terms, electricity consumption in the transport sector grows the most. The share of electricity in this sector increases dramatically from around 1% in 2020 to almost 50% in 2100.

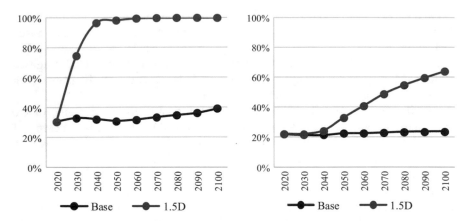

Fig. 4 Share of decarbonised electricity in total global annual generation (left) and share of electricity in global annual final energy consumption (right)

5.2.2 Energy-Service Demands and Investment Needs in the Power Sector

Although more extensive deployment of energy efficient technologies reduces electricity demand in the decarbonisation scenarios in the short term, vast electrification of end-users leads to considerably higher electricity demand in these scenarios over the longer term. However, the level of electricity demand and the required investments in the power sector tightly depend on the level of energy-service demands (Fig. 5). In other words, reducing energy-service demands will significantly offset the increase in electricity demand and consequently reduce the level of investments in the power sector. Over the period 2020–2100, the 1.5D-DEM requires considerably higher investment levels, largely for renewables, compared to the 1.5D. In contrast, in the 1.5D-VES, the cumulative investment need is lower than in the 1.5D. In relative terms, China experiences the highest reduction in the cumulative investment levels (17%) in the 1.5D-VES compared with the 1.5D.

5.3 The Relative Roles of Mitigation Measures

Shifting from the Base to the 1.5D needs a wide range of supply-side and demand-side mitigation options, with the major contributor being renewables (excluding BECCS), accounting for a 30% share in cumulative global CO_2 emissions reductions over the period 2020–2100. Energy-service demands reductions play a substantial role with 20% contribution, while nuclear contributes 17%, BECCS 16%, fossil CCS 12%, fuel switching 4% and efficiency 1% (Fig. 6). The low contribution of efficiency improvement is due to the fact that energy efficiency

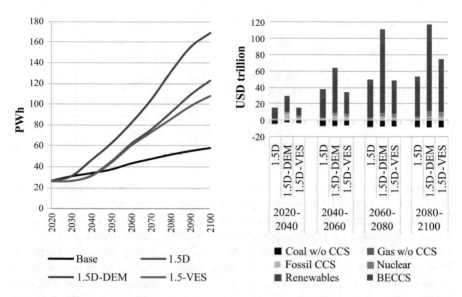

Fig. 5 Global final electricity demand by scenario (left) and cumulative additional investment in the decarbonisation scenarios relative to the Base (right)

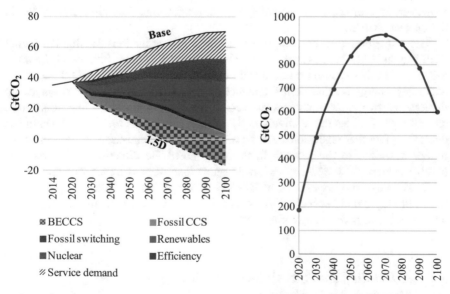

Fig. 6 Global energy-related CO_2 emissions reductions from the Base to the 1.5D by mitigation measures (left) and global cumulative energy-related CO_2 emissions in the 1.5D

Table 3 GtCO$_2$ cumulative CO$_2$ emissions reductions over the period 2020–2100 by mitigation measure and scenario

	1.5D	1.5D-DEM	1-5D-VES
Service-demand	20	0	23
Efficiency	1	4	1
Renewables	30	45	29
Nuclear	17	17	17
Fossil switching	4	3	3
Fossil CCS	12	15	11
Biomass CCS	16	16	16

measures, which are cost-saving over the long run, are already deployed in the Base, as these would be an integral part of a least-cost pathway.

Despite the rapid decarbonisation of the global energy system which reaches carbon-neutrality by 2065 (the power sector reaches carbon-neutrality by 2040), meeting the 1.5 °C target results in an "overshoot" of the energy sector carbon budget that must be counterbalanced by a significant deployment of negative emissions technologies, especially after 2060 (Fig. 6). This sheds light on the key role of BECCS in this context. In fact, more than 300 GtCO$_2$ of net negative emissions resulted from deployment of BECCS bring temperature back to the target level in 2100.

It is important to note that although BECCS is the most mature negative emissions technology, there are other options that deliver negative emissions such as direct air capture, biochar and lime-soda process (McGlashan et al. 2012). However, as their costs and potential are quite uncertain, they are not included in the current chapter.

The mitigation role of energy-service demands reductions is slightly higher (23%) in the 1.5D-VES (Table 3). However, the relative contributions of the other measures in this scenario do not notably differ from those in the 1.5D. In contrast, excluding energy-service demand reduction option in the 1.5D-DEM scales up deployment of the technological measures, especially renewables that contributes around 45%. Besides the relatively high technical potentials for renewables, this is due to the fact that renewables can be directly used by end-users, which provides higher flexibility to the system. In the 1.5D-DEM the direct-use of renewables (excl. Bioenergy) by end-use sectors is more than 40% compared to that in the 1.5D. Although the importance of BECCS is already recognized, its contribution does not vary among scenarios. As stated, this is mainly due to the limited availability of biomass for energy purposes.

5.4 A Detailed Look at the Role of Energy-Service Demands Reductions

The level of energy-service demands reductions depends on the costs of alternative technologies and fuels available to meet the service demands, costs of the other

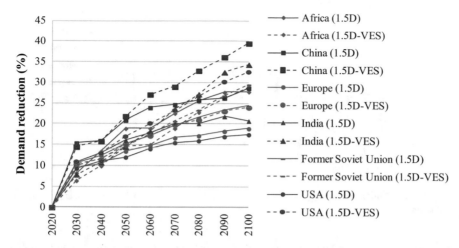

Fig. 7 Energy-service demand reduction level for six regions under the 1.5D and 1.5D-VES scenarios compared to the Base

production input factors (i.e. capital-labour) and the size of the elasticity of substitution. In order to provide more insights into this mitigation measure, Fig. 7 depicts energy-service demand reduction levels for six different regions under the 1.5D and the 1.5D-VES scenarios relative to the Base.

Pursuing the limitation of warming to below 1.5 °C reduces the flexibility in mitigation measures almost completely. In fact, the urgency of putting ambitious mitigation strategies into place and limited availability of cost-effective technological options in the near future impose a "price-shock", leading to significantly higher energy-service prices over the next 10–15 years. As a result, energy-service demands experience a notable reduction over the period 2020–2030. In both scenarios, relative energy-service demands reductions increase over time towards the end of the century. In most of the regions, the level of energy-service demands reductions in both scenarios is almost the same in the short to medium term. However, increasing energy-service prices and the higher elasticity of substitutions in the 1.5D-VES, favours higher reductions in energy-service demands in this scenario in the long-term. This is especially true for developing regions such as China and India, which experience a rapid economic growth.

It is noteworthy that although in the 1.5D-VES scenario, India has the highest elasticity of substitution in 2100, in this scenario, as well as in the 1.5D, China represents the most significant energy-service demands reductions over the whole century. This elevates the importance of the role that energy-service demands in China will play in reaching the 1.5 °C objective.

A look on the sectoral level is necessary for a more rigorous evaluation of the role of energy-service demands reductions. For this purpose, emissions from secondary energy carriers, such as electricity, are accounted for in each end-use sector. As presented in Fig. 8, energy-service demands reductions play the most critical

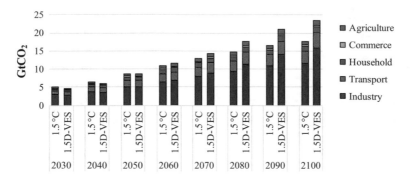

Fig. 8 Contribution of energy-service demand reduction to global overall CO_2 emissions reduction

role in decarbonising the industry, followed by the transport sector. A reasonable explanation for this issue is the relatively expensive abatement opportunities in these two sectors. Generally speaking, the challenges of decoupling expanding demands for the industry and transport sectors from CO_2 emissions will require considerable changes in their current structure and processes, which may dramatically increase the prices of energy services. In contrast, the other sectors provide relatively cheap mitigation measures as their decarbonisation are mainly characterised by the decarbonisation of the electricity consumption.

5.5 Marginal Abatement Costs and GDP Losses

Analysis of marginal costs of CO_2 emissions abatement indicates that reducing energy-service demands has a significant impact on the direct costs of CO_2 emissions. Figure 9 shows that in the 1.5D, the marginal abatement costs rise sharply, especially in the second half of the century, reaching almost 4.5$/kg in 2100, which is beyond reasonable levels (Koljonen and Lehtilä 2012). In the 1.5D-DEM, however, marginal abatement costs increase even more rapidly through 2100 to a final price of 10.2$/Kg. It can be concluded that price-induced energy-service demands reductions play a crucial role in ensuring a more cost-effective transition to the 1.5 °C ambition.

Suffice to say that achieving an ambitious decarbonisation target cost-effectively not only requires technological options but also needs reductions in energy-service demands. However, for a more comprehensive and integrated assessment of the mitigation role of energy-service demand reduction it is of crucial importance to analyse macroeconomic implications of the relevant scenarios.

A widely-used measure in this context is GDP loss representing the difference in GDP between the Base and each of the decarbonisation scenarios. In fact, in the mitigation scenarios GDP would be lower than in the Base because of higher

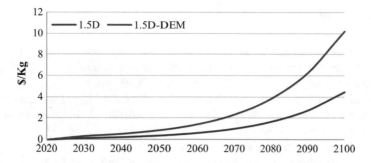

Fig. 9 Marginal CO_2 emissions abatement costs under the decarbonisation scenarios

energy and emissions mitigation costs, as well as changes in resource allocation. Some of the losses, however, can be recovered by reducing energy-service demands.

Figure 10 depicts regional GDP losses in the 1.5D and 1.5D-VES scenarios. It can be observed that if economies become more flexible to replace energy-service demands with other production inputs, they will be less affected by ambitious global mitigation targets. This implies that besides decarbonisation of the energy mix, achieving the 1.5 °C target requires a considerable decoupling between energy consumption and economic growth. In relative terms, China suffers the most from the implemented decarbonisation target with GDP losses of 6.6% and 7.7% by 2100 in the 1.5D-VES and the 1.5D, respectively.

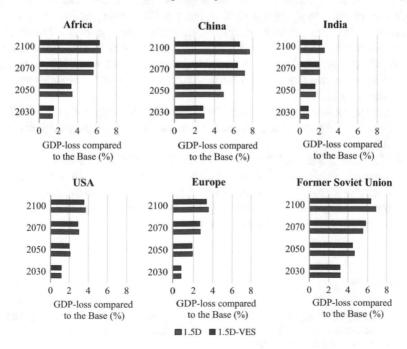

Fig. 10 GDP-Losses compare to the Base by region and scenario

6 Conclusion

The following key messages emerge from the proposed scenario analysis. First, the pace and scale of efforts needed to achieve carbon neutrality by 2065 and the considerable deployment of BECCS in the post-2040 period in the 1.5 °C scenario emphasize that there would be almost no room for delay. Rapid and strong policy action as well as increased effort and sustained international collaboration is needed to support aggressive deployment of a mixture of supply-side and demand-side mitigation measures in all nations in the world.

Second, decarbonising the power sector and substituting fossil fuels in less flexible end-use sectors with the decarbonised electricity is found to be a cost-effective strategy for all regions. However, a rapid and strong growth in electricity demand may bring new challenges to the power sector. In this regard, reducing energy-service demands plays a critical role in offsetting the additional pressure on the power sector.

Third, uncertainty concerning the size and function of the elasticity of substitution can have some impact on mitigation pathways and the associated costs. Nevertheless, this uncertainty does not weaken the key insight that reducing energy-service demands notably facilitates the transition towards the 1.5 °C target. Furthermore, this measure plays a vital role in recovering the negative macroeconomic impacts of the implemented mitigation policies. In other words, although reaching the 1.5 °C target without reducing energy-service demands is technically feasible, it comes with huge economic impacts. This implies that besides decarbonisation of the energy mix, achieving the 1.5 °C target requires a considerable decoupling between energy demand and economic growth. Therefore, a cost-effective mitigation strategy should benefit from policy instruments that aim at modifying consumer behaviour to further support the transition. This is especially the case for countries like China, whose economy requires deep changes under the decarbonisation policy.

Moreover, policy makers should be aware of the importance of flexibility mechanisms in the energy supply sector to support the large deployment of variable renewable sources.

Finally, due to the need for negative emissions, BECCS is found to be a necessary option to reach the 1.5 °C goal. However, deployment of large-scale BECCS technologies is limited to the availability of sustainable and sufficiently large biomass supply. Therefore, an effective system is essential to support efficient production and consumption of biomass for energy purposes.

References

Bairam E (1991) Elasticity of substitution, technical progress and returns to scale in branches of Soviet industry: a new CES production function approach. J Appl Econometrics 6:91–96. https://doi.org/10.1002/jae.3950060108s

Dessens O, Anandarajah G, Gambhir A (2016) Limiting global warming to 2 °C: what do the latest mitigation studies tell us about costs, technologies and other impacts? Energy Strategy Rev 13–14:67–76. https://doi.org/10.1016/j.esr.2016.08.004

Fujimori S, Kainuma M, Masui T et al (2014) The effectiveness of energy service demand reduction: a scenario analysis of global climate change mitigation. Energy Policy 75:379–391. https://doi.org/10.1016/j.enpol.2014.09.015

IEA (2017) Energy Technology Perspectives

IPCC (2014) Climate Change 2014: synthesis report. Contribution of Working Groups I, II and III to the Fifth Assessment Report of the IPCC

Kesicki F, Anandarajah G (2011) The role of energy-service demand reduction in global climate change mitigation: combining energy modelling and decomposition analysis. Energy Policy 39:7224–7233. https://doi.org/10.1016/j.enpol.2011.08.043

Koljonen T, Lehtilä A (2012) The impact of residential, commercial, and transport energy demand uncertainties in Asia on climate change mitigation. Energy Econ 34:S410–S420. https://doi.org/10.1016/j.eneco.2012.05.003

Kypreos S, Lehtila A (2015) Decomposing TIAM-MACRO to assess climatic change mitigation. Environ Model Assess 20:571–581. https://doi.org/10.1007/s10666-015-9451-9

Loulou R, Labriet M (2008) ETSAP-TIAM: The TIMES integrated assessment model Part I: model structure. CMS 5:7–40. https://doi.org/10.1007/s10287-007-0046-z

Lovell AK (1973) Estimation and prediction with CES and VES production functions. Int Econ Rev 14:676–692

McGlashan NR, Workman MHW, Caldecott B, Shah N (2012) Negative emissions technologies. Grantham Inst Clim Change Briefing Pap No 8:1–27

Mousavi B, Lopez NSA, Chiu ASF et al (2017) Driving forces of Iran's CO_2 emissions from energy consumption: an LMDI decomposition approach. Appl Energy 206:804–814. https://doi.org/10.1016/j.apenergy.2017.08.199

Pye S, Usher W, Strachan N (2014) The uncertain but critical role of demand reduction in meeting long-term energy decarbonisation targets. Energy Policy 73:575–586. https://doi.org/10.1016/j.enpol.2014.05.025

Remme U, Blesl M (2006) Documentation of the TIMES-MACRO model

Resch G, Held A, Faber T et al (2008) Potentials and prospects for renewable energies at global scale. Energy Policy 36:4048–4056. https://doi.org/10.1016/j.enpol.2008.06.029

Rogelj J, den Elzen M, Höhne N et al (2016) Paris agreement climate proposals need a boost to keep warming well below 2 °C. Nature 534:631–639. https://doi.org/10.1038/nature18307

Roskamp KW (1977) Labor productivity and the elasticity of factor substitution in west germany industries. Rev Econ Stat 59:366–371

World Bank (2017) World bank country and lending groups, income data https://datahelpdesk.worldbank.org/knowledgebase/articles/906519-world-bank-country-and-lending-groups

Zellner A, Ryu H (1998) Alternative functional forms for production, cost and returns to scale functions. Appl Econometrics 13:101–127

The Role of Population, Affluence, Technological Development and Diet in a Below 2 °C World

Kenneth Karlsson, Jørgen Nørgård, Juan Gea Bermúdez,
Olexandr Balyk, Mathis Wackernagel, James Glynn
and Amit Kanudia

Key messages

- Focusing only on technological development is likely not to be sufficient to limit the increase in global mean temperature to well below 2 °C.
- Increasing educational attainment, family planning and less meat intensive diets will reduce environmental impacts and increase the probability of stabilising temperature increase well below 2 °C.
- Attaining an equitable distribution of economic growth between the regions of the world can have positive and negative impacts on how challenging it will be to reach a well below 2 °C world.

K. Karlsson (✉) · J. Nørgård · J. G. Bermúdez · O. Balyk
Technical University of Denmark, Kongens Lyngby, Denmark
e-mail: keka@dtu.dk

J. Nørgård
e-mail: jsn@dtu.dk

J. G. Bermúdez
e-mail: jgeab@dtu.dk

O. Balyk
e-mail: obal@dtu.dk

M. Wackernagel
Global Footprint Network, Geneva, Switzerland
e-mail: mathis@foootprintnetwork.org

J. Glynn
MaREI Centre, Environmental Research Institute, University College Cork,
Cork, Ireland
e-mail: james.glynn@ucc.ie

A. Kanudia
Kanors, Delhi, India
e-mail: amit.kanudia@gmail.com

- The global TIMES integrated assessment model, TIAM-World, is combined with two other modelling tools to assess the socio-technical challenges towards a well below 2 °C world.

1 Introduction

From ancient time people have realized that local environments could be degraded in the case where large numbers of people living in close proximity could cause excessive pollution and local shortages of resources like land, water, wood and minerals. Such local crowding problems resulted in people migrating to other locations. More general concerns for overpopulation were expressed in Europe in the 1700s (Lütken 1758; Malthus 1798). During the 1800s and 1900s these regional and national problems in Europe were resulted in warfare and disease, as well as by migrating and colonizing other continents (Nørgård and Xue 2017).

Over recent centuries world population has grown near continuously, now reaching 7.6 billion and adding about 80 million people every year. Such growth in population will likely push environmental damages—including climate change impacts—in an upward direction and through a rebound effect counteract the possible environmental benefits through relative decoupling from improvements achieved through modifications in technology and affluence (Nørgård and Xue 2017).

In 2005, the United Nation's Ecosystem Assessment identified population growth as a principal driver of environmental change, along with economic growth and technological evolution (Engelman 2011).

Economic activity is also closely linked to environmental impacts. More demand for goods, services and food increase production and demand for land and raw materials. The increase in demand is not only from increasing population, but also from increased discretional material consumption as regions become increasingly wealthy. Market economies have a built-in positive feedback loop: More demand leads to more production and investment in production capacity, which creates more work, increasing demand for labour in amount or higher skilled labour, which in turn increases the income in society which again increase demand (Keynesian income multiplier) (Jespersen and Chick 2016). So even without increasing population, environmental impact can arise from the growing economy. Improved productivity can on the other hand reduce the environmental impacts from economic activity, when more is produced per unit of production because of improved skills, technology and efficiency.

Technologies dependent on fossil fuels are the main reason for the increasing concentration of CO_2 in the atmosphere and as well as other environmental impacts, but technology is also part of the solution for reducing these impacts. More efficient technologies reduce the demand for energy per produced unit and renewable low carbon technologies can remove or reduce CO_2 emission. More efficient production can decouple environmental impacts from economic activity, but this decoupling needs to be absolute to reduce environmental impacts in gross terms, rather than relative decoupling which can decrease the environmental intensity of economic activity, but increase environmental damage in gross terms overall.

The discussion on mitigation of climate change has had considerably more focus on technology supply side development, rather than a focus on the demand side of economic activity and maybe less on population. Therefore, this chapter proposes to mix different methodologies and explore a diverse set of scenarios to better understand the interdependencies between these three drivers of development.

2 Methodology

The proposed approach to analyse the impact of population, consumption and technology development links three methodologies (Fig. 1): IPAT, Ecological Footprint and Integrated Assessment Modelling (IAM). The first two methodologies have been built into the IPAT(D) model, while the latter is applied by soft-linking to the TIMES Integrated Assessment Model (TIAM-World).

The IPAT(D) model can be run from https://ipat.tokni.com/. The user can choose different combinations of input assumptions and view the results as global mean temperature increase and Ecological Footprint, to quickly assess the sustainability of the scenario created without having to run a full IAM model such as TIAM-World. The impacts are tracked on the 16 regions of TIAM-World (Table 1) from 2015 until 2100. The historical data used in the model to illustrate the past evolution of the parameters is based on Worldbank (2017), NASA (2017), The Global Footprint (2017) and Kanors-EMR (2017).

The required inputs to the model are population, affluence (represented by the GDP per capita), diet (CO_2e/person/year), climate sensitivity, and technology development represented by the carbon intensity of the economy in different technology rich climate mitigation scenarios.

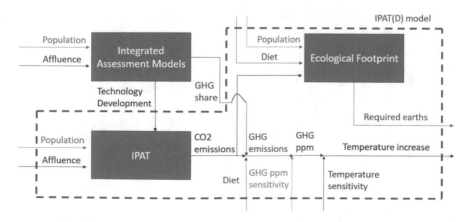

Fig. 1 Methodology structure

Table 1 TIAM-WORLD regions

AFR	Africa	JPN	Japan
AUS	Australia	MEA	Middle East
CAC	Central Asia & Caucasus	MEX	Mexico
CAN	Canada	ODA	Other Developing Asia
CHI	China	OEE	Other East Europe
CSA	Central and South America	RUS	Russia
EUR	Europe 27+	SKO	South Korea
IND	India	USA	United States of America

Source Kanors-EMR (2017)

In the IPAT(D) model, the technology development scenarios are obtained from TIAM-World runs where carbon intensity per unit of GDP is a model output and the temperature increase is calculated assuming linearity (transient climate response to emissions) between the anthropogenic temperature anomaly and cumulative CO_2 in the atmosphere (Matthews et al. 2009).

In the IPAT(D) model, the diet (D) model variant calculates the impact of shifting diets on factors such as process emissions and ecological footprint. The amount of calories and the composition of the diet differ from region to region. This affects the land-use requirements and related emissions of greenhouse gases. Dietary impacts on overall greenhouse gas emissions can make an important difference in a highly populated world to the remaining carbon budget and non-CO_2 gases. The data used is the specific energy content consumed per capita, kcal/day/person in 2011 as reference (FAOSTAT 2017) (Fig. 2 left axis), the Green House Gas (GHG) intensity of each food product, CO_2e/kcal (Shrinkthatfootprint 2017), and land-use intensity of each food product, ha/kcal, (Ranganathan et al. 2016). The first two data sources are combined to obtain the GHG intensity of each regional

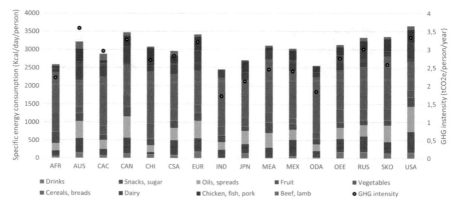

Fig. 2 Specific energy content consumed per capita and day in 2011 (left axis) and GHG intensity for food production in 2011 (right axis). *Source* FAOSTAT (2017), Shrinkthatfootprint (2017)

diet (CO_2e/person/year) (Fig. 2 right axis). This parameter is multiplied by the population and is divided by the GHG emissions to calculate the share of GHG linked to diet of each region. These shares are assumed constant over time based on 2011 data.

When a region shifts towards a lower emission intensity diet, the ratio between the GHG intensity of the new diet and the original one is calculated and applied to adjust the GHG emissions linked to the new diet. The land-use intensity of the diets is applied in a similar way. The analysis of diet shifting can also be useful for Energy System Models to estimate energy demands and land-use change. For this chapter however, the diet data is not fed back as an input to TIAM-World, and its impact was only studied in the IPAT(D) model.

2.1 IPAT Methodology

The equation $I = P \times A \times T$ calculates ecological impact (I) as a function of population (P), affluence (A) and technological emissions intensity factor (T) (Ehrlich and Holdren 1971). It highlights the role of population and affluence in climate change mitigation. In other words, focusing only on mitigation via technology adoption may not be enough to tackle climate change if no measures are taken for demand side reduction measures influencing population and affluence.

One of the limitations of the IPAT equation is that it does not consider any inter-dependency between the population, affluence and technology, whereas in reality they are quite interrelated (Boserup 1981; Simon 1980, 1981). As the IPAT model does not check for realistic combinations of P, A and T it is up to the user to create combinations that can fit into a realistic storyline. On the other hand, there are no "wrong" combinations, as the future has turned out to be difficult to predict in hindsight. The population factor masks a lot of information that can influence demand, such as age, gender, culture etc. Affluence (in this methodology measured as GDP/capita in each region) is also a rough representation of prosperity, where demand depends on region, consumer groups, level of income and much more. The technology parameter is the most aggregated factor as it includes all measures that can reduce GHG emissions driven by technological substitution and deployment. By isolating the impact on GHG emissions from diet the technological impact factor becomes a little less aggregated.

Therefore, when using the IPAT methodology one has to be aware of these limitations, but also acknowledge that many of the same limitations influence many energy system models and IAM's. In this chapter, the technology rich IAM TIAM-World is used to estimate the technology parameter, whereas the projections for population and affluence are taken from external transdisciplinary sources including the Shared Socioeconomic Pathways (SSPs) described in a further section.

2.2   What the Ecological Footprint Measures

Ecological Footprint accounting is driven by one key question: How much of the planet's (or a region's) regenerative capacity does a defined activity require from nature? The utilized regenerative capacity represents all the mutually exclusive, biologically productive spaces that are required to provide the demanded ecosystem services, e.g. food, timber provision, waste assimilation, space occupied for infrastructure, etc.

Since regeneration of ecosystems varies over time with technology, management practices, and climate, both the amounts demanded and the areas regeneration of these ecosystem services need to be specified with a reference date. The ability of ecosystems to provide for these resources—its renewable capacity—is called "bio-capacity". The human demand on this bio-capacity is called "Ecological Footprint".

"Activities" can refer to the entire consumption metabolism of humanity, of a particular population (a city), of a production process, or of something as small and discreet as producing one kg of durum wheat spaghetti. The method and its evolution is described in (Rees and Wackernagel 1996; Wackernagel et al. 2002, 2014).

Figure 3 shows the ecological footprint and bio-capacity in the regions represented in TIAM-World. Some regions have surplus bio-capacity while others are exceeding theirs. The carbon footprint used herein represents the amount of afforestation (hectares) required to absorb the carbon emissions of a given year.

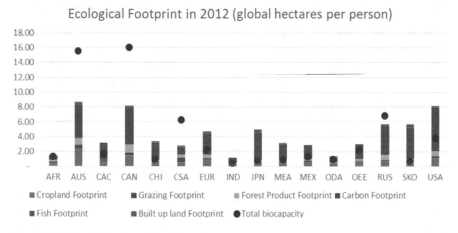

Fig. 3 Ecological footprint in 2012 and regional bio-capacity. *Source* The Global Footprint Network 2017

2.3   The Integrated Assessment Model TIAM-World

The TIMES Integrated Assessment Model (TIAM) is a multi-regional and inter-temporal partial equilibrium model of the entire energy/emission system of the world, based on the TIMES paradigm (Loulou and Labriet 2008; Loulou 2008). Several variants of the model exist that differ with regard to, e.g. regional aggregation, sectoral details, etc. TIAM-World is the model variant used in this study.

TIAM-World represents the energy system of the World divided in 16 regions. The model contains explicit detailed descriptions of more than one thousand technologies and one hundred commodities in each region, logically interrelated in a Reference Energy System, the chain of processes that transform, transport, distribute and convert energy into services from primary resources in place to the energy services demanded by end-users. The model also integrates a climate module permitting the computation and modelling of globally averaged temperature-change related to concentrations and radiative forcing.

TIAM-World is used to calculate different technology development scenarios that serve as input for the IPAT(D) model. This is done by using consistent assumptions for population and economic development from the SSP narratives and then running the model with different projections of CO_2 prices to force technology change. Based on this a large number of technology scenarios was created with different CO_2 trajectories and GHG shares, and thereby different climate impact. Carbon capture and storage (CCS) is not included in these TIAM-World runs, as the overlap with the ecological footprint methodology on removal of CO_2 from the atmosphere has not yet been solved.

3   Input Scenarios to the IPAT(D) Model

The IPAT(D) model includes a database with the scenarios listed below for population, affluence, technological development, GHG emissions and diet. All combinations of these parameters can be tested in the model.

3.1   Population and Affluence Scenarios

The population and affluence scenarios are based on the Shared Socio-economic Pathways (SSPs) data (Moss et al. 2010; Kriegler et al. 2012; O'Neill et al. 2014) illustrating possible pathways for change in population and economic development (Tables 2 and 3).

Each SSP consists of quantitative projections of GDP, Population and Urbanisation along with resource and technology constraints consistent with the underlying qualitative narrative. SSP1 "Sustainability—taking the green road" is a

Table 2 Population scenarios

SSP1	Low fertility in current low and medium income countries, medium fertility in current rich OECD countries
SSP2	Medium fertility in all countries
SSP3	High fertility in current low and medium income countries, low fertility in current rich OECD countries
SSP4	High fertility in current low income countries, low fertility in current medium income and rich countries and medium fertility in rich OECD countries
SSP5	Low fertility in current low and medium income countries, high fertility in current rich OECD countries
Low fertility	Low fertility, immediately fall to <1.5 birth per woman in all countries
No change	No change in population from 2015 and forward

Table 3 Affluence scenarios

SSP1	High growth in current low and medium income countries, medium growth in current high income countries
SSP2	Medium uneven economic growth in all countries
SSP3	Low economic growth in all countries
SSP4	Low economic growth in current low income countries, medium growth in other
SSP5	High economic growth in all countries
No change in affluence	No change from 2015, constant in each region on the level of 2015
Economic crisis	Crisis until 2050, stronger in high income countries, and then slow recovery
Equality	All regions reach by 2100 the level of USA in 2015
Catching up	Continuing historic growth rates until 2050, after 2050 the current low income countries catch up with current high income countries
Shifting power	Low economic growth in current high income countries and high growth in current low income countries

scenario in which the world shifts gradually, but pervasively towards a more sustainable path, with a focus on inclusive development within environmental limits. SSP2, "Middle of the Road" is a world in which social, economic and technological trends do not significantly change from historical rates of change. SSP3, "Regional Rivalry—A Rocky Road" is characterised by resurgent nationalism, competitiveness, fragmentation and little cooperation on environmental policy with a focus on energy and food security within nation development goals. SSP4, "Inequality—A Road Divided", is a scenario in which there is increasingly inequality in economic opportunity and political power, leading to social stratification across the world and within countries. Social cohesion degrades and conflict becomes common. SSP5, "Fossil-fuelled Development—Taking the Highway", relies heavily on the open market and innovation to drive rapid technological change and development of human capital towards sustainable development.

3.2 Diet Scenarios

Dietary impacts can be altered for each region in the model, but for simplicity sake, only three alternatives are used in this study (Table 4).

3.3 Technology Scenarios and Associated GHG Share

Three scenarios are chosen from a large number of technology scenarios from TIAM-World driven by the different SSP scenarios, to represent the potential technological development (Table 5).

From each scenario in TIAM-World, CO_2 emissions per GDP has been calculated by year and region, and is used in the IPAT(D) model to scale the GHG emissions with change of regional GDP, which is further scaled with the change in regional population.

Table 4 Diet scenarios

IND diet	India's diet in all countries, transition over 20 years
USA diet	USA or EU diet in all countries, transition over 20 years
No change	No change from today

Table 5 Technology scenarios

Business as usual	No policies or targets implemented, competitive markets secure global cost minimized solution
Strong technology development	90% non-fossil power, 50% non-fossil primary energy in 2050. More than 90% non-fossil primary energy in 2100
Radical technology development	100% non-fossil power, 85% non-fossil primary energy in 2050. 95% non-fossil primary energy in 2100

4 IPAT(D) Results from Modelled "Futures"

4.1 The Middle of the Road Scenario

In the middle of the road base case (Fig. 4), SSP2 assumptions are selected for population and affluence ("Middle of the Road" scenario), which gives the level of demand. This is combined with continuing existing diet, and a "Strong Technology Development".

This base scenario has a very strong focus on investments in renewable energy. Population peaks at 9 billion and over time when GHG emissions are reduced, the footprint gets below one Earth by 2090. However, temperature increase does not stay below 1.5 °C and the income gap between rich and poor countries increases as well.

4.2 The Effect of Population

To illustrate the impact from population development two futures with different population scenarios are compared: SSP3 (high fertility (12 billion people by 2100) versus "Low fertility" (3 billion people by 2100) (Fig. 5).

Both above futures assume a radical low carbon technology development, a high economic growth and no change in diet. When comparing the results of the ecological footprint of these futures, the required number of Earths are 0.7 higher in the

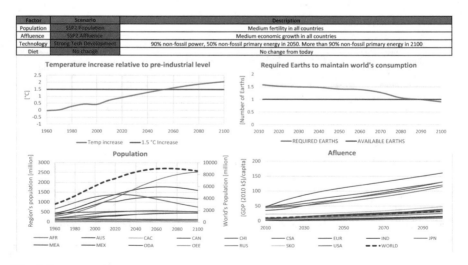

Fig. 4 Base scenario

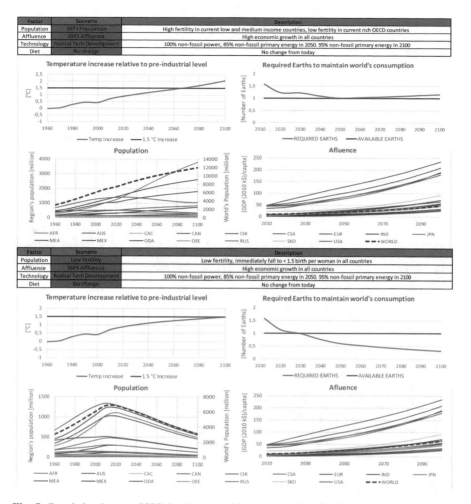

Fig. 5 Population impact. SSP3 development (above) versus low fertility (below)

high population scenario, while the temperature increase in the "Low fertility" scenario is 0.5 °C lower. Moreover, the environmental impacts would have been considerably higher if the technological development had been less optimistic.

4.3 The Effect of Economic Development

Two futures with completely different economic growth scenarios (Fig. 6) illustrate the importance of economic development per capita.

The first future assumes a "Catching up" development where low income countries develop considerably to converge to the levels of high income countries

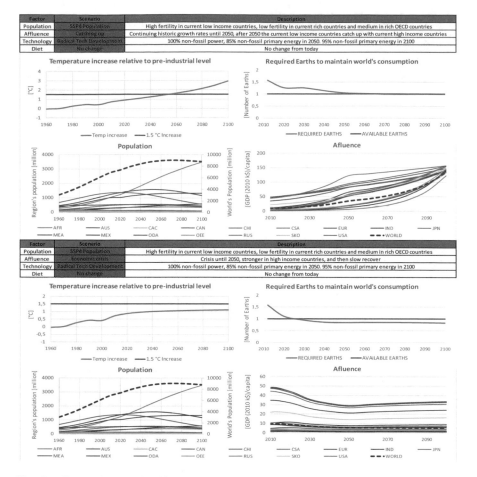

Fig. 6 Affluence impact. Catching up scenario (above) versus economic crisis (below)

by the end of the century, whereas the second illustrating the impact of a global economic stagnation. They both assume the SSP4 population scenario, a very optimistic technology development, and no change in diet. The results show that the "economic crisis" future would achieve the climate targets, while in the "Catching up" scenario the temperature increase would reach 3 °C.

In Fig. 5 the same high economic growth rate was applied to all countries, resulting in an increased gap between rich and poor countries in the future. Continued global development inequality may make it easier to limit ecological impacts. If the low income countries of today, who also are projected to have the largest share of the global population in the future, achieve the same level of wealth per capita as the current high income countries, like in the "Catching up" scenario (Fig. 6), then the challenge becomes much more difficult due to increased cumulative emissions.

4.4   The Effect of Technology

Is technological development then the solution? Most of the previous scenarios, assume a radical technology development, and because of that, the results may not appear to be critical in terms of temperature increase and footprint. However, what would happen if GDP grows considerably ("Shifting power" scenario)? Would the technological development still be enough to stabilise temperature increase? This is what is investigated in Fig. 7, which compares the climate impact of three scenarios with high population and high increase in affluence per capita, no change in diet, with different technology developments. The results show that, even in the best case, the expected temperature increase is more than 3 °C, and the needed Earths are 1.4 by 2100. The other two technology scenarios are much worse. Therefore, focusing on technological development alone is most likely not sufficient to combat climate change when having high growth in global GDP.

4.5   The Effect of Diet

Diet can also play an important role in the transition towards a sustainable future. In Fig. 8 a future with SSP1 assumptions for population and affluence (GDP/capita), and a strong technology development is combined with three different diets: USA or EU diet adopted worldwide, no change or India's diet adopted worldwide. The results show that diet has a stronger impact on the ecological footprint than on temperature increase. The impact is dependent on population, affluence and the energy system configuration, and will therefore vary considerably in scenarios with different assumptions for these parameters.

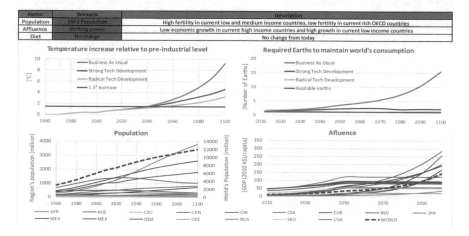

Fig. 7 Technology development impact

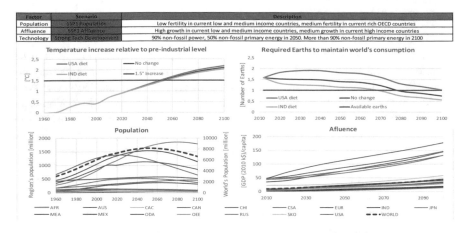

Fig. 8 Diet impact on climate and global footprint

4.6 A Future with a Stagnating Economy and Decreasing Population

As an illustrative scenario of the importance of economic and population development, (Fig. 9) shows what could happen if income per capita stays at the same level as today until 2100 and population decreases to 3 billion in the same period. This would lead to a decrease in global GDP down to one third of today by 2100.

With such a decrease in GDP the world could continue using the same fossil fuel mix as today and still slow global temperature increase to around 1.5 °C by 2100 while bio-capacity surplus would be increased at the same time. Reversing the

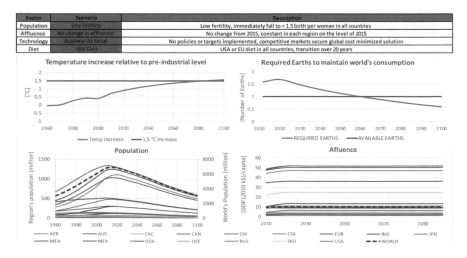

Fig. 9 Future with fossil fuels achieving the climate targets by 2100

outcome of this experiment is interesting: using fossil fuels at same shares as today requires that, the increase in income per capita stop and the global population is reduced to 3 billion to stabilise global temperature. However, this future is still far from equitable for all regions of the world as the income inequality continues.

4.7 A Path Towards an Economically Converging and Well Below 2 °C World

Another future (Fig. 10) considers a path towards a well below 2 °C and equitable world where GDP per capita is assumed to converge to the same level by 2100 in all regions. Population decreases following the low fertility scenario, technology improves radically, and everyone adopts India's low meat diet to minimize their climate impact. The output is a peaking temperature increase of 1.7 °C by 2100 and a very low resource pressure on the Earth. This future shows that an equal world is possible while avoiding undesired environmental consequences. However, it requires redistribution of wealth and strong focus on family planning.

4.8 A Possible but Unfair Well Below 2 °C World

As mentioned earlier: technological development alone is likely not enough to reach the ambitious Paris Agreement goals, but on the other hand, without a fast transition to a net zero carbon energy system, the chance to stay well below 2 °C temperature increase is low.

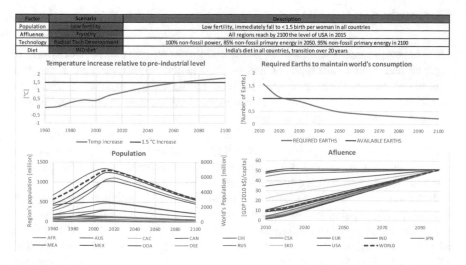

Fig. 10 Optimistic future: a climate friendly and economically equal world

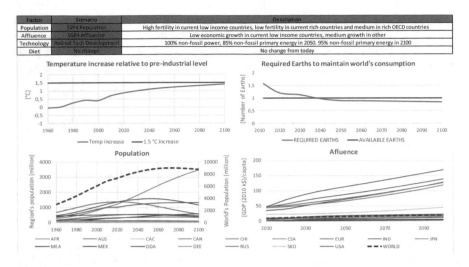

Fig. 11 Possible pathway: a stabilised low temperature but un-equal world

Figure 11 shows a future with the SSP4 storyline, which assumes a continued high fertility in low-income countries and low fertility in high income countries, with low economic growth in low income countries and medium growth in others, a radical technology development, and no change in diets. In this scenario, the results show that the temperature increase can stay below 1.5 °C and that the ecological footprint will not exceed one Earth. Although positive when focussing on environmental impacts, this scenario implies increased inequality with increasing pressure on bio-capacity in the low-income countries.

5   Conclusion

This chapter presents a combination of methodologies to give a broad assessment of regional and global development and environmental impact through a range of interrelated factors. The presented IPAT(D) model's purpose is to illustrate the importance of not only thinking of technological development in climate mitigation but also considering population and the effects of the distribution of affluence between regions. IPAT(D) can, besides from increasing awareness in general about dependencies between population, diet, affluence and technology, also be used as a tool for creating storylines for IAM model input drivers, by testing different combinations before running fully integrated assessment models.

IPAT(D) illustrates especially the importance of population development. A taboo on the population control is widespread among decision makers, who wish to appear neutral on this sensitive issue of people's choice of family size. However, all political decisions have indirect impacts on fertility rates through tax systems, education, health care, social security, etc. Lowering fertility rates would gradually facilitate improved standards of living in general and provide environmental benefit through the reduced ecological footprint of fewer children and all their future descendants.

Rapid technological development is also a key factor in achieving well below 2 °C scenarios. Only in the scenarios with radical low carbon technology development, can the target be met, unless a global long lasting economic recession takes hold.

Regarding economic development, the IPAT(D) scenarios illustrate how the global GHG emissions increase if the low income countries catch-up with the high income countries, because of the size of their projected population. High economic growth in low income countries will result in a high global GDP and thereby high demand for energy and food following historical trends. This means it is much easier to reach climate targets if economic inequality between regions of the world continues, which is an unpleasant outcome under these model assumptions. Therefore, aiming to develop and support a fair and more equal pathway to a sustainable world, then limiting economic growth in the current high income countries and decreasing fertility in current low income countries would probably be needed, where all countries converge on a lower level of material consumption than today.

Diet also has an important role to play in the transition towards a well below 2 °C and low ecological footprint world. The European and American diet, which contains relatively high levels of carbon intensive animal protein, is not sustainable if developing nations continue to adopt it. Therefore, achieving a lower ecological footprint would require developed regions, such as the US and EU, to change towards a lower carbon intensity diet and the rest of the world not to adopt the current European and American diet. If the world population stabilizes at 10–12 billion then the average diet probably needs to look more like what is seen in India today—an almost vegetarian diet.

A better assessment of non-CO_2 greenhouse gases would certainly be welcome in further work. Those emissions are driven mostly by land-use/agriculture and these emissions are not detailed in TIAM-World. The share of CO_2, CH_4 and N_2O in total emissions depends a lot on the assumptions on the type of agriculture and on the mitigation options available for this sector. Different demographic developments will have an impact on the structure of agriculture and land-use. This is partly handled in the Ecological footprint method represented by bio-capacity and footprint, but GHG related to land-use change are not evaluated in detail here either. There are no easy solutions to this, other than linking the proposed modelling system with a dedicated land-use model.

References

Boserup E (1981) Population and technological change. University of Chicago Press, Chicago
Ehrlich P, Holdren J (1971) Impact of population growth. Science (New Science) 171 (3977):1212–1217

Engelman R (2011) An end to population growth: why family-planning is key to a sustainable future. Solutions Sustain Desirable Future 2(3). http://www.thesolutionsjournal.com

FAOSTAT (2017) Food and agriculture organization of the united nations data. Available at: http://www.fao.org/faostat/en/#data

Jespersen J, Chick V (2016) John maynard keynes (1883–1946). Faccarello IG, Kurz HD (red), Handbook on the history of economic analysis: great economists since petty and boisguilbert, vol. 1. Edward Elgar Publishing, Incorporated, Cheltenham, UK, s. 468–483

Kanors-EMR (2017) TIAM-WORLD. Available at: http://www.kanors-emr.org/models/tiam-w

Kriegler E et al (2012) The need for and use of socio-economic scenarios for climate change analysis: a new approach based on shared socio-economic pathways. Global Environ Change 22:807–822

Loulou R (2008) ETSAP-TIAM: the TIMES integrated assessment model Part II: Mathematical formulation. Comput Manag Sci, Spec Issue Managing Energy Environ 5(1–2):41–66

Loulou R, Labriet M (2008) ETSAP-TIAM: the TIMES integrated assessment model Part I: model structure. Comput Manage Sci, 5(1):7–40. https://EconPapers.repec.org/RePEc:spr:comgts:v:5:y:2008:i:1:p:7–40

Lütken OD (1758) An enquiry into the proposition that the number of people is the happiness of the realm, or the greater the number of subjects, the more flourishing the state. Danmarks og Norges Oeconomiske Magazin

Malthus T (1798) An Essay on the principle of population. Available at: http://www.esp.org/books/malthus/population/malthus.pdf

Matthews HD, Gillett NP, Stott PA, Zickfeld K (2009) The proportionality of global warming to cumulative carbon emissions. Nature 459(7248):829–832

Moss RH, Edmonds JA, Hibbard KA, Manning MR, Rose SK, Van Vuuren DP, Carter TR, Emori S, Kainuma M, Kram T, Meehl GA, Mitchel JFB, Nakicenovic N, Riahi K, Smith SJ, Stouffer RJ, Thomson AM, Weyant JP, Wilbanks TJ (2010) The next generation of scenarios for climate change research and assessment. Nature 463(7282):747–756

NASA (2017) Data. Available at: http://climate.nasa.gov/vital-signs/global-temperature/

Nørgård J, Xue J (2017) From green growth towards a sustainable real economy, real world economics review, #80. http://www.paecon.net/PAEReview/issue80/NorgaardXue80.pdf

O'Neill BC et al (2014) A new scenario framework for climate change research: the concept of shared socioeconomic pathways. Clim Change 122:387–400

Ranganathan J et al (2016) Shifting diets for a sustainable food future. Available at: http://www.wri.org/sites/default/files/Shifting_Diets_for_a_Sustainable_Food_Future_1.pdf

Rees W, Wackernagel M (1996) Our ecological footprint: reducing human impact on the earth. New Society Publishers, Gabriola Island, BC

Shrinkthatfootprint (2017) Life cycle assessment data. Available at: http://shrinkthatfootprint.com/food-carbon-footprint-diet

Simon J (1980) Resource, population, environment: an oversupply of false bad news. Science 208:1431–1437

Simon J (1981) Environmental disruption or environmental improvement? Social Sci Q 62(1):30–43

The Global Footprint Network (2017) Data. Available at: http://www.footprintnetwork.org/licenses/public-data-package-free-edition-copy/

Wackernagel M et al (2002) Tracking the ecological overshoot of the human economy. Proc Nat Acad Sci 99:9266–9271

Wackernagel M et al (2014) Ecological footprint accounts, handbook of sustainable development (second revised edition). Edward Elgar Publishing, Cheltenham, Glos, UK

Worldbank (2017) Data. Available at: http://data.worldbank.org/

Part II
The Diversity of the National Energy Transitions in Europe

A Scandinavian Transition Towards a Carbon-Neutral Energy System

Pernille Seljom and Eva Rosenberg

Key messages

- A carbon neutral transport sector with no import of biofuels to Scandinavia requires use of hydrogen.
- Hydrogen production without access to CO_2 storage requires extensive investments in wind power and photovoltaics.
- Solid decision support from energy system models requires an appropriate representation of weather dependent renewable electricity generation and heat demand.
- The entire energy system, as included in TIMES models, should be considered when designing policy instruments to reach carbon neutrality.

1 Introduction

The Scandinavian countries, Denmark, Norway and Sweden, have ambitious climate targets that are in accordance to a well-below 2 °C warming of the global temperature from a preindustrial level. Scandinavia has a high share of renewables in the electricity generation mix, and a reduction of greenhouse gas (GHG) emissions will require extensive adaptions in other parts of the energy system, such as the building, transport and industry sectors. This study provides a cost-optimal adaptation of the Scandinavian energy system to reach a carbon neutral energy system by 2050, which is to a minimum extent dependent on

P. Seljom (✉) · E. Rosenberg
Institute for Energy Technology (IFE), Kjeller, Norway
e-mail: Pernille.Seljom@ife.no

E. Rosenberg
e-mail: Eva.Rosenberg@ife.no

© Springer International Publishing AG, part of Springer Nature 2018
G. Giannakidis et al. (eds.), *Limiting Global Warming to Well Below 2 °C: Energy System Modelling and Policy Development*, Lecture Notes in Energy 64,
https://doi.org/10.1007/978-3-319-74424-7_7

external markets. The Scandinavian region is assumed to be self-sufficient with biomass resources, have no Carbon Capture and Storage (CCS) in the power sector and no investments in hydrogen production from natural gas steam methane reforming (SMR). The analysis is conducted with a stochastic TIMES model (The Integrated MARKAL EFOM System) (Loulou et al. 2005a, b, c, 2008; Loulou and Labriet 2008) that provides investments that explicitly takes into account the weather dependent uncertainty related to renewable electricity generation and heat demand.

1.1 Scandinavian Characteristics and Energy Systems

The electricity mix in Scandinavia is unique. Denmark has a large share of electricity generation from Combined Heat and Power (CHP) and wind power, at 47 and 51% respectively in 2015 (Eurostat 2017). The electricity generation in Norway and Sweden is also distinctive, since the two countries have the largest hydro production among the EU countries, with a 96 and 47% hydropower share in the respective electricity generation mix in 2015 (Eurostat 2017) and have jointly 70% of the European hydro storage capacity.

The electricity market in the Scandinavian countries is highly integrated with each other and the European electricity market, and the spot price is the same in specific areas of the regions (Fig. 1). With flexible CHP plants, hydro reservoirs and transmission capacity to Europe, the Scandinavian region is well suited to integrate a larger share of intermittent electricity generation and to electrify a larger part of the end-use sectors such as buildings, transport and industry.

Sweden has twice as many inhabitants as Norway and Denmark, while the total GHG emissions are similar among the Scandinavian countries. This is primarily caused by structural differences of the energy sectors, manufacturing activities and a large agricultural sector in Denmark. The amount of combustion fuels and the use of CHP plants are approximately the same in Denmark and Sweden, but the type of fuels used differ. The use of coal and natural gas, relative to the consumption of solid biomass, is larger in Denmark compared to Sweden. This is reflected in the larger GHG emissions in the energy industry in Denmark compared to Sweden (Fig. 2). The emissions the in the energy industry is highest in Norway among the Scandinavian countries, due to the oil and gas production activity.

The energy consumption per capita is highest in Norway, followed by Sweden, and is about half in Denmark compared to Norway. In 2015, both Sweden and Norway were net electricity exporters, while Denmark was a net electricity importer. For the same year, the electricity generation was over five times higher in Sweden and Norway compared to Denmark.

The future Scandinavian energy consumption, electricity trade and electricity generation is uncertain due to various outcomes of the electricity market and climate parameters. For example, the inflow to the hydropower plants depend on melted snow and precipitation and the energy consumption is highly dependent on

Fig. 1 Illustration of Scandinavian spot price areas and external transmission connections (existing and under construction)

the outdoor temperature due to the cold climate. Key information that is relevant to this section, indicating the population and the energy systems characteristics of the Scandinavian countries are summarised in Table 1.

1.2 Scandinavian Climate Policies

A common feature of the Scandinavian countries is that they have ambitious climate targets that are in accordance with a well-below 2 °C warming temperature target.

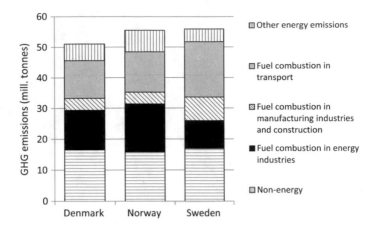

Fig. 2 2015 GHG emissions in Denmark, Norway and Sweden (Eurostat 2017)

Table 1 Key information of the Scandinavian countries (Eurostat 2017)

Country		Denmark	Norway	Sweden
Population	mill.	5.7	5.2	9.7
Gross inland energy consumption	TWh	195	350	530
Gross inland fossil fuel consumption	TWh	129	205	152
Final electricity consumption	TWh	31	111	125
Electricity generation	TWh	28	145	159
Net electricity export	TWh	−6	15	23

The Danish Climate law was adopted in 2014, with an objective of fossil fuel independency by 2050 (Danish Energy Agency 2017b). It is in relation to the climate law, established an independent council, that report to the Danish government annually. There are also ambitious local climate targets in Denmark, where for example Copenhagen, is planning to be the first carbon-neutral capital by 2025.

The Norwegian long-term target is a reduction of 80–95% of the GHG emissions compared to in 1990 by 2050 (The Norwegian Ministry of Climate and Environment 2017). Further, the Norwegian emissions need to be reduced by at least 40%, compared to a 1990 level, by 2030. The Norwegian climate act was adopted in March 2017 and will be in force from 1 January 2018. Climate targets of 2030 and 2050 will be established by law and annual updates on GHG emissions and projections has to be reported by the government.

Sweden has committed to zero GHG emissions into the atmosphere by 2045. The Swedish climate act will enter into force on 1 January 2018 (The Government of Sweden 2017). The purpose of the climate act is to ensure that the climate policy of the government considers the national climate goals, and propose measures on how to reach the climate goal. The government is required to present a climate report every year, and the government is required to draw up a climate policy action plan on how to achieve the climate goals every fourth year.

1.3 Previous Work on the Nordic Region

The TIMES modelling tool is used in the Nordic Energy Technology Perspective (NETP) study to analyse how the Nordic countries (Denmark, Finland, Iceland, Norway and Sweden) can achieve their ambitious goals, even below the global 2 °C Scenario (IEA 2013). This was a joint project with International Energy Agency (IEA), the Nordic research council and various Nordic research institutions. The study states that the energy related CO_2 emissions in the Nordic region needs to be lowered by 70% within 2050, compared to 1990, to be consistent with the 2 °C scenario of the global Energy Technology Perspective 2012 (IEA 2012). In a carbon neutral scenario, it is assumed that the CO_2 emissions need to be reduced with 85% by 2050 within the Nordic countries.

A conclusion of NETP is that it is possible to achieve a near decarbonisation of the Nordic energy system with a considerably increase in wind power production, nuclear power production and use of biomass. Passenger cars will to a large extent be electrified (by batteries or hydrogen) and road freight, aviation and shipping will use biofuels. To cover the large use of biofuels in the transport sector, the Nordic region will be a net importer of bioenergy. This is despite NETP present a net export of electricity from the region, at 50–100 TWh in the long-term. An update and strengthening of the NETP study is presented in (IEA 2016). In this report, there is a decrease in nuclear power generation, an increase in wind power production and the Nordic region is a net exporter of electricity with about 50 TWh/year in a carbon neutral energy system. Biofuels are primarily used to supply the long-distance passenger road traffic, heavy-duty road, marine freight and aviation.

In NETP, the energy demand decreases considerable, despite an increase in building area and industry production, due to more efficient heating of buildings and more energy efficient industrial processes. Another important assumption of NETP is the availability of Carbon Capture and Storage (CCS). However, the future availability of CCS is highly uncertain due to technical, market and legal issues. Also, the high biofuel import to the Nordic countries in the NETP is questionable in regards of sustainability. The large biomass resources in the Nordic countries make it difficult to justify a dependency of biomass import to the region. This study differs from the NETP studies regarding the regional coverage (Finland and Iceland is not included), the future energy demand, and availability of imported bioenergy and CCS. The presented analysis does not assume considerable reduction in energy demand, implementation of CCS, or import of bio products to the region. The CCS restriction excludes also hydrogen production from natural gas as a part of the solution in a carbon neutral scenario. Furthermore, natural gas based hydrogen production, by steam methane reforming, in Norway requires that there will be a market for hydrogen outside of Scandinavia due to the extensive investment requirements.

2   Methodology and Assumptions

2.1   TIMES Model Structure and Assumptions

TIMES is a bottom-up optimisation modelling framework that provides a detailed techno-economic description of resources, energy carriers, conversion technologies and energy demand from a social welfare perspective. The model minimizes the total discounted cost of the energy system to meet the demand for energy services. The model decisions are made with full knowledge of future events and assume free competition with no market imperfections. To provide a macroeconomic solution, current policy instruments are excluded in this study, including taxes and subsidies. Further, the annual discount rate is set to 4%.

The model periods are every fifth year within the time horizon from 2010 to 2050. To consider seasonal and daily variations in energy supply and demand, each model period is represented by 12 2-h steps for a representative day of four seasons, giving 48 time-slices in total. While investments are made for each mode period, the operational decisions are optimised on the two-hourly daily level to satisfy the energy demand at least cost. The operational decisions include the activity level of each installed capacity type such as the electricity generation, fuel use and electricity trade.

The model is regionally divided into the spot price areas, with two regions in Denmark, five regions in Norway and four regions in Sweden (Fig. 1). The model has endogenous investment options in new capacity expansions to Europe, but the capacities within the Scandinavian model region are fixed.

The model includes a set of technologies to transform energy sources to final demand, including conversion processes such as electricity and heat generation technologies and demand technologies such as boilers and vehicles. The characterisation of the energy technologies, such as cost data and efficiencies, are exogenous input to the model and are inter alia based on Lind and Rosenberg (2013) and NVE (2015). Future energy demand of heat, transport and non-substitutable electricity are exogenous input to the model and are based on reference energy projections for Denmark (Danish Energy Agency 2017a), Norway (Rosenberg and Espegren 2014) and Sweden (Swedish Energy Agency 2012, 2013).

2.2 Stochastic Modelling Approach Applied to Weather Dependent Parameters

This study applies a two-stage stochastic model (Kall and Wallace 1994; Higle 2005) to provide cost-optimal investments that explicit consider the weather dependent short-term uncertainty of the following stochastic parameters: Photovoltaics (PV) production, wind production, hydro production, heat demand in buildings and the electricity prices outside Scandinavia. Short-term uncertainty is characterised by uncertainty that are periodically recurring and that affect operational decisions only.

Each uncertain parameter is represented by 21 discrete realisations, called scenarios, which are assigned equal probabilities to occur. The scenarios are generated by random sampling from large data sets that are derived from historical data. Further, the scenarios are adjustments to match the statistical mean of the uncertain parameters. Each scenario represents a given outcome of the uncertain parameters. This study applies the same discrete scenarios as used in (Seljom et al. 2017), and applies the similar stochastic model formulation as the following papers (Seljom and Tomasgard 2015, 2017; Seljom et al. 2017).

To illustrate typical scenario characteristics, a selection of the model input are shown through a 25/75 quantile, minimum, maximum and median of the 21 stochastic scenarios (Figs. 3 and 4). Figure 3 clearly shows that the availability factors vary significantly between the scenarios and time of the day. Figure 4 demonstrates that the heat demand varies significantly by time of day and by scenario.

Figure 5. illustrates a scenario tree with the information structure of the two-stage stochastic model. In the first stage, the realisation of the operational scenarios is unknown and investments in new capacity for the entire model horizon, from 2010 to 2050, are made. In the second stage, starting at the branching point of

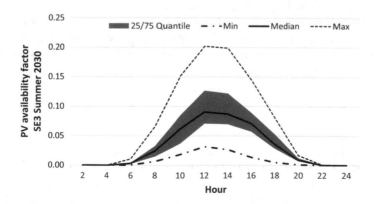

Fig. 3 Model input on PV availability for SE3 (Swedish spot price region with highest population) in summer 2030

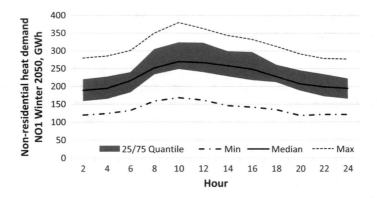

Fig. 4 Model input on heat demand for NO1 (Norwegian spot price region highest population) in winter 2050

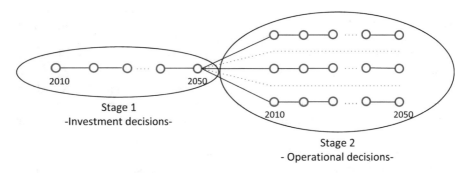

Fig. 5 Illustration of a two-stage scenario tree with short-term uncertainty

the scenario tree, the outcomes of the different scenarios are known, and operational decisions are made for each of the scenarios for the entire model horizon. Consequently, the investments are identical for all scenarios, whereas operational decisions are scenario dependent. The second stage decisions represent possible operational situation that can occur in the presence of short-term uncertainty related to renewable energy production and energy consumption. To consider the different operational situations in the optimisation, the TIMES model minimize the investment costs and the average of the operational costs for all scenarios. This gives investment decisions that recognize the expected operational cost, and that are feasible for all the model specified realisations of the uncertain parameters. To the author's knowledge, this is the first study that uses this stochastic approach in an analysis of the transition to a carbon neutral energy system.

Unlike a stochastic approach, a simplified deterministic model has only one operational scenario. Consequently, the investment decisions in a deterministic model do not take into account a range of operational situations that can occur.

2.3   Four Model Cases Based on Climate and Energy Prices

Four different model cases are analysed, with two different climate targets on climate neutrality under two different assumptions on the future prices of biomass, fossil fuels and electricity outside Scandinavia (Tables 2 and 3). Note that the model cases and stochastic scenarios are two different types of model input. The stochastic scenarios describe the short-term uncertainty of electricity supply and heat demand whereas the model cases represent different long-term realisation of climate targets and energy prices. The model input on the stochastic scenarios is therefore identical in the four model cases.

The first climate target, *Reference (REF)*, includes no restrictions on the use of fossil fuels towards 2050. The second climate target, *Climate Neutral (CN)*, involves a gradual phase out of fossil fuel consumption to meet the future energy demand towards 2050. The phase out is modelled by gradually constraining the CO_2 emissions from fossil fuels used in energy related processes, excluding the emission related to fossil fuel consumption in industrial processes. All model cases limit energy efficiency measures to the implementation of new buildings standards and the use of more efficient vehicles, have no CO_2 storage, and import no bio products to the Scandinavian region. The motivation behind these assumptions is elaborated in Sect. 1.3.

Two different price assumptions are included to provide robust model insights in light of the uncertain development of the future energy market. The two price assumptions have different assumptions with regard to long-term energy prices. The first price assumption, *P1*, assumes an increase in the long-term energy prices according to Table 3. The second price, *P2*, assumes constant energy prices towards 2050 where the energy prices of 2015 in Table 3 are used as a model input for all future model periods. Note that the electricity price in the Scandinavian region is endogenous whereas the electricity price outside Scandinavia is exogenous.

Table 2 Overview of model cases

Model case	Reference		Climate neutral	
	REF_1	REF_2	CN_1	CN_2
Climate target	No	No	−80% in 2030 −100% in 2050	
Energy prices	Increasing	Constant	Increasing	Constant
CCS	No use of CCS			
Hydrogen from SMR	No hydrogen from natural gas SMR			
Bioenergy import	No import of bio products to Scandinavia			

Table 3 Overview of energy price assumptions for increasing prices (P1)

	2015	2030	2040	2050
Fossil fuels (EUR/MWh)				
Coal	10	11	12	13
Natural gas	25	32	34	36
Oil	67	79	83	86
Diesel	76	89	92	95
Gasoline	73	85	89	92
Biomass (EUR/MWh)				
Pellet	40–67	47–79	49–82	50–85
Straw	22	26	27	27
Chips	27–39	32–45	33–48	39–55
Bio gas	84–121	84–121	84–121	84–121
Bio diesel	111–187	120–204	123–208	125–212
Electricity (EUR/MWh)				
Finland (FI)	31	121	165	208
Germany (DE)	35	126	171	214
Lithuania (LI)	35	126	171	214
Poland (PO)	35	126	171	214
The Netherlands (NL)	45	129	172	215
United Kingdom (UK)	68	139	179	222

3 Analysis of the Transition to a Carbon Neutral Energy System

This section presents the impact of a transition to a carbon neutral Scandinavian energy system towards 2050, implying no use of fossil fuels to meet the energy demand. Each model case proposes one investment strategy towards 2050 that is feasible for 21 different operational situations that can given the short-term uncertainty of electricity generation and heat demand. A model result is therefore 21 different outcomes of the operational decisions towards 2050. The analysis focuses on the electricity, transport and building sector, since these sectors are exposed to large changes with a phase out of fossil fuels. If not otherwise specified, this section reports the average of the 21 operational decisions.

3.1 How to Satisfy the Increasing Electricity Demand

The transition to a carbon neutral Scandinavian energy system requires an extensive electrification of the end-use sectors such as transport and buildings. This gives a significant increase in the electricity consumption, higher electricity prices and investments in new electricity generation capacity.

The expected value, the maximum and the minimum among the stochastic scenarios of electricity generation, consumption and net export are indicated (Table 4). For a given year, the range of each of these represents possible outcomes of the electricity sector given the short-term uncertainty of electricity generation, heat demand and electricity prices outside Scandinavia.

Carbon neutrality increases the electricity generation with 35% for price assumption 1, *P1*, and with 33% for price assumption 2, *P2*, in 2050. The additional electricity generation is primarily used to cover the increased electricity demand within Scandinavia. A significant part of the electricity consumption is used to produce hydrogen to the transport sector. The share of the electricity supplied to water electrolysis, related to the total increased electricity consumption, is 86% and 80% in 2050 for *P1* and *P2* respectively. This implies that the decarbonisation of the transport sector has a significant impact on the cost-optimal investments in new generation capacity.

Investments in wind power and PV highly increase with a carbon neutrality climate target (Fig. 6). This gives a higher share of intermittent renewables in the electricity generation mix, increasing the value of flexible solutions, such as the hydro reservoirs in Norway and Sweden, to ensure that the energy demand is met at all times. Since the impacts of climate neutrality on investments are similar for both price assumptions, these results are robust for different developments of the future energy market. For *P1*, the wind power capacity is increased with 519% in 2030 and 621% in 2050 with climate neutrality. Further, PV is not considered cost-optimal when fossil fuels are a part of the future energy system, but constitutes a large share of the electricity capacity when fossil fuels are phased out. In 2050 the PV capacity is 55 GW and 32 GW and the PV production is 52 TWh and 30 TWh for *P1* and *P2* respectively.

If the European electricity prices are kept at the current price level, it is to some extent more cost-optimal to import electricity than to build new generation capacity in Scandinavia. However, a significant net import of electricity to Scandinavia is most probably not a long-term solution. This is because decarbonisation policies in Europe can increase the general investments and price level in Europe, and that the

Table 4 Electricity balance for all model instances in 2030 and 2050

Minimum/Average/Maximum, TWh					
	Model case	REF_1	CN_1	REF_2	CN_2
2030	Generation	293/**333**/371	370/**412**/437	284/**330**/375	361/**406**/444
	Consumption	303/**318**/336	393/**400**/411	310/**316**/328	397/**404**/419
	Net export	−28/**0**/19	−41/−**7**/15	−47/−**3**/43	−58/−**18**/32
	Exp. loss	15	16	19	19
2050	Generation	334/**361**/381	452/**486**/505	313/**349**/391	430/**464**/491
	Consumption	322/**345**/363	464/**471**/488	333/**338**/355	467/**475**/489
	Net export	7/**0**/2	−36/−**7**/1	−42/−**6**/34	−58/−**32**/-4
	Exp. loss	16	17	23	21

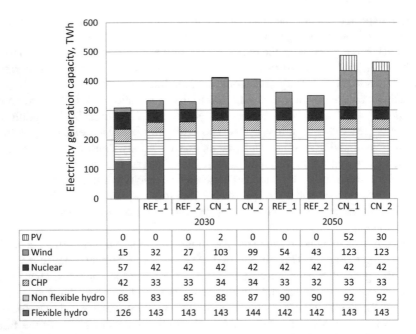

	REF_1	REF_2	CN_1	CN_2	REF_1	REF_2	CN_1	CN_2
			2030				2050	
▥ PV	0	0	0	2	0	0	52	30
▦ Wind	15	32	27	103	99	54	43	123	123
■ Nuclear	57	42	42	42	42	42	42	42	42
▨ CHP	42	33	33	34	34	33	32	33	33
▤ Non flexible hydro	68	83	85	88	87	90	90	92	92
■ Flexible hydro	126	143	143	143	144	142	142	143	143

Fig. 6 Electricity generation capacity in 2010, 2030 and 2050 for all model cases

Scandinavia region has better renewable resources compared too many other European countries. A net import of electricity to Scandinavia can therefore indicate a sub-optimal distribution of renewable electricity generation among the North European countries.

3.2 No Fossil Fuels in the Transport Sector

The fuel mix composition in the transport sector is significantly changed in a carbon neutral energy system (Fig. 7). In the carbon neutral transition, the fossil fuels are gradually replaced with more biofuels, electricity and hydrogen. It is also a decline in total fuel consumption with the climate neutral policy since electric and hydrogen vehicles are more fuel efficient than fossil fuelled vehicles. Due to the assumptions on no import of biofuels to the Scandinavian region, and the limited use of electricity in heavy vehicles, hydrogen is in the long-term the dominant fuel used in a carbon-neutral transport sector. It is cost-optimal to invest in a combination of both base load hydrogen production, with no flexibility at lower investment costs, and flexible hydrogen production with higher capital costs. In 2050, the share of flexible production capacity of the total hydrogen capacity is 53% and 69% for *CN_1* and *CN_2* respectively. Based on these observations, it is clear that flexible hydrogen

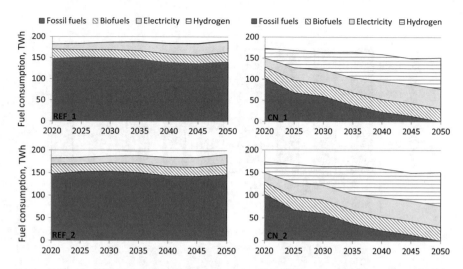

Fig. 7 Transport fuel consumption from 2020 to 2050 for all model cases

production, that is adjustable to the electricity price, can facilitate a large integration of intermittent renewable electricity generation into the energy system.

3.3 Biomass, Heat Pumps and Efficient Electric Heating in the Building Sector

The heat supply from biomass technologies and heat pumps increase, whereas the heat supplied with low-efficient electricity based heating decreases with climate neutrality (Fig. 8). For example for price assumption 1, *P1*, there is a 122% increase in biomass and a 29% increase in heat supply by heat pump in 2050. The heat is used to satisfy the heating demand in residential and commercial buildings.

Note that the y-axis starts at 75 TWh to better illustrate the differences between the technologies beyond district heat (DH) supply. The technology named *Elc* includes both direct electric heating and electric boilers that have considerably lower efficiency than heat pumps.

3.4 Impact of Using a Stochastic Model

This section compares investment decisions obtained from using a stochastic and a deterministic modelling of the uncertain parameters. The two model types are similar, except for the representation of the uncertain parameters, where the average value of the uncertain parameters is used as a model input in the deterministic

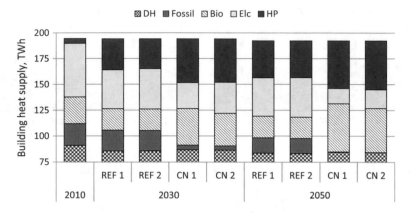

Fig. 8 Building heat supply in 2010, 2030 and 2050 by technology

approach. This implies that the stochastic model, with a more realistic representation of the energy sector, requires higher computational effort than its deterministic equivalent. A deterministic modelling approach cannot be justified if the model
results differ significantly from the stochastic approach since it can cause misleading model insights.

The modelling approach affects primarily the electricity and building sector
(Figs. 9 and 10). In general, the stochastic approach invests in significant more
capacity than the deterministic approach, and consequently the deterministic
approach underestimate investment needs that are required to meet the future
energy demand under carbon neutrality. This is caused by stochastic modelling,

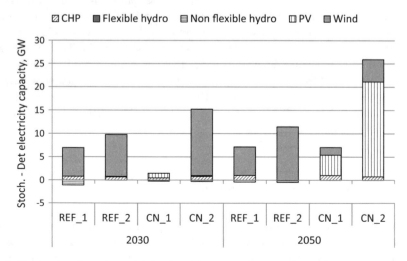

Fig. 9 Stochastic minus deterministic electricity generation capacity in 2030 and 2050 for all
model cases

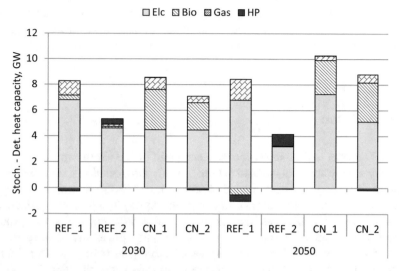

Fig. 10 Stochastic minus deterministic buildings heat capacity in 2030 and 2050 for all model cases

unlike the deterministic approach, explicitly models a range of operational situations which can occur. This includes for example periods with poor wind conditions, low hydro inflow and high European electricity prices, and periods with excess power supply in Scandinavia and limited export opportunities.

In the electricity sector, the stochastic approach gives higher investments than the deterministic approach, ranging from 4% in *CN_1* to 21% in *CN_2* in 2050. It is the investments in wind power and PV that differs most with the modelling approach. For example, for *CN_2* in 2050, the PV capacity is 173% higher for the stochastic modelling approach. Note that it is not given that a stochastic modelling approach has higher investments in electricity generation capacity. For example in (Seljom and Tomasgard 2015, 2017; Seljom et al. 2017), the stochastic approach has significantly lower investments in wind power capacity. However, none of these listed papers restrict the future use of fossil fuels, whereas a fossil fuel phase out is the core assumption of this study.

In the building sector, the stochastic approach has more heating capacity than the deterministic approach, ranging from 6% in *REF_2* to 14% in *REF_2* in 2050. For all model cases, the stochastic methodology invests in more low efficient electricity technologies whereas the influence on biofuels, gas and HP depends on period and model case. It is particularly the installed capacity for direct electric heating and electric boilers that is affected by the representation of the uncertain parameters. This can be explained by the fact that the stochastic approach, with scenario dependent electricity prices, invests in additional low-cost electricity-based heating, to meet the heat demand in periods with low electricity prices. Although the difference in heat capacity between the two approaches is relative small compared to the total installed capacity, it has a significant impact on the model results for some model cases. For example, for *REF_1* in 2050, the stochastic low-cost electricity heating (Elc) is 150% higher than the capacity obtained from the deterministic model.

4 Conclusion

A TIMES model with a stochastic representation of the short-term uncertainty of electricity supply and heat demand is used to study the transition to a carbon neutral Scandinavian energy system towards 2050. The stochastic approach is valuable since it provides cost-optimal investments in the energy system that considers a set of operational situations and that explicitly value flexibility. For this study, a simplified deterministic approach gives significantly lower investments in wind power, PV and low-efficient electricity based heating in buildings. This confirms that the method used to represent short-term uncertainty can significantly affect the model results, and a simplified representation of this uncertainty can give misleading model insights. Consequently, an appropriate representation of short-term uncertainty is an important prerequisite to provide reasonable policy recommendations from energy system models.

This study presents a novel analysis since it uses a different set of model assumptions than previous studies addressing a Scandinavia transition towards carbon neutrality. Here, the carbon neutrality strategy does not rely on import of biofuels to Scandinavia and it is assumed no CO_2 storage. With these assumptions, a phase-out of fossil fuels requires major changes to the Scandinavian energy system. It involves a major electrification of the end-use sectors, in particular the transport sector. The electricity capacity needs to be about double the size compared to the current capacity level, and the share of intermittent wind power and PV in the electricity generation mix increases considerably. A large share of the increased electricity generation is used to produce hydrogen to the transport sector from water electrolysis, and hydrogen is the dominant fuel in the transport sector.

The results emphasise the importance of considering the entire energy system when designing policy to reach carbon neutrality. This is because the required investments in electricity capacity are highly dependent on the degree of electrification in the transport and building sectors. Further, the degree of electrification depends on the future availability, competiveness and sustainability of biofuels.

References

Danish Energy Agency (2017a) Basisfremskrivning 2017. Available at https://ens.dk/sites/ens.dk/files/Forsyning/bf2017_hovedpublikation_13_mar_final.pdf

Danish Energy Agency. (2017b) Dansk klimapolitik (Danish Climate Policy). Available at https://ens.dk/ansvarsomraader/energi-klimapolitik/fakta-om-dansk-energi-klimapolitik/dansk-klimapolitik

Eurostat (2017) Environment and energy. Available at http://ec.europa.eu/eurostat/data/database

Higle LJ (2005) Stochastic programming: optimization when uncertainty matters. Tutorials in Operational Research, New Orleans, INFORMS

IEA (2012) Energy technology perspectives 2012: pathways to a clean energy system. France, Paris

IEA (2013) Nordic energy technology perspectives: pathways to a carbon neutral energy future. Paris, France, IEA and Nordic Energy Research. Available at http://www.iea.org/media/etp/nordic/NETP.pdf

IEA (2016) Nordic energy technology perspectives 2016—cities, flexibility and pathways to carbon-neutrality. IEA and Nordic Energy Research, Paris, France

Kall, P, Wallace SW (1994) Stochastic programming. John Wiley & Sons, Chichester

Lind A, Rosenberg E (2013) TIMES-Norway model documentation. Kjeller, Norway, Institute for Energy Technology

Loulou R (2008) ETSAP-TIAM: the TIMES integrated assessment model. part II: mathematical formulation. CMS 5(1–2):41–66

Loulou R, Labriet M (2008) ETSAP-TIAM: the TIMES integrated assessment model Part I: model structure. CMS 5(1–2):7–40

Loulou R, Lehtila A, Kanudia A, Remme U, Goldstein G (2005a) Documentation for the TIMES model—part III. Energy technology systems analysis programme. Available at http://iea-etsap.org/docs/TIMESDoc-GAMS.pdf

Loulou R, Remme U, Kanudia A, Lehtila A, Goldstein G (2005b) Documentation for the TIMES model—Part I. Energy technology systems analysis programme. Available at http://iea-etsap.org/docs/TIMESDoc-Details.pdf

Loulou R, Remme U, Kanudia A, Lehtila A, Goldstein G (2005c) Documentation for the TIMES Model—Part II. Energy technology systems analysis programme. Available at http://iea-etsap.org/docs/TIMESDoc-Details.pdf

NVE (2015) Kostnader i energisektoren. Kraft, varme og effektivisering. The norwegian water resources and energy directorate. Oslo, Norway. Available at http://publikasjoner.nve.no/rapport/2015/rapport2015_02a.pdf

Rosenberg E, K. A. Espegren (2014) CenSES energiframskrivinger mot 2050. Available at https://www.ife.no/no/publications/2014/ensys/censes-energiframskrivinger-mot-2050

Seljom P, Lindberg KB, Tomasgard A, Doorman G, Sartori I (2017) The impact of zero energy buildings on the scandinavian energy system. Energy 118(Supplement C): 284–296

Seljom P, Tomasgard A (2015) Short-term uncertainty in long-term energy system models—A case study of wind power in Denmark. Energy Econ 49(Supplement C): 157–167

Seljom P, Tomasgard A (2017) The impact of policy actions and future energy prices on the cost-optimal development of the energy system in Norway and Sweden. Energy Policy 106 (Supplement C): 85–102

Swedish Energy Agency (2012) Färdplan 2050. Available at http://www.energimyndigheten.se/Global/F%C3%A4rdplan%202050%20Bost%C3%A4der%20och%20lokaler.pdf

Swedish Energy Agency (2013) Långtidsprognos 2012. Available at http://www.energimyndigheten.se/Global/Statistik/Prognoser/L%C3%A5ngsiktsprognos%202012.pdf

The Government of Sweden (2017) The climate policy framework. Available at http://www.government.se/articles/2017/06/the-climate-policy-framework/

The Norwegian Ministry of Climate and Environment (2017) Climate Act (Klimaloven), Prop. 77 L (2016–2017). Available at https://www.regjeringen.no/no/dokumenter/prop.-77-l-20162017/id2546463/

Net-Zero CO_2-Emission Pathways for Sweden by Cost-Efficient Use of Forestry Residues

Anna Krook-Riekkola and Erik Sandberg

Key messages

- Electricity, district heating and space heating can be close-to-zero emissions by as early as 2025; carbon neutrality of transport and industry sectors is more challenging.
- The focus by decision makers on the use of biofuels within the transport sector must be expanded to all sectors and include carbon capture and storage.
- Using existing biomass more efficiently, integrating biofuel production with the pulp and paper industry to use residues, and using waste heat for district heating are promising measures.
- Competitiveness of biofuels in transport would require a carbon price that is flexible with respect to fluctuations in oil price.
- TIMES-Sweden was efficiently used to explore how Sweden can convert to a carbon-neutral energy system by 2045.

1 Introducing the Swedish Case

In July 2016, the Swedish Government's Cross-Party Committee on Environmental Objectives tabled a National Climate Policy Framework and a Climate and Clean Air Strategy. The Strategy included immediate and long-term climate targets as well as strategies on how to improve urban air quality. Sweden defines its *climate target* as achieving net-zero greenhouse-gas (GHG) emissions by 2045, which beside domestic

A. Krook-Riekkola (✉) · E. Sandberg
Energy Science Unit, Luleå University of Technology, 971 87 Luleå, Sweden
e-mail: anna.krook-riekkola@ltu.se

E. Sandberg
e-mail: erik.sandberg@ltu.se

© Springer International Publishing AG, part of Springer Nature 2018
G. Giannakidis et al. (eds.), *Limiting Global Warming to Well Below 2 °C: Energy System Modelling and Policy Development*, Lecture Notes in Energy 64,
https://doi.org/10.1007/978-3-319-74424-7_8

reductions can be achieved by (i) GHG emission reductions abroad, (ii) bio-energy with carbon capture and storage (BECCS), and (iii) increased carbon dioxide (CO_2) uptake in land use, land-use change and forestry (LULUCF). The three reduction measures mentioned above are essentially in place to compensate for emissions with no recognised reduction alternatives, such as GHGs from agriculture and some industrial processes, which are estimated to account for 15% of GHGs. Consequently, Sweden has committed to an 85% reduction of domestic GHGs by 2045 in comparison with its 1990 levels (SOU 2016:47, 2016). Furthermore, when there are alternatives to fossil fuels, the Swedish Government has translated the 85% reduction in domestic GHGs to a net-zero CO_2-emission by 2045.

Sweden is rich in natural resources. This offers both advantages and challenges in meeting large CO_2 reductions. Indeed, among European Union (EU) countries, Sweden has the largest share of renewable energy sources in primary supply (Eurostat 2017). In 2015, the country's share of fossil fuels in its final energy consumption for buildings was less than 3%, while less than 3% of generated electricity was derived from fossil fuels (SEA 2017). However, the abundance of natural resources has also resulted in fossil fuel dependant industry—the most pronounced being the iron and steel, cement, and copper industries. The high consumption of fossil fuels by these industries generates nearly 30% of domestic GHGs. Sweden also has a large pulp and paper industry, but this relies on biomass for its energy demand.

Next to hydro and wind power, biomass plays an important role in the Swedish energy system. The country's main source of biomass is forestry residues and forest industry by-products, where the potential feedstock is linked to the outtake of logs used for non-energy purposes. Börjesson (2015) assumes the potential of biomass will increase by 144 PJ by 2030 and 216 PJ by 2050, respectively, compared with current levels of use. Nonetheless, this potential biomass increase cannot replace today's use of fossil fuels as a primary energy source, which was 584 PJ in 2015, for example. It would be important, therefore, to identify how and where biomass is cost-efficiently used so that Sweden's net-zero CO_2-emission target can be met. This can be assessed with energy system optimisation models (ESOMs).

In this context, the aim of the chapter is to identify cost-efficient strategies for the Swedish energy system in its quest to achieve net-zero CO_2 emission. Different net-zero CO_2-emission pathways until 2050 are assessed by using TIMES-Sweden, an ESOM of the comprehensive Swedish energy system.

2 TIMES-Sweden to Explore Future Net-Zero CO_2-Emission Pathways

TIMES-Sweden is built on the TIMES (The Integrated MARKAL-EFOM System) model platform. TIMES is an energy-economic model generator for developing comprehensive energy system optimisation models on a local, national or multi-regional level, and provides a technology-rich basis for estimating energy dynamics over a long-term, multi-period time horizon (Loulou et al. 2016a, b). The Pan-European and Swedish TIMES models have the same main structure, as described in RES2020 (2009) and Simoes et al. (2013), but TIMES-Sweden has been enhanced to describe the Swedish situation more specifically (Krook-Riekkola 2015).

TIMES-Sweden is structured in accordance with Eurostat energy statistics (see Eurostat 2004) and is divided into seven main sectors: Supply, Electricity and District Heating, Industry, Agriculture (also includes fishery and forestry), Commercial, Residential and Transport.

The marginal prices of each final energy commodity (delivered energy) are a result of the cost-optimisation, while globally imported energy products (coal, natural gas and oil) and the price of different biomass fractions (branches and tops, agriculture waste, sawdust and other industry by-products) are exogenous parameters.

2.1 Key Scenario Assumptions

The central scenario (*REFCLIM*) considers that the World is aiming to keep the rise in global temperatures to "well below two-degrees Celsius above pre-industrial levels", in line with the Paris Agreement (UN 2015). This premise has an impact on the assumed CO$_2$ price within the EU Emissions Trading System (EU ETS), as well as on the assumed price of oil, gas and coal products (Tables 1 and 2). *REFCLIM* price projection is based on the PRIMES modelling runs from Autumn 2015 (retrieved through personal communication with Swedish Environmental Protection Agency). Two alternative baseline scenarios, *REFPRIMES* and *REFLOW,* are proposed with less ambitious development of the EU ETS targets and therefore lower projected EU ETS carbon prices. *REFPRIMES* consider a strengthening of the EU ETS target as proposed by the European Commission (2013) at that time, but is not sufficient to be in line to meet a 2 degree target. The *REFLOW* price projection represents a scenario with a weak climate treaty, in which the carbon price is assumed to be constant beyond 2025.

Table 1 Projected EU ETS carbon prices (EUR$_{2005}$/tCO$_2$)

	EU ETS price projection								
	2010	2015	2020	2025	2030	2035	2040	2045	2050
Severe climate treaty (REFCLIM)	5.3	8.6	17.3	23.0	34.9	72.8	145	246	265
PRIMES reference price (REFPRIMES)	5.3	8.6	17.3	23.0	30.5	42.1	70.9	96.8	100
Weak climate treaty (REFLOW)	5.3	8.6	17.3	23.0	23.0	23.0	23.0	23.0	23.0

Table 2 Projected fossil fuel import prices (EUR_{2005}/GJ)

	Import price projection						Source of statistics	
	2010	2015	2020	2030	2040	2050	2010	2015 and later
Crude oil	10.1	7.2	10.2	10.4	10.4	10.4	IEA (2012)	IEA (2015)
Natural gas	6.4	6.4	6.7	6.9	6.9	6.9	IEA (2012)	Statista (2015)
Coal	2.7	2.1	2.3	2.3	2.3	2.3	IEA (2012)	Assumed constant beyond 2020

The existing demand in 2000, 2005 and 2010 are based on statistics from Eurostat, Statistics Sweden and the Swedish Energy Agency. Thereafter, projection drivers are applied to calculate the future demand; those drivers are specified for each demand segment and attained by soft-linking the TIMES-Sweden model with a national computable general equilibrium model of Sweden (EMEC) (Krook-Riekkola et al. 2013, 2017a). Demand projections for heating water and space are based on assumptions about an increased population and fewer individuals per household (i.e. more dwellings are required), together with an assumption about dwellings having improved insulation (resulting in a lower heating demand per square metre) in line with relevant building code regulations (Boverket 2014). The demand for transportation is aligned with the assumptions underlying the work by the Swedish Environmental Protection Agency (SEPA 2015). Base-year data were updated according to current biofuel plants (various sources, including Grahn and Hansson 2015) and the present stock of vehicles (Transport Analysis 2015). Thus, the model incorporates the recent trend towards an increased use of biofuel as well as the shift from petrol to diesel vehicles. Furthermore, in line with the Swedish Transport Administration's 2014 recommendations for large public infrastructure investments, we applied a socio-economic discount rate of 3.5% (SIKA 2014).

Sectors not included in the EU ETS have carbon, environmental and energy taxes based on levels dictated by the Swedish Tax Agency from time to time (Swedish Tax Agency 2008a, b, 2012). For the period modelled in our study, we assumed the 2015 tax levels be kept intact for all remaining years, with the exception of biofuels and natural gas for transportation which were exempt from all taxes in 2015 in order to facilitate fair competition with mature fuels. In the model we assumed these fuels would gain critical momentum from 2020, i.e. (i) should bear their external costs, and (ii) are in the model subjected to taxes from 2020 onwards. The present Swedish Electricity Certificate System is modelled endogenously by defining quotas in line with Swedish Energy Agency directives (SEA 2015), as applied to final electricity consumption in the Agriculture, Commercial and Residential sectors.

Existing biomass mainly consists of forest residues and some biogas. In this study, we assume an increase of biomass from forestry residues, energy forestry and energy crops from today's (2015) levels by 144 PJ by 2030, and by 216 PJ by 2050 (Börjesson 2015). Furthermore, this assumed level of biomass increase is still based on an active forest industry under the 1993 Swedish Forest Act with two overarching equal objectives on production and environment, which the Swedish Energy Protection Agency narrows down as follows: "[F]orests and forest lands should be used effectively and responsibly so they produce sustainable yields" (SEPA 2017). Also, since the potential increase of biomass has the same underlying assumptions as the assumed LULUCF in the definition of the Swedish GHG target, using the biomass will not prevent the Swedish GHG target from being met.

2.2 Model Improvements for Assessing Net-Zero CO$_2$ Emissions

This study represents the first time that the TIMES-Sweden Model has been used to assess climate policy up to 2050. Previously, the model did not include enough options to achieve net-zero CO$_2$ emissions. We have therefore updated it to encompass more biomass and electricity options as well as more carbon, capture and storage or use (CCS/U) technologies. More specifically, the technology database was expanded with respect to biofuel plants, electric and biofuel vehicles, and with CCS options within the cement, iron and steel and pulp and paper industry. Clearly, considerable uncertainties remain in the latter data, due to technologies still requiring demonstration in industrial environments (Hoenig et al. 2009, 2012). To deal with those uncertainties, special attention has been paid to the differences in energy flows when adding carbon capture to existing industrial sites, in other words, to identify the additional energy needed when adding carbon capture. Biomass details and options incorporated in TIMES-Sweden is largely documented in Krook-Riekkola et al. (2017b), although the new biomass reference energy system and industrial processes were not implemented in TIMES-Sweden at the time of the analysis presented in this chapter. Main sources of CCS information and how they were interpreted are as follows:

- Techno-economic cost assumptions on CCS/U within the cement industry: (IEAGHG 2013a), covering costs for both retrofitting of CCS/U technologies to already existing cement plants and installations of new cement plants combined with CCS/U technologies. Post-combustion capture, full oxyfuel combustion, and partial oxyfuel combustion are considered as promising technologies. Techno-economic cost assumptions on CCS/U within the iron and steel industry: IEAGHG (2013b), covering only. CCS options applied to the blast furnace are covered. The existing production of ore-based steel in Sweden is done with blast furnaces, and applying CCS technologies to alternative means of steel production seems therefore unlikely, i.e. are not considered in our model.
- Techno-economic assumptions of SSC/U within the pulp and paper industry: Pettersson and Harvey (2012). The study covers many different alternative configurations of the steam- and power-production units of an average Swedish pulp mill, including CCS technology alternatives for regular boilers as well as

integrated gasification combined-cycle power plants for extra power generation. However, the considered technology alternatives are only applicable to stand alone pulp mills, i.e. they are only implemented as such in our model.

The implemented CCS technology alternatives have a mitigation rate ranging between 1500 and 3100 kt of captured CO_2 per million tonnes of produced pulp.

3 Four Distinct Scenarios

Several climate scenarios (CT) are defined to assess the impact of different intermediate sector targets on the Swedish energy system (Table 3). The main difference between the CT scenarios is the CO_2 target (overall domestic target and/or sector target; magnitude of the target). The impact of climate targets is assessed by comparing scenarios with baseline scenarios without climate target. The main impact assessment will take place between climate scenarios and the REFCLIM baseline scenario, since they all apply the same REFCLIM EU ETS price projections.

In addition, several sensitivity analyses were performed. Two elements seemed to have the biggest impact on the results: the projected demand for transport, and fossil fuel prices. In order to reflect that one important action to reduce CO_2-emission, not captured within the model, is to reduce the demand for road transportations, a parallel scenario analysis was performed in which lower transport demand projections were implemented. In the main scenarios fossil prices are low (assuming a low-carbon world with low competition of fossil fuels), alternative scenarios with higher fossil fuel prices have been applied to reflect the uncertainty in fuel price projection. In addition, we have performed sensitivity analysis in which we have assumed (i) a faster cost and technical development of electric cars, (ii) lower potential of biomass and (iii) further restriction on CCS/U.

4 Climate Mitigation Pathways

To shed further light on our findings, we focus the scenario analysis on describing the differences in climate mitigation actions in the various economic sectors investigated as well as on how biomass has been allocated between sectors.

Table 3 Main scenarios described based on CO_2 price assumption and CO_2 target

Three baseline scenarios without specific climate targets

Baseline scenario	Assumed CO_2 price/tax		CO_2 reduction target by year and magnitude	
	EU ETS	Carbon tax	2030	2045
RefClim	REFCLIM	Current level	–	–
RefPRIMES	REFPRIMES	Current level	–	–
RefLow	REFLOW	Current level	–	–

Climate Target (CT) scenario	Assumed CO_2 price/tax		CO_2 reduction target by year and magnitude	
	EU ETS	Carbon tax	2030	2045
CT	REFCLIM	Current level	NETS + ETS $\leq$ −40%	NETS + ETS $\leq$ −100%
CT + NETS40			NETS + ETS $\leq$ −40% NETS $\leq$ −40%	NETS + ETS $\leq$ −100% NETS $\leq$ −40%
CT + NETS50			NETS + ETS <= −40% NETS $\leq$ −50%	NETS + ETS $\leq$ −100% NETS $\leq$ −50%
CT + NETS60			NETS + ETS $\leq$ −40% NETS $\leq$ −60%	NETS + ETS $\leq$ −100% NETS $\leq$ −60%
CT + NETS70			NETS + ETS $\leq$ −40% NETS $\leq$ −70%	NETS + ETS $\leq$ −100% NETS $\leq$ −70%

Five main climate target scenarios with an overall net-zero CO_2-emission target for 2045, with different sector targets. Sector emissions are divided into ETS and NETS, where *ETS* and *NETS* respectively represent emissions from activities included and not included by the EU ETS. *Reduction* refers to a reduction from 2005 levels

4.1 Differences in Climate Mitigation Actions in Various Economic Sectors

The resulting total CO_2 emissions from energy supply, energy conversion and energy use, including process emissions from energy-intensive industries, are also reduced over time in the *REFCLIM* baseline scenario without climate targets (represented by the dashed line in Figs. 1 and 2). During the first reduction period (2015–2025), the reduction initially takes place in the Electricity and District Heating sector and thereafter in the transport sector. In the second reduction period (2035–2045), the reduction takes place first in the Transport sector and finally in the

Industry sectors. A similar pattern is seen in the *REFPRIMES* baseline scenario. Consequently, the EU ETS price in line with *REFPRIMES* is high enough to make invest in some energy-efficiency improvements and fossil-fuel-free options in the Electricity and District Heating and in the Industry sector cost-efficient, i.e. impose emission abatements. Nevertheless, even in the scenario with high EU ETS prices, emissions are only reduced by 50% by 2045 compared with 2005. However, in the *REFLOW* baseline scenario, CO_2 emissions first decrease due to increase use of electric and biofuel vehicles, but is increased when the transport demand increase, i.e. an emission reduction is not seen if the carbon price is not increased over time.

When instead imposing an overall net-zero CO_2-emission target (represented by the stacked area in Fig. 1), biomass-combined Heat and Power (biomass-CHP) with CCS/U are introduced to get equal amount of negative emissions, resulting in an emissions reduction by 86%. The significantly largest reduction of CO_2 emission in the scenario without a climate target (*REFCLIM*) takes place in the Transport sector, where final consumption of diesel and gasoline is reduced with 36% between 2015 and 2030 despite an increase in transport demand. An effect of gasoline and diesel being replaced with electricity and biofuels is reduced emissions from refineries. Additional reduction takes place in the Industry sector (when the iron and steel industry switches to electricity wherever possible, while CCS/U is introduced in the cement industry that does not have any electrification options) and from decreased use of waste incineration in the Electricity and District Heating sector. The resulting decrease of waste incineration should be seen in relation to the assumptions made in the model: the compensation for municipal solid waste incineration is considered constant at today's level in the model. In other words, its relative value compared to the increased carbon prices is not high enough to make waste incineration profitable compared to other electricity generating technologies.

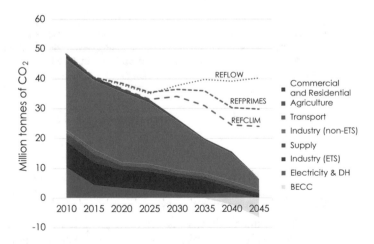

Fig. 1 Sector CO_2-emission pathway in scenario *Climate Target* (staked area). For comparison, the emission pathways in the three baseline scenarios (*REFCLIM, REFPRIMES* and *REFLOW*) without climate targets is included (dashed line)

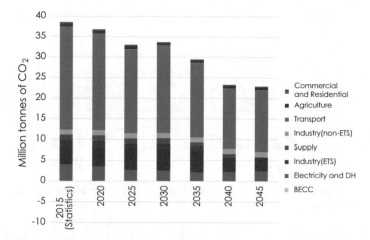

Fig. 2 CO$_2$ emission per sector and over time in the *REFCLIM* baseline scenario

In reality, however, the receiving compensation will increase (and waste incineration will be used to generate electricity and heat) until alternative waste treatments are less costly. Finally, some reductions are due to biofuels replacing diesel for vehicles used within the agriculture sector and biogas replacing natural gas for buildings (space heating and cooking) and for transportation (cars and busses). However, when the original fuel demand within the Agriculture sector and the natural gas use within buildings and transportation are low, these are minor emission reductions compared with the overall emission reduction.

When one introduces an additional climate target on the NETS sectors (CT + NETSXX scenarios in Fig. 3), the overall CO$_2$ emissions are reduced at an earlier stage in comparison with only having the overall climate target, starting in year 2030. Moreover, this reduction mainly takes place in the Transport sector (Fig. 3, left). With the most ambitious target (CT + NETS70), the emission

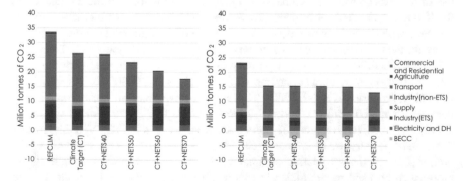

Fig. 3 CO$_2$ emission per sector and over time in different scenarios for year 2030 (left) and year 2040 (right)

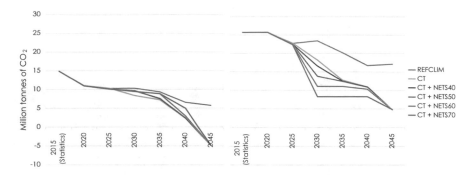

Fig. 4 CO_2 emissions from ETS (left) and non-ETS (right) sectors, main scenarios that all share the same CO_2 prices

reductions in the Transport sector are sufficient to also meet the overall target in 2040 without using negative emission reduction from BECC technologies (Fig. 3, right). By 2045, the net emissions are the same among all scenarios with climate targets and BECC is also used in the CT + NETS70 scenario. Further differences can be seen in the use of biomass (Sect. 4.2).

Introducing specific emissions reduction targets for the sectors not included in EU ETS (NETS) increases the competition of biomass, resulting in lower biomass use and corresponding higher use of fossil fuel within the ETS sectors. As a result, the CO_2 emissions from the ETS sectors are higher in scenarios with NETS specific emission reduction targets (Fig. 4). Nevertheless, the net emission decreases when one introduces another CO_2-emission target on the NETS.

4.2 Differences in Biomass Use

The resulting final demand of biomass, in each of the scenarios, follows the intuitive use of biomass between the scenarios (Fig. 5, right)—the tougher the climate target the higher the demand for biomass. However, the primary use of biomass does not correlate with the final energy demand. In the end of the modelled period (2040 and beyond), the two scenarios with the toughest CO_2-emission targets on NETS sectors (CT + NETS60/70) have a lower primary use of biomass compared with the three other scenarios with climate targets (CT and CT + NETS40/50), (Fig. 5, left). Explicitly, scenario CT + NETS60/70 has lower primary use but higher final energy demand compared with CT and CT + NETS40/50. The explanation is that more resource-efficient (and expensive) biofuels are used in CT + NETS60/70. The further decrease of primary use of biomass, seen in CT + NETS70, is a consequence of the higher share of district heating compared with the other scenarios, which allows biomass to be used more efficiently (district heating becomes a by-product when producing biofuels).

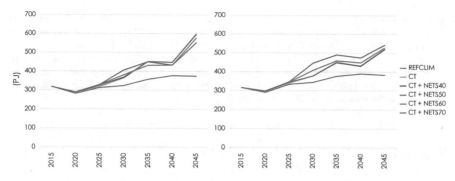

Fig. 5 Annual primary use of biomass (left) and final energy demand (right) for each of the main scenarios, in PJ (includes black liquor, municipal waste and sludge). Scenarios CT, CT+NETS40 and CT+NETS50 follow each other most of the years, i.e. are overlapping

5 Conclusion

The Swedish Government's climate goal is for the country to be carbon-neutral by 2045. The energy pathways calculated from TIMES-Sweden show pathways on how Sweden can convert to a carbon-neutral energy system by 2045. Electricity, district heating and space heating can be close-to-zero emissions by as early as 2025, apart from some fossil-fuel consumption during peak periods. The challenge lies in the transport sector and in energy-intensive industries. For the transport sector, it is critical to advance technologies relating to fossil-fuel-free vehicles and sustainable biofuel production, especially in terms of freight transportation. However, it is more difficult to attain carbon-neutrality in some sectors, even if energy efficiency and fossil-fuel-free energy commodities are introduced. A case in point is the cement industry, which releases CO$_2$ in certain processes. Despite this, net-zero CO$_2$ emissions are achieved by introducing carbon capture and storage/use in the cement industry in combination with BECC to achieve negative emission (by capturing carbon emissions derived from biomass) and at the same time generating electricity and heat.

A key feature for making it possible to meet net-zero targets in 2045 is the use of forestry residues—whether derived from logging, from the sawmill industry, or from pulp and paper. The primary use of biomass to meet the specific demand for energy related service and goods will be lower if choosing advanced biofuels with high efficiency both in respect of resources use in the production step and when being used in the vehicles, in combination by using waste heat from the biofuel production for district heating.

Finally, the resulting system cost of achieving net-zero CO$_2$ emissions is significant lower when one reduces the demand for transport. Thus, it is also important to introduce measures that reduce transport needs and/or increase the number of people per vehicle.

Acknowledgements This study forms part of a delivery to the Swedish Government's Cross-Party Committee on Environmental Objectives and the conclusion from the modelling runs fed into a climate policy framework for Sweden, i.e. SOU 2016:21 and SOU 2016:47. In addition, part of the study has been documented in Swedish (Krook-Riekkola 2016). Financial support from the Swedish Energy Agency is gratefully acknowledged, as are constructive comments on the analysis from Eva Jernbäcker, Karl-Anders Stigzelius, Jonas Forsberg and Erik Furusjö. Any errors remain the sole responsibility of the authors.

References

Börjesson P (2015) Biomassapotential från svenskt skogs- och jordbruk – Uppdaterade uppskattningar [Biomass potential from Swedish forestry and agriculture]. Secretariat of the Cross-Party Committee on Environmental Objectives, Stockholm

Boverket [Swedish National Board of Housing, Building and Planning] (2014) Boverkets föreskrifter om ändring i verkets byggregler (2011:6) – Föreskrifter och allmänna råd [Changes to the Board's building regulations (2011: 6)—regulations and general advice]. Resolved on 17 June 2014. Available via https://rinfo.boverket.se/BBR/PDF/BFS2014-3-BBR-21-rattelseblad. pdf

European Commission (2013) EU energy, transport and GHG emissions: trends to 2050—reference scenario 2013. Directorate-General for Energy, Directorate-General for Climate Action and Directorate-General for Mobility and Transport, European Commission, Brussels

Eurostat (2004) Energy statistics manual: a joint report by Eurostat, IEA and OECD. Available via http://epp.eurostat.ec.europa.eu/portal/page/portal/energy/introduction

Eurostat (2017) Share of renewables in gross final energy consumption, 2015 and 2020. Available via http://ec.europa.eu/eurostat/statistics-explained/index.php/File:Share_of_renewables_in_gross_final_energy_consumption,_2015_and_2020_(%25)_YB17.png

Grahn M, Hansson J (2015) Prospects for domestic biofuels for transport in Sweden 2030 based on current production and future plans. Wiley Interdisc Rev Energy Environ 4(3):290–306. https://doi.org/10.1002/wene.138

Hoenig V, Hoppe H, Koring K, Lemke J (2009) ECRA [European Cement Research Academy] CCS project—report about phase II. Available via https://ecra-online.org/research/ccs/

Hoenig V, Hoppe H, Koring K, Lemke J (2012) ECRA [European Cement Research Academy] CCS project—report on phase III. Available via https://ecra-online.org/research/ccs/

IEA/International Energy Agency (2012) World energy outlook 2012. IEA, Paris

IEA/International Energy Agency (2015) Medium-term oil market report 2015. Nov 2015. IEA, Paris

IEAGHG/International Energy Agency Greenhouse Gas Research and Development Programme (2013a) Deployment of CCS in the cement industry, 2013/19, December 2013. IEAGHG, Cheltenham

IEAGHG/International Energy Agency Greenhouse Gas Research and Development Programme (2013b) Iron and steel CCS study (Techno-Economics Integrated Steel Mill), 2013/04, July 2013. IEAGHG, Cheltenham

Krook-Riekkola A (2015) National energy system modelling for supporting energy and climate policy decision-making: the case of Sweden. Ph.D. Thesis, Department of Energy and Environment, Chalmers University of Technology, Sweden

Krook-Riekkola A (2016) Bilaga 12: Klimatmålsanalys med TIMES-Sweden: Övergripande klimatmål 2045 i kombination med sektormål 2030 [Modeling Swedish climate targets using TIMES-Sweden]. In: SOU/Statens Offentliga Utredningar (ed) En klimat- och luftvårdsstrategi för Sverige [A climate policy framework and a climate and clean air strategy for Sweden]. The Cross-Party Committee on Environmental Objectives, Swedish Government Official Reports Series, SOU 47(Part II), Stockholm

Krook-Riekkola A, Berg C, Ahlgren EO and Söderholm P (2013) Challenges in soft-linking: the case of EMEC and TIMES-Sweden. National Institute of Economic Re-search, Working Paper No.133

Krook-Riekkola A, Berg C, Ahlgren E, Söderholm P (2017a) Challenges in top-down and bottom-up soft-linking: lessons from linking a Swedish energy system model with a CGE model. Energy 141:803–817. https://doi.org/10.1016/j.energy.2017.09.107

Krook-Riekkola A, Wetterlund E, Sandberg E (2017b) Biomassa, systemmodeller och målkonflikter [Biomass, system models and conflicting targets]. Fjärrsyn report 2017:407. Available via http://www.energiforsk.se/program/fjarrsynfutureheat/rapporter/biomassa-system modeller-och-malkonflikter-2017-407

Loulou R, Goldstein G, Kanudia A, Lehtilä A, Remne U (2016a) Documentation for the TIMES model-part 1: TIMES concepts and theory. Available via http://iea-etsap.org/index.php/ documentation

Loulou R, Lehtilä A, Kanudia A, Remne U, Goldstein G (2016b) Documentation for the TIMES model, part II: comprehensive reference manual. Available via http://iea-etsap.org/index.php/ documentation

Pettersson K, Harvey S (2012) Comparison of black liquor gasification with other pulping biorefinery concepts—systems analysis of economic performance and CO$_2$ emissions. Energy 37:136–153. Available via http://doi.org/10.1016/j.energy.2011.10.020. Accessed 18 Nov 2016

RES2020 (2009) The pan European TIMES model for RES2020 model description and definitions of scenarios. Monitoring and evaluation of the RES directives implementation in EU27 and policy recommendations for 2020. A project funded under the intelligent energy for Europe programme. Project no: EIE/06/170/SI2.442662. Available via http://www.cres.gr/res2020/

SEA/Swedish Energy Agency (2015) Kvotplikten varierar årligen. Nedan finns de fastställda kvoterna för åren 2003-2035. [Electricity certificates: quotas decide for the period 2003–2035]. Available via http://www.energimyndigheten.se. Accessed 11 Oct 2015

SEA/Swedish Energy Agency (2017) Energy in Sweden: facts and figures 2017. Spreadsheet with statistics. Available via http://www.energimyndigheten.se/en/facts-and-figures/publications/

SEPA/Swedish Environmental Protection Agency (2015) Sammanställning trafikarbete Kontrollstation 2015 och Färdplan 2050 [Summary National transportation 'Control Station 2015' and 'Roadmap 2050']. Excel spread sheet. Personal Communication

SEPA/Swedish Environmental Protection Agency (2017) Report for Sweden on assessment of projected progress, March 2017. In accordance with Articles 13 and 14 under Regulation (EU) No. 525/2013 of the European Parliament and of the Council Decision: a mechanism for monitoring and reporting greenhouse gas emissions and for reporting other information at national and Union level relevant to climate change and repealing Decision No. 280/2004/EC

SIKA (2014) Samhällsekonomiska principer och kalkylvärden för transportsektorn: ASEK 5.1 [Socio-economic principles and calculation values for the transport sector]. Version 2014-04-01. Swedish Transport Administration

Simoes S, Nijs W, Ruiz P, Sgobbi A, Radu D, Bolat P, Thiel C, Peteves S (2013) The JRC-EU-TIMES model—assessing the long-term role of the SET plan energy technologies. JRC scientific and policy reports. JRC85804, EUR 26292 EN. ISBN 978-92-79-34506-7. Available via http://publications.jrc.ec.europa.eu/repository/handle/JRC85804

SOU/Statens Offentliga Utredningar 2016:21 (2016) Ett klimatpolitiskt ramverk - Delbetänkande av Miljömålsberedningen [A climate policy framework for Sweden], The Cross-Party Committee on Environmental Objectives, Swedish Government Official Reports Series (SOU), Stockholm

SOU/Statens Offentliga Utredningar 2016:47 (2016) En klimat- och luftvårdsstrategi för Sverige [A climate policy framework and a climate and clean air strategy for Sweden], The Cross-Party Committee on Environmental Objectives, Swedish Government Official Reports Series (SOU), Stockholm

Statista (2015) Natural gas prices in the US and in Europe from 2013 to 2025 (in US Dollars per million British thermal units). Available via www.statista.com. Accessed April 2015

Swedish Tax Agency (2008a) Skattesatser 2009. Tax rates in 2009. Available via www. skatteverket.se

Swedish Tax Agency (2008b) Punktskatter, SKV 505 utgåva 21. [Selective purchase taxes, SKV 505 Edition 21]. Available via www.skatteverket.se

Swedish Tax Agency (2012) Ändrade skattesatser på bränslen och el fr.o.m. 1 januari 2012. [Changes in fuel and electricity tax levels from January 2012]. Available via www. skatteverket.se

Transport Analysis (2015) Road traffic—vehicle statistics. Available via https://www.trafa.se/en/ road-traffic/ Accessed January 2015

UN/United Nations (2015) Paris agreement. Authentic texts of the Paris agreement. Available via http://unfccc.int/paris_agreement/items/9485.php

A Long-Term Strategy to Decarbonise the Danish Inland Passenger Transport Sector

Jacopo Tattini, Eamonn Mulholland, Giada Venturini,
Mohammad Ahanchian, Maurizio Gargiulo, Olexandr Balyk
and Kenneth Karlsson

Key messages

- Danish decarbonisation target is in line with an increase in global temperatures of 1.75–2 °C
- Early ban on Internal Combustion Engines (ICE) allows the largest decrease in cumulative emissions from Danish transport sector
- Early ban on ICE generates the highest tax revenue for the exchequer
- The innovative modelling framework that links a national optimization energy system model with a private car simulation model provides consumer realism to study the decarbonisation of the inland transport sector.

J. Tattini (✉) · G. Venturini · M. Ahanchian · O. Balyk · K. Karlsson
Technical University of Denmark, Kongens Lyngby, Denmark
e-mail: jactat@dtu.dk

G. Venturini
e-mail: give@dtu.dk

M. Ahanchian
e-mail: moah@dtu.dk

O. Balyk
e-mail: obal@dtu.dk

K. Karlsson
e-mail: keka@dtu.dk

E. Mulholland
MaREI Centre, Environmental Research Institute, University College Cork, Cork, Ireland
e-mail: eamonn.mulholland@umail.ucc.ie

M. Gargiulo
E4SMA, Turin, Italy
e-mail: gargiulo.maurizio@gmail.com

© Springer International Publishing AG, part of Springer Nature 2018
G. Giannakidis et al. (eds.), *Limiting Global Warming to Well Below 2 °C: Energy System Modelling and Policy Development*, Lecture Notes in Energy 64,
https://doi.org/10.1007/978-3-319-74424-7_9

137

1 Introduction

The 21st meeting of the Conference of Parties (COP21) witnessed an agreement to pursue efforts to limit temperature increase to 1.5 °C above pre-industrial levels (UNFCCC 2016) through Intended National Determined Contributions (INDCs). Limiting global temperature rise to 1.5 °C above pre-industrial levels relates to a total carbon budget of between 400 and 850 $GtCO_2$ eq (as of 2011) with the respective probability of achievement varying between >66 and >33% (IPCC 2014). While each signatory of the COP21 agreement will play a varied role in adhering to these carbon budgets, there has yet to be an agreement for the equitable sharing of national carbon budgets. This chapter creates a range of provisional carbon budgets for Denmark and focuses on the potential of policies aimed at the inland transport sector compliance with these budgets. Denmark is chosen as a case study following the ambitious target set by the Danish government to decarbonise the entire energy system by 2050 (The Danish Government 2011). Furthermore, the inland transport sector is given focus considering that its share amounted to 28% of the total energy consumption in 2015 (Eurostat 2017). So far, attempts to encourage renewables within the transport sector have been largely offset by an increase in transport activity and a lack of alternatives available. Significant levels of policy intervention are required to reduce the transport sector reliance on fossil fuels. The study aims at determining the contribution of policies to decarbonise the inland passenger transport sector and to calculate national cumulative greenhouse gas (GHG) emissions, which are compared to a range of carbon budgets necessary to contribute to limit global temperature rise. This chapter aims at answering the following research questions:

1. How much GHG emissions reduction can be achieved in Denmark through policies focusing on inland passenger transport?
2. Will the cumulative GHG emissions up to 2050 exceed the carbon budget available for Denmark to maintain the average global temperature rise well below 2 °C?

An innovative modelling framework is adopted, which links a techno-economic energy systems optimisation model of Denmark—TIMES-DKMS—with a hybrid techno-economic and socio-economic simulation of the Danish private car sector—the Danish Car Stock Model (DCSM)—to provide realistic answers to the research questions underlying this study. The transport sector within TIMES-DKMS features endogenous modal shift. DCSM represents the heterogeneous nature of the private car sector. A variety of policy packages aimed at reaching an ambitious decarbonisation of the inland transport sector are implemented iteratively in both TIMES-DKMS and the supporting simulation models.

2 Methodology

This study is carried out with an original modelling framework, which integrates TIMES-DKMS—the national energy system model of Denmark equipped with modal shift add-on (Tattini et al. 2018a)—with DCSM—a consumer choice model of the private transport sector accompanied by a sectoral simulation model of the private car sector.

2.1 TIMES-DKMS

TIMES-DKMS is built on the TIMES (The Integrated MARKAL EFOM System) model generator, developed by the Energy Technology Systems Analysis Program (ETSAP)—a technology collaboration programme of the International Energy Agency (IEA). It is a partial equilibrium, linear optimisation model, which determines a least-cost solution for the energy system, subject to certain constraints. TIMES performs a simultaneous optimisation of operation and investments across the represented energy system over the modelling horizon. TIMES is based on the bottom-up approach, as it requires a database of technologies characterised by a high technical, economic and environmental detail. Loulou et al. (2016) provide a detailed description of TIMES.

TIMES-DKMS is a multi-regional TIMES model, covering the entire Danish energy system. It is geographically aggregated into two regions, with technological and economic projections to 2050. TIMES-DKMS is composed of five sectors: supply, power and heat, transport, industry and residential (Balyk et al. 2018). Within the scope of this study, we focus on inland passenger transport, which includes private car, bus, coach, rail (metro, train, S-train), 2-wheeler (motorcycle and moped) and non-motorized modes (bike and walk). Within the inland passenger transport sector, TIMES-DKMS determines modal shares endogenously. The mode- and length-specific transport service demands are merged into length-only specific transport service demands, thus enabling competition between modes. Modal competition is based on both the levelised costs of the modes and on new parameters in the TIMES framework: speed and infrastructure requirements. Modal speeds are complemented by a constraint on the total travel time budget (TTB), historically observed for the Danish transport sector (Transport DTU 2016). The TTB ensures the competitiveness of faster yet more expensive modes in a cost-optimisation modelling framework. Infrastructure accounts for the cost of adapting the existing transport networks to demand increases and possible significant modal shift. Infrastructure requirements regulate modal shift, as this may end up in infrastructure saturation, subsequently requiring additional infrastructure capacity, which implies a cost (Tattini et al. 2018a). Moreover, constraints on the maximal and minimal modal shares and on the rate of shift derived from the Danish National Travel Survey are included in TIMES-DKMS to guarantee the realism of

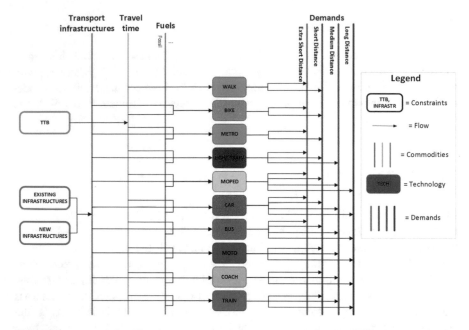

Fig. 1 Structure of the inland passenger transport sector in TIMES-DKMS for incorporating modal shift

the shift. Tattini et al. (2018a) provide a detailed description of TIMES-DKMS. Figure 1 provides a schematic description of the structure of TIMES-DKMS.

TIMES-DKMS outputs the least-cost decarbonisation pathway that meets all the constraints included in the model. However, the description of the private car sector in TIMES-DKMS is purely techno-economic, and does not account for heterogeneity within the private car market, thus suggesting a solution that may not be technically feasible (Mulholland et al. 2017a; Daly et al. 2011).

2.2 Danish Car Stock Model—DCSM

DCSM is a simulation model composed of two core components; a socio-economic consumer choice model and a techno-economic CarSTOCK model. DCSM checks the feasibility of the vehicle portfolio deployment pattern identified by TIMES-DKMS and introduces the necessary adjustments (as described in Sect. 2.3).

2.2.1 Consumer Choice Model

The consumer choice model estimates the influence of various policies on the Danish private vehicle market via a simulation market share algorithm, which has also been employed in the CIMS hybrid energy-economy model (Rivers and Jaccard 2005). This algorithm uses the tangible costs (investment cost, maintenance costs, fuel cost and vehicle-related taxes) along with a monetised representation of the intangible costs (model availability, range anxiety, and refuelling infrastructure) faced by the consumers to calculate the market share of a technology in a specific year when competing against a set of technologies. Heterogeneity of private vehicle preferences are accounted for through splitting transport users into 18 segments, divided geographically (urban/rural), by driving profile (Modest Driver, Average Driver, Frequent Driver) and by adoption propensity (Early Adopter, Early Majority, Late Majority), inspired by McCollum et al. (2017). Five technologies split into three categories are represented in the model—gasoline internal combustion engine (ICE), diesel ICE, natural gas (NG) ICE, battery electric vehicle (BEV) and plug-in hybrid electric vehicle (PHEV) disaggregated into the classes small, medium, and large (for ICEs, based off engine size) and into short, medium, and long range for BEVs (<125, 125–175, >175 km respectively). Mulholland et al. (2017b) provide a further description of the market segmentation and of the tangible and intangible costs for Denmark. The consumer choice model generates the market shares of the vehicle stock and outputs this result to the CarSTOCK model to determine the impact of policy measures on aggregate stock.

2.2.2 CarSTOCK Model

CarSTOCK is a bottom-up model that uses the outputs of the consumer choice model to create stock projections and analyse the net effect of policy measures in Denmark (Mulholland et al. 2017b). The CarSTOCK model draws upon detailed Danish statistics (FDM 2017), relating to the composition of private car sales, average mileage, efficiency, and life-time of vehicles with a disaggregation of vintage, fuel type and engine size (vehicle range in the case of BEVs). Using these inputs, it determines the long-term evolution of the private car stock, energy use and related CO_2 emissions to 2050 based off the ASIF methodology developed by Schipper et al. (2000). Total vehicle stock resulting from TIMES-DKMS is fed to CarSTOCK, which combines private car profiles and market shares (from the consumer choice model) to calculate the market shares of each car type with respect to total car stock. The CarSTOCK model has a detailed disaggregation of private car technologies into technology type (in line with those described in the Consumer Choice Model) and 30 vintage categories to represent the evolution of the car fleet.

2.3 Multi-model Approach

Integrating models has become an increasingly common approach in the field of energy system modelling (Mulholland et al. 2017a; Merven et al. 2012; Schäfer and Jacoby 2005). In the modelling framework used here, the policy measures are run in both the consumer choice model and in TIMES-DKMS in parallel. TIMES-DKMS first determines the optimal technology investments to meet the exogenous end-use demands at the least overall systems cost. Then, DCSM checks the technical feasibility of the solution obtained with TIMES-DKMS for the private passenger transport sector. If the solution is not feasible, capacity constraints bounding the stock of specific car technologies are added in TIMES-DKMS to comply with the realistic car shares projections calculated by the CarSTOCK model. A new solution is obtained with TIMES-DKMS, which is again verified in DCSM. Data exchange between the two models is iterated until there is convergence between the results (Fig. 2).

To ensure consistency within the model framework, the private vehicle costs in TIMES-DKMS and DCSM are harmonised for 2015 (Fig. 3). The road infrastructure cost is omitted from Fig. 3, as it is identical for all car types. Upon including the intangible costs in DCSM, the merit order of the car technologies changes compared to an analysis limited to tangible costs. This suggests that DCSM offers a more comprehensive view on the characteristics of cars perceived by consumers. Therefore, the multi-model approach employed in this study benefits from the models' respective strengths: the holistic representation of the integrated Danish energy system and the behaviourally-detailed insight of the Danish car consumer choice.

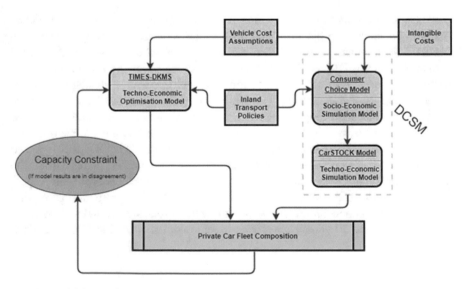

Fig. 2 Model integration between TIMES-DKMS and DCSM

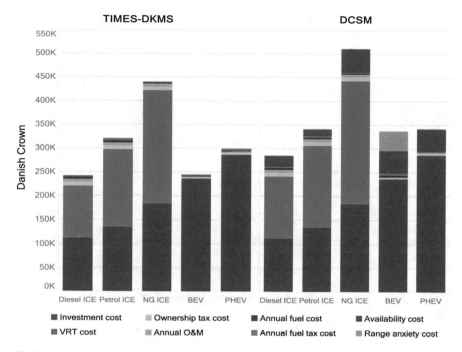

Fig. 3 Comparison of tangible and intangible costs in 2015 in TIMES-DKMS and DCSM

2.4 Carbon Budget for Denmark

This study allocates a carbon budget for Denmark based on population ('equity') and emissions ('inertia'), following the approach proposed by Raupach et al. (2014). To establish the carbon budget for Denmark, the global carbon budgets required to limit global temperature rise to varying levels with varying probabilities of achievement are taken from the 5th Assessment Report by IPCC (2014), which uses a base year of 2011. Denmark's national share is calculated using emission data from UN (2017a) and population data from UN (2017b). This national budget is brought up to a base year of 2015 using emissions data from UN (2017a). Land use and land use change and forestry (LULUCF) related emissions are subtracted using data from CDIAC (2016), resulting in the range of carbon budgets for the Danish energy system presented in Table 1.

2.5 Scenario Definition

In this study, we analyse the potential reduction of GHG emissions in Denmark enabled by alternative developments of the vehicle registration tax (VRT), the fuel

Table 1 Carbon budgets for the Danish energy system from 2015 corresponding to different levels of global temperature rise and levels of confidence [MtCO2 eq]

Temperature rise/confidence level	66%	50%	33%
4 °C target	3438	4031	4562
3 °C target	2090	2499	2958
2.5 °C target	1375	1733	2065
2 °C target	660	967	1171
1.5 °C target	48	201	507

Table 2 Description of scenarios for the policy analysis

Scenario	Description
Ref	Reference scenario, only 2020 targets included
Fuel tax	The tax paid on electricity used for transport, equal to 245.8 DKK/GJ in 2015, is derogated from 2020 onwards
VRT	The Vehicle Registration Tax (VRT) is derogated for all electric, hybrid and hydrogen vehicles from 2020 onwards
Fuel tax and VRT	Combination of the scenarios Fuel Tax and VRT
ICE bans	A ban on the purchase of new ICE cars is introduced from the year 2025, 2030, 2035 or 2040 (i.e. four scenario variants)

tax and from banning the sale of ICE vehicles in different years (Table 2). Denmark taxes cars through a *VRT* based on the capital cost and fuel efficiency of the vehicle, through a *circulation tax* based on the efficiency and weight of the vehicle and through *fuel taxes*. The *VRT* scenario assesses the effect of the derogation of the VRT for BEV and PHEV from 2020 onwards. In the *Fuel Tax* scenario, the tax on electricity used in transport is lifted from 2020 onwards, while keeping all other fuel taxes constant. In the *Fuel Tax and VRT* scenario, we examine the combined effect of the VRT derogation with removing the fuel tax on electricity from 2020. Some countries are currently discussing banning the sales of ICE vehicles in the near future (International Energy Agency 2017), which justifies the interest in analysing the effects of banning ICE cars sales.

All policy scenarios are consistent with Denmark's target of becoming independent from fossil fuels by 2050 (Danish Energy Agency 2015). This constraint is set on all sectors represented in TIMES-DKMS, with the exception of inland transport, for which the policies under assessment are the only option to reach the decarbonisation. Moreover, short- and medium-term targets complying with the European objective of minimum 10% renewable energy share in transport by 2020 and a 39% GHG emission reduction in 2030 with respect to 2005 levels (European Commission 2016) are applied. The policy scenarios are compared against a reference scenario (*Ref*), which maintains only the short-term targets for 2020, i.e. minimum 10% renewable energy share in transport and 50% electricity production from wind power.

3 Results: Focus on Technologies

3.1 Evolution of the Car Stock

The *Ref* scenario is characterised by a minor penetration of EVs, which represent 2.5% of the total car stock in 2050 (Fig. 4). The policies modelled boost the penetration of EVs with different degrees of effectiveness. The derogation of the tax on electricity for transport does not foster a strong penetration of EVs, while a derogation of the VRT on EVs enables a significant electrification of the car stock by 2050 (24% of car stock). The combined effect of fuel tax and VRT derogation accelerates the process of electrification of the car stock (32% in 2050). Setting a ban on the sale/import of vehicles run solely by an ICE strongly promotes the total electrification of the car stock. In the *ICE_Ban_2040* scenario, 93% of the stock is electric in 2050, while in *ICE_Ban_2035* scenario the entire stock becomes electric in 2050. The complete electrification of the car stock is anticipated by 2045 in the *ICE_Ban_2030* scenario and by 2040 for *ICE_Ban_2025*. Among EVs, PHEV

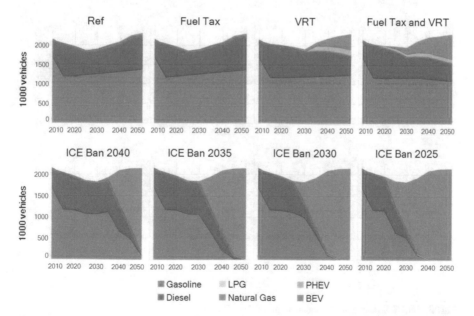

Fig. 4 Evolution of the car stock and car technologies in time across scenarios in TIMES-DKMS

technology only reaches a significant share of the total stock in the VRT scenario (6.5% in 2050), which demonstrates that the major barrier to the wide deployment of this technology is its high investment cost.

3.2 Modal Shares for Inland Passenger Transport

In most policy scenarios, the car stock decreases over the period 2020–2030 due to the increase in the average cost of ICE vehicles to fulfil the more stringent EU fuel standards concurrent with stagnant alternative fueled vehicle (AFV) costs. These have not decreased enough as to become a widely accepted technology, prompting a modal shift to public buses, which is a cheaper option (Fig. 5). After 2035, BEVs achieve a significant cost reduction due to the decrease in battery costs and cars gain again a higher modal share at the expense of public buses. In 2050, across all policy scenarios, bike, coach, metro, S-train and train modes increase their market share with respect to 2010, at the expense of transport by bus, car, 2-wheeler and walk. In particular, bike and metro transport witness the highest increment of use with respect to 2010.

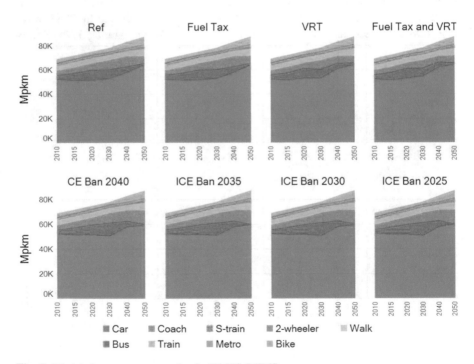

Fig. 5 Modal shares across scenarios in TIMES-DKMS

3.3 Fuel Mix for Inland Passenger Transport

The combined consumption of diesel and bio-diesel increases until 2020 in the *Ref* scenario, and then decreases by 2050 following improvements in fuel-economy and fuel switching, predominantly to electricity (Fig. 6). The share of bio-diesel in the blend gradually increases until reaching 72% in 2050. The combined consumption of gasoline and bio-ethanol decreases over time, while the share of bio-ethanol in the blend increases from 5.8% in 2015 to 45.4% in 2050. Moreover, electricity acquires a higher importance as fuel, constituting 5.3% of the total inland passenger transport fuel consumption in 2050. The drop in fuel consumption in 2030 is a consequence of multiple factors: a shift away from cars towards buses (characterised by a lower relative energy-intensity), an electrification of the car stock and efficiency improvements.

The fuel consumption varies across scenarios, due to changes in modal shares and technology shares within the car stock (Fig. 7).

In all the policy scenarios, with the exception of *Fuel_Tax*, the total fuel consumption reduces in 2050 with respect to the *Ref* scenario. Placing a ban on ICE vehicles causes reduction in fuel consumption due to the switch to electric vehicles (EVs), which have a significantly higher fuel economy than their ICE counterpart. It should also be noted that the private car sector has a major impact on total fuel consumption from the perspective of the inland passenger transport sector, illustrated by the similarities between the variations in car stock (Fig. 4) and fuel consumption (Fig. 7).

3.4 GHG Emissions

The annual GHG emissions from inland passenger transport sector undergo a significant decrease over time across all scenarios (Fig. 8). GHG emissions in 2050 drop by 37.3% in the *Ref* scenario with respect to 2010, due to a penetration of biofuels, EVs, and increases in the average efficiency of vehicles. These reductions are achieved despite the overall increase of transport activity over the same period.

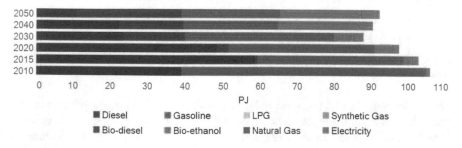

Fig. 6 Fuel consumption from inland passenger transport in Ref scenario

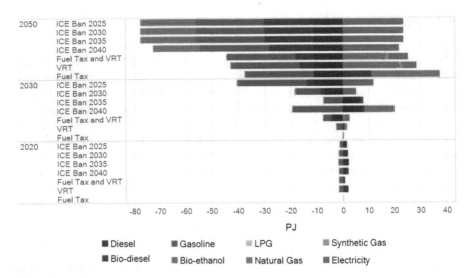

Fig. 7 Difference in fuel consumption from inland passenger transport across policy scenarios with respect to the Ref scenario

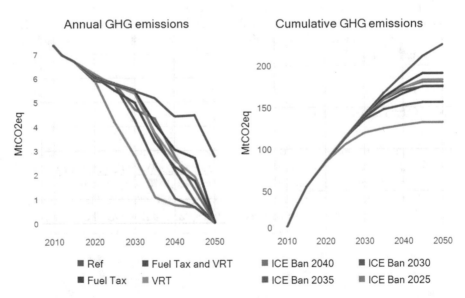

Fig. 8 Annual (left side) and cumulative (right side) GHG emissions from inland passenger transport sector

The implementation of transport policies enables the achievement of more ambitious decarbonisation targets. The inland passenger transport sector is completely decarbonised by 2050 in the *Fuel_Tax* scenario and all scenarios that include an ICE ban. The greatest cumulative reduction in GHG emissions is achieved through

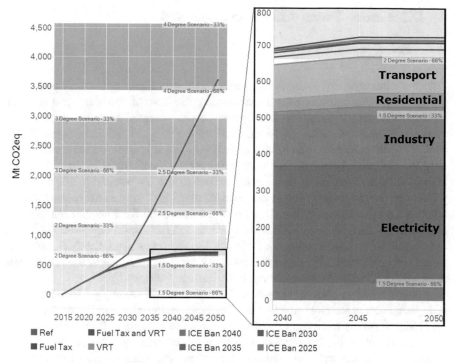

Fig. 9 Cumulative GHG emissions from the entire Danish energy system

an early ban placed on ICE vehicles (in 2025 and in 2030), while taxation-focused policy has a similar effect to that of later bans (in 2035 and 2040).

Figure 9 extends the focus of the analysis from inland passenger transport to the entire Danish energy system, showing the cumulative GHG emissions of the energy system over the modelled time horizon. In the *Ref* scenario, the cumulative GHG emissions diverge from the policy scenarios from 2025, and in particular, the steepness of cumulative GHG emissions increases after 2030 due to the adoption of coal-fired plants for power generation. In the policy scenarios, GHG emissions gradually decrease over time, to comply with the Danish environmental target of becoming fossil-free by 2050. Granted a fossil-free energy system is achieved in all sectors excluding inland passenger transport, the policy scenarios indicate that cumulative GHG emissions from the entire Danish energy system in 2050 are in line with a national contribution of an increase in global temperatures of 1.75–2 °C (excluding the possibility of negative emissions in the second half of the century). Figure 9 also shows the contribution of each energy sector towards national cumulative GHG emissions for the *ICE_Ban_2025* scenario, with the marginal emissions for the inland passenger transport scenarios shown above these contributions. The electricity sector is accountable for half of all emissions over the time-frame 2015–2050, matching those from the residential, industry, and transport sectors combined.

4 Discussion: Focus on Lessons Learned

4.1 Policy Insights

This study has analysed a range of regulatory measures focused on inland passenger transport while simultaneously decarbonising the rest of the energy system at least-cost. A central focus has been given to the potential of these measures to minimise cumulative GHG emissions to adhere to national carbon budgets. While evaluating the potential outcome of transport policies, it is important to consider not only their effectiveness, but also their efficiency, which can be evaluated as difference in actualised tax revenue with respect to the *Ref* scenario. The effect of policies on the tax revenue shows that *Fuel Tax and VRT* implies the highest loss of revenue for the exchequer (Fig. 10). The 6.2% reduction for *Fuel Tax and VRT* is explained by the uptake of BEV and PHEV from 2020, upon which no VRT and tax on electricity consumption are imposed. On the other hand, the *ICE_Ban* scenarios enforced from 2035 onwards benefit the tax revenues, due to the penetration of taxed AFV when their investment costs have not dropped yet.

Although from an environmental and tax revenue perspective, the *ICE_Ban_2025* is the most effective of all policies analysed, the different degrees of feasibility of policy instigation should be considered, stemming from their different timing, method of implementation, and public acceptability. Changes to taxation schemes require several government consultations while the introduction of a ban of ICEs presents a challenge in terms of negotiations (on timing and exceptions) with the automotive industry, let alone the preferences of consumers. While identifying the early ban on the sale of ICE as a suitable policy to decarbonise the Danish inland passenger transport sector, we recognise the lack of comprehensiveness of the policy measures analysed, e.g. measures affecting modal shift have not been addressed (Tattini et al. 2018b).

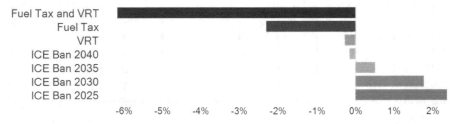

Fig. 10 Actualised cumulative change in tax revenue with respect to the Ref scenario

4.2 Methodology Insights

The adopted model framework improves the representation of the transport sector compared to traditional bottom-up energy systems optimisation models. Modal shift provides an additional option to explore decarbonisation pathways, and realistic consumer preferences in the private car sector are accounted through to the integration of the DCSM. However, there are still some limitations that future research may address. One limitation relates to the fact that TIMES-DKMS results do not take into account that even with a ban on ICE vehicles in 2025, there would still be some ICE vehicles circulating in 2050 according to DCSM (without any incentive for early scrapping). Modal shares are determined only via a suitably constrained socio-economic optimisation. A possible way to overcome this shortcoming consists of integrating consumers' heterogeneity into the model (differentiating their travel habits, perceptions and thus preferences) and determining modal shares resulting from a set of decisions taken by diverse consumers. Moreover, the level-of-service attributes characterising the modes should go beyond speed, to include also other relevant ones, e.g. waiting time and transfer time (Tattini et al. 2018b). Finally, while this study has calculated potential national carbon budgets based on a combination of equity and inertia sharing, these budgets will not be fully effective unless there is a global agreement on the method for allocating national and regional carbon budgets.

5 Conclusion

This study developed an innovative multi-model approach for Denmark that integrated an energy systems optimisation model (TIMES-DKMS) with a simulation model of the private car sector (DCSM) to assess the influence of various policy measures on the decarbonisation of the inland transport sector of Denmark. The multi-model approach developed combines the strengths of both modelling methods and provides a greater degree of consumer realism to the analysis of the private car sector. The analysis of potential contribution of seven policy measures towards the decarbonisation of the Danish inland transport sector revealed that a ban on the sale of ICE cars in 2025 enables the largest decrease in GHG emissions, i.e. 41% reduction of cumulative GHG emissions from the inland passenger transport sector with respect to the reference scenario. Moreover, the ICE ban in 2025 generates the highest tax revenue for the exchequer among the scenarios analysed. Regulatory measures focused on the derogation of tax have a lower relative effect on cumulative GHG emissions reduction and have a net negative impact on tax revenue when compared against the baseline. Nonetheless, all scenarios have a significant level of decarbonisation by 2050, with a complete decarbonisation of inland passenger transport in all scenarios where a ban on the sale of ICEs was imposed, and a greater than 90% reduction relative to 2015 in policies focused on tax derogation.

A broader analysis focusing on the entire energy system revealed that neither a total derogation of VRT and fuel tax for EVs nor an early ban on the sale of ICE vehicles would not contribute to maintaining the increase of global temperature limited to 1.5 °C.

Acknowledgements The work presented in this paper is a result of the research activities within the COMETS (Co-Management of Energy and Transport Sector) project (COMETS 4106-00033A), which has received funding from The Innovation Fund Denmark.

References

Balyk O, Andersen SK, Dockweiler S et al (2018) TIMES-DK: technology-rich multi-sectoral optimisation model of the Danish energy system. Energy Strategy Rev [under review]

CDIAC (2016) Global carbon project. US Department of Energy. http://cdiac.ornl.gov/GCP/. Accessed 10 Aug 2017

Daly H, Lavigne D, Chiodi A et al (2011) An integrated energy systems and stock modelling approach for modelling future private car energy demand. Paper presented at the 60th semi-annual ETSAP meeting, Stanford University, California, USA, 6â€"13 July 2011

Danish Energy Agency (2015) Denmarkâ€™s energy and climate outlook 2015. https://ens.dk/sites/ens.dk/files/Analyser/danish_energy_and_climate_outlook_2015.pdf. Accessed 12 Sep 2017

European Commission (2016) Regulation of the European parliament and of the council. http://eur-lex.europa.eu/legal-content/EN/TXT/HTML/?uri=CELEX:52016PC0482&from=en. Accessed 12 Aug 2017

Eurostat (2017) Energy balances 2017. Available at: http://ec.europa.eu/eurostat/web/energy/data/energy-balances. Accessed 25 Sep 2017

FDM (2017) Bildatabasen. http://www.fdm.dk/bildatabasen. Accessed 18 Jun 2017

International Energy Agency (2017) Global EV outlook 2017. https://www.iea.org/publications/freepublications/publication/GlobalEVOutlook2017.pdf. Accessed 11 Aug 2017

IPCC (2014) Climate Change 2014: synthesis report. In: Pachauri RK, Meyer LA (eds) Contribution of working groups I, II and III to the fifth assessment report of the Intergovernmental Panel on Climate Change. Geneva, Switzerland. http://www.ipcc.ch/report/ar5/syr/. Accessed 18 Jun 2017

Loulou R, LehtilÄ¤ A, Kanudia A et al (2016) Documentation for the TIMES modelâ€"part I: TIMES concepts and theory. Energy Systems Technology Analysis Programme (ETSAP). International Energy Agency (IEA). https://iea-etsap.org/docs/Documentation_for_the_TIMES_Model-Part-I_July-2016.pdf

McCollum DL, Wilson C, Pettifor H et al (2017) Improving the behavioral realism of global integrated assessment models: an application to consumersâ€™ vehicle choices. Transport Res Part D Transport Environ 55:322â€"342. https://doi.org/10.1016/j.trd.2016.04.003

Merven B, Stone A, Hughes A et al (2012) Quantifying the energy needs of the transport sector for South Africa: a bottom-up model. http://www.sanea.org.za/CalendarOfEvents/2013/SANEALecturesCT/Aug14/AdrianStone-EnergyResearchCentreUniversityofCapeTown.pdf

Mulholland E, Rogan F, Ó Gallachóir B (2017a) From technology pathways to policy roadmaps to enabling measuresâ€"a multi-model approach. Energy 138:1030â€"1041. https://doi.org/10.1016/j.energy.2017.07.116

Mulholland E, Tattini J, Ramea K et al (2017b) The cost of electrifying private transportâ€"evidence from an empirical consumer choice model of Ireland and Denmark. Transport Res Part D Transport Environ [Under Review]

Raupach MR, Davis SJ, Peters GP et al (2014) Sharing a quota on cumulative carbon emissions. Nat Clim Change 4(10):873â€"879. https://doi.org/10.1038/nclimate2384

Rivers N, Jaccard M (2005) Combining top-down and bottom-up approaches to energy-economy modeling using discrete choice methods. Energy J 26(1):83â€"106. https://doi.org/10.2307/41323052

Schäfer A, Jacoby HD (2005) Technology detail in a multisector CGE model: transport under climate policy. Energy Econ 27(1):1â€"24. https://doi.org/10.1016/j.eneco.2004.10.005

Schipper L, Marie-Lilliu C, Gorham R (2000) Flexing the link between transport and greenhouse gas emissions: a Path for the World Bank. International Energy Agency, Paris, p 16

Tattini J, Gargiulo M, Karlsson K (2018a) Reaching carbon neutral transport sector in Denmarkâ€"evidence from the incorporation of modal shift into the TIMES energy system modeling framework. Energy Policy 113:571â€"583. https://doi.org/10.1016/j.enpol.2017.11.013

Tattini J, Ramea K, Gargiulo M et al (2018b) Improving the representation of modal choice into bottom-up optimization energy system modelsâ€"the MoCho-TIMES model. Appl Energy 212:265282. https://doi.org/10.1016/j.apenergy.2017.12.050

The Danish Government (2011) Our future energy. 2011. https://stateofgreen.com/files/download/387. Accessed 8 Sep 2017

Transport DTU (2016) Danish national travel survey, dataset TU0615v1, from May 2006 to December 2015. Accessed 10 Sep 2017

UN (2017a) Carbon dioxide emissions. http://data.un.org/Data.aspx?q=emissions&d=MDG&f=seriesRowID%3a749#MDG. Accessed 16 Sep 2017

UN (2017b) World population prospects 2017. https://esa.un.org/unpd/wpp/DataQuery/. Accessed 11 Sep 2017

UNFCCC (2016) Report of the conference of the parties on its twenty-first session, 30 Novâ€"13 Dec 2015. Part two: action taken by the conference of the parties at its twenty-first session, Paris

Challenges and Opportunities for the Swiss Energy System in Meeting Stringent Climate Mitigation Targets

Evangelos Panos and Ramachandran Kannan

Key messages

- Electrification, efficiency, active participation of consumers in energy supply and demand side management are key pillars for achieving the fast and deep emissions reduction required to go beyond NDC.
- New business models involving microgrids, smart grids, virtual power plants, storage and power-to-gas technologies emerge and create new opportunities for low carbon development.
- The transition to a low carbon energy system requires immediate action, long-term price signals, effective implementation of carbon pricing mechanisms, regulations and legislation for supporting new emerging and exponential technologies.
- The use of the TIMES modelling framework in assessing deep decarbonisation policy scenarios enables an integrated assessment of the challenges, opportunities and trade-offs across the whole energy system in a consistent way that considers complex interdependencies between the energy actors.

E. Panos (✉) · R. Kannan
Laboratory for Energy Systems Analysis, Energy Economics Group,
Paul Scherrer Institut, Villigen, Switzerland
e-mail: evangelos.panos@psi.ch

R. Kannan
e-mail: kannan.ramachandran@psi.ch

© Springer International Publishing AG, part of Springer Nature 2018
G. Giannakidis et al. (eds.), *Limiting Global Warming to Well Below 2 °C: Energy System Modelling and Policy Development*, Lecture Notes in Energy 64,
https://doi.org/10.1007/978-3-319-74424-7_10

155

1 Introduction: Zero Emission Challenges in a Nuclear Phase-Out Context

In its Nationally Determined Contribution (NDC) to the Paris Agreement on Climate Change, Switzerland has committed to reducing its greenhouse gas (GHG) emissions by 50% in 2030, compared to 1990 levels. Proportionally, this reduction is to be achieved by 60% domestically and by 40% with the use of international credits (UNFCCC 2015). The Swiss government has formulated an indicative goal to reduce emissions in 2050 by 70–85% compared to 1990, including the use of international credits, as well as the vision to reduce per capita emissions in Switzerland to 1–1.5 t-CO_2eq in long-term (UNFCCC 2015). Given that in Switzerland the CO_2 emissions alone account for about 80% of the total GHG, a CO_2 tax on heating and process fuels was introduced in 2008, which is 96 CHF/t-CO_2 (83 EUR/t-CO_2) in 2018. Around two-thirds of the revenues from the CO_2 levy are redistributed through health insurers and the old-age insurance system; the remaining is used to finance building renovation programmes. From 1990 to 2014, the Kyoto protocol GHG emissions in the country declined by about 10% (BAFU 2016a, b).

The highest-emitting sectors are the transport and buildings, representing almost 2/3 of all emissions. The Swiss electricity sector is already almost CO_2-free as electricity is mainly generated from hydropower (59%), nuclear (33%), and renewables (5%) (BFE 2015). The new Swiss energy strategy aims at gradually phasing out the existing nuclear power (safety is the sole criterion) and promoting energy efficiency and renewable energy (BFE 2017).

Much of the additional sustainable renewable energy potential is on roof-top solar PV (Bauer and Hirschberg 2017). Wind conditions are less beneficial for wind power than in other countries, and wind projects are challenged by social opposition. Expansion of hydropower depends on political and social boundary conditions, though unexploited sustainable potentials are limited. The realisation of additional biomass (mainly manure) faces challenges regarding logistics and costs. Geothermal energy is also a controversial issue in the country, due to induced seismic activity in the recent past (Stauffacher et al. 2015). Finally, carbon capture and storage (CCS) is surrounded by uncertainties in costs and geological storage, additionally to issues related to public perception and the absence of a legal framework (Sutter et al. 2013).

In this context, the phase-out of nuclear energy could risk the ambition of the Swiss climate change mitigation policy and impose challenges for energy stakeholders and policymakers. A decentralised energy system built around small-scale renewable projects would imply that utilities need to move into new service-oriented business models beyond the old commodity-based model of cost-effective supply. Energy consumers would be increasingly interested in managing their energy use patterns and balancing their electricity and heating needs real-time. These developments imply a profound technological shift towards digitalisation, and policymakers have to develop a better understanding of the opportunities, challenges and risks that arise from it.

The limited availability of domestic renewable energy resources, the need for new market designs and legal frameworks, and the social impedance or acceptance of low carbon technologies constitute the climate change mitigation a formidable challenge for Switzerland. This chapter presents a techno-economic feasibility of scaling-up climate change mitigation efforts in a developed country with an innovative economy. It creates a link between national policies and global climate change mitigation efforts, and it assesses national barriers challenges and opportunities that are often overlooked in studies focusing on global deep decarbonisation pathways. The insights provided could be useful to a range of similar developed European countries, regarding resource availability or policies, such as Belgium, Netherlands, Sweden, Germany, Denmark, Austria, Norway, Ireland and the Czech Republic.

Assessment of CO_2 reduction policies for Switzerland, in the context of a nuclear phase-out, has been performed in the past with the modelling frameworks of the International Energy Agency's—Energy Technology System Analysis Programme (ETSAP), such as the Swiss-MARKAL energy systems model (Kypreos 1999; Schulz et al. 2008; Weidmann et al. 2012), the Swiss-TIMES energy systems model (Kannan and Turton 2016), the CROSS-border Swiss-TIMES electricity model (Pattupara and Kannan 2016); and non-ETSAP tools such as the CITE Computable General Equilibrium model (Bretschger and Zhang 2017), a systems dynamic model for the electricity supply security (Osorio and van Ackere 2016), the MERGE-ETL integrated assessment model (Marcucci and Turton 2012) and Prognos modelling framework (Prognos 2012). Although these studies demonstrated the technological feasibility of meeting stringent climate change mitigation targets, none of them assessed a scenario targeting below 2 °C global warming.

To this end, we employ an enhanced version of the Swiss-TIMES energy systems model (STEM) with a more detailed representation of the electricity sector, given the central role of electricity in meeting ambitious climate change mitigation goals. The model includes a range of features suitable for an in-depth analysis of decarbonisation pathways, such as: (a) early capacity retirement mechanisms to evaluate stranded assets; (b) electricity grid topology to assess challenges in the electricity infrastructure; (c) representation of the stochastic variability of renewable sources; and (d) modelling of ancillary services markets.

2   Methodology

2.1   The Swiss TIMES Energy Systems Model

The Swiss TIMES energy systems model (Kannan and Turton 2014) has a long-term horizon (2010–2100) with 288 hourly intra-annual timeslices (four seasons and three typical days per season). It covers the whole Swiss energy system with a broad suite of energy and emission commodities, technologies and infrastructure, from resource supply to energy conversion and usage in 17 energy demand sectors (Fig. 1).

The electricity sector of the model has been enhanced, due to the central role of electricity in the decarbonisation. There is a representation of four electricity grid levels, from very high to low voltage, to which different power plant and storage options are connected (Panos and Kannan 2016). Technical operating constraints of the hydrothermal power plants are approximated via a linearised formulation of the unit commitment problem (Panos and Lehtilä 2016). The power plants can be retired before the end of their technical lifetime when they have higher fixed or operating costs than the investment cost in new technology (Lehtilä and Noble 2011). The stochastic variability of electricity supply and demand is calculated from a bootstrapped sample of weather and consumption data in Switzerland over the last 15 years (Fuchs et al. 2017). The concept of the stochastic residual load duration curve (RLDC) is then employed to assess the needs in storage and dispatchable generation capacity (Lehtilä et al. 2014). The requirements in capacity for primary

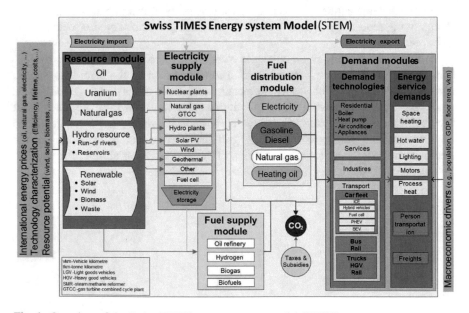

Fig. 1 Overview of the Swiss TIMES energy systems model (STEM)

and secondary reserve are endogenously modelled via ancillary services markets (Fuchs et al. 2017; Panos and Kannan 2016; Welsch et al. 2015).

The model includes an ad hoc representation of the electricity grid transmission topology with fifteen grid nodes. Seven Swiss regions are represented as a single node each. Each of the four existing nuclear power plants and each one of the four neighbouring countries is also represented as a single node. A power flow model (Schlecht and Weigt 2014) is employed to aggregate the detailed electricity transmission grid into the 15 nodes and 319 bi-directional lines included in the STEM model (Fuchs et al. 2017; Lehtilä and Giannakidis 2013).

Because of these enhancements, the transport sector is not endogenously modelled to reduce the model's equation matrix. The demand for transport fuels is exogenously provided based on the Swiss energy strategy scenarios (Mathys and Justen 2016; Prognos 2012). The rationale for excluding the transport sector from this study is that mobility choices of the individuals are not always cost-optimal. They are also based on other non-cost related factors (e.g. comfort), which are not adequately represented in the TIMES modelling framework and are subject to additional research (Daly et al. 2015).

2.2 Definition of the Baseline and Low Carbon Scenarios

The chapter assesses two scenarios, differentiated by the level of the climate change mitigation effort and the exogenously provided fuel consumption in transport. The Baseline scenario is considered to be compatible with the Swiss commitments, while the Low Carbon scenario is a deep decarbonisation scenario. The definition of the Low Carbon scenario is based on the post-2015 CO_2 budget that limits the warming to 1.5 °C.

The STEM model considers energy-related CO_2 emissions. We assume that the CO_2 emissions from industrial processes decrease according to the CO_2 emissions from fuel combustion in both scenarios, based on the reduction in the future cement production (Prognos 2012) and the implementation of emission savings measures (Zuberi and Patel 2017). Since we use the post-2015 global CO_2 emissions budget, we do not impose additional assumptions regarding the abatement of the non-CO_2 Kyoto protocol gases.

2.2.1 The Baseline Scenario

The energy service demands are derived from the macroeconomic developments and the efficiency measures assumed in the Politische Massnahmen (POM) scenario of the Swiss Energy Strategy (Prognos 2012). The main assumptions are summarised in Table 1. The Baseline scenario implements the Swiss commitments to UNFCCC, and it can be considered as the "NDC" scenario. Hence, based on domestic measures only, the scenario achieves a 30% reduction in the CO_2

emissions by 2030 from 1990 (UNFCCC 2015). In 2050, the CO_2 emissions decline by at least 42% from 1990, equivalent to the minimum target of 70% with the same ratio between domestic measures and international credits as in 2030.

2.2.2 The Low Carbon Scenario

The energy service demands, technology assumptions and resource availability are the same as in the Baseline scenario. The fuel consumption in transport is derived from the Neue Energie Politik (NEP) scenario of the Swiss energy strategy (Prognos 2012), which is compatible with the long-term Swiss target of 1–1.5 t-CO_2 per capita.

Table 1 Main assumptions in the "Baseline" scenario

	2010	2030	2050
Economy/demography			
Real GDP (billion CHF_{2010})	546.6	670.5	800.7
Population (million)	7.9	8.8	9.0
Space heating area (million m^2)	708.8	863.2	937.5
Renewable energy potentials			
Hydropower (TWh_e)			39.0
Solar PV rooftop (TWh_e)			19.2
Wind (TWh_e)			4.3
Geothermal (TWh_e)			4.4
Biomass (PJ_{th})			104.0
Imported fuel and CO_2 emissions trading scheme prices			
Swiss border gas price (CHF_{2010}/GJ)	7.9	11.0	13.2
Swiss border diesel price (CHF_{2010}/GJ)	16.4	21.0	25.4
CO_2 price (ETS, CHF_{2010}/t-CO_2)	15.0	48.0	59.0
Specific investment costs of key technologies (depending on size and application)			
Solar PV(CHF_{2010}/kW)		1400–2200	1000–1400
Heat pumps (CHF_{2010}/kW)		400–2700	400–2200
CHP gas CHF_{2010}/kW$_e$)		1200–6100	1200–5900
Efficiency measures			
Buildings: Labelling, renovation, oil-heating replacement, new building construction codes			
Industry/commercial: Efficiency incentives in industrial processes, ORC, best available technologies			
Power sector			
Nuclear phase-out by 2034, CO_2 capture and storage is not available, electricity net imports are allowed, grid expansion limited to the already announced plans for 2025 and beyond by the Swissgrid			

Sources Prognos (2012), Bauer and Hirschberg (2017), IEA (2016), Panos and Kannan (2016) and own estimations
Exchange rate: 1 CHF2010 = 0.96 USD2010 = 0.72 EUR2010

The emission reduction trajectory is based on the post-2015 Swiss CO_2 budget, derived from a per capita allocation of the global post-2015 CO_2 budget by using the medium fertility population projections (UN 2017b). Starting from a global post-2015 CO_2 budget of 250 GtC that limits warming to 1.5 °C with 66% probability by assuming adaptive mitigation of non-CO_2 climate drivers (Millar et al. 2017), the Swiss post-2015 budget is about 860 Mt CO_2. The trajectory imposes CO_2 emission reduction targets relative to 1990 of 50% in 2030, 70% in 2040 and 85% in 2050, which have to be achieved domestically and without carbon dioxide removal technologies. The post-2050 Swiss CO_2 budget is then 93 Mt CO_2, and it requires zero emissions by 2085 and cumulative negative emissions of 8 Mt CO_2 thereafter.

The scenario achieves 0.7 t-CO_2 per capita in Switzerland by the year of 2050, which is below than the world average and close to the OECD and EU average in the Beyond 2 °C Scenario (B2DS) of the IEA (IEA 2017). It could be argued that the Low Carbon scenario falls within the Paris Agreement range of ambition.

3 Results and Policy Recommendations

Moving beyond NDC requires fast and deep CO_2 emissions reductions. The emissions peak in 2020 in both scenarios, but the cumulative carbon budget over the period to 2050 is about 33% lower in the Low Carbon scenario than in Baseline (Fig. 2). In the Low Carbon scenario, emissions decline at rates exceeding past increases and accelerating to reach 5.6%/yr. after 2030. The transport and buildings sectors lead the way in the shift from the Baseline to the Low Carbon scenario, accounting for one-third each of the incremental cumulative abatement over the period of 2015–2050. Increased electrification in the end-use sectors sees the power sector absorb increased demand in the Low Carbon scenario while delivering about one-fifth of the additional abatement due to the deployment of new renewable technologies. Energy conservation also has an essential role to play, contributing to 19% of the mitigation in 2050.

3.1 Transforming the Electricity System

Consumers are turned into prosumers facilitated by digitalisation. While in the Baseline scenario the existing nuclear generation is mainly replaced by large gas power plants, in the Low Carbon scenario the electricity sector undergoes a profound restructuring towards renewable generation. In this scenario, renewables contribute about 90% to the total electricity supply by 2050 (Fig. 3). Decentralised electricity supply sources (combined heat and power CHP and photovoltaic PV) and prosumers, i.e. consumers that are also producers, are important in both scenarios. A large part (up to 85%) of decentralised electricity generated in the

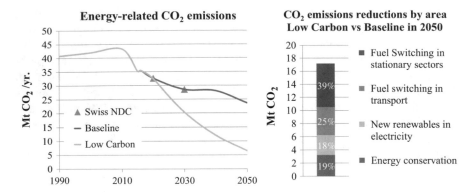

Fig. 2 Direct CO_2 emissions from fuel combustion (excluding international aviation) and emissions reductions from baseline by technology area

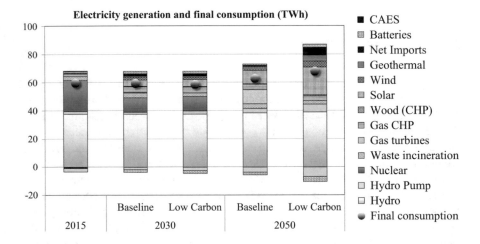

Fig. 3 Electricity generation mix (negative values denote charging of electricity storage)

residential sector is consumed on-site. The opposite holds in the commercial sector, driven by the larger installations.

Digitalisation helps in integrating variable and distributed energy sources in the future electricity markets. Better data and rigorous analysis of consumption and production patterns are essential. However, regulatory frameworks are needed regarding data collection and storage, connectivity and privacy. Similar concerns hold in other countries too, such as Denmark, Sweden and Germany.

High voltage networks need to support distributed generation. The share of distributed generation in end-use sectors increases from 27% in the Baseline scenario to 41% in the Low Carbon scenario by 2050. New connection requests at the lower grid levels lead to structural congestion in the upper levels. In order to eliminate congestion, the transmission grid has to be expanded beyond the announced plans from the Swiss Transmission System Operator (von Kupsch 2015). Since grid expansion requires long run-in times to be implemented and must achieve acceptance by public and stakeholders, early dialogue, planning and action are necessary. Similar issues are also faced by other European countries, such as Germany.

Microgrids and virtual power plants are the new business models for flexibility. The future Swiss energy system needs a profound technological shift towards flexibility. The uptake of variable renewable energy in the Low Carbon scenario increases the requirement for secondary reserve capacity to 700 MW in 2050, from 400 MW in 2015, with the peak demand for reserve occurring in summer instead of winter. Hydropower remains the main reserve provider, but the opening of the ancillary markets to smaller units owned by consumers is required to meet the increased demand. Flexible CHP units and storage increase their role in grid balancing in the Low Carbon scenario and provide by at most 200 MW of reserve.

New business models emerge involving the evolution of microgrids and virtual power plants. Utilities could be smart energy integrators that operate the distribution of electricity but no longer own generation units. Alternatively, utilities could deal directly with customers and sell services such as heat, lighting and cooling. An internet of energy could appear as new software platforms need to be developed to remotely, securely, and automatically dispatch generation and storage units in a web-connected system. This opportunity encapsulates the challenge for policy-makers to create regulatory frameworks that ensure privacy and security. Most of the developed European countries face similar challenges today.

Storage helps in achieving ambitious climate change mitigation goals. Accelerated deployment of storage, driven by the hedging against the increased variability of supply and arbitrage with the hourly electricity prices, is a crucial enabler in supporting the transition to a low-carbon power system. The lion's share in the future stationary electricity storage is in batteries (Fig. 4). About 6 GW of additional capacity in batteries is required in the Low Carbon scenario by 2050. Batteries complement pump storage when hydro storage is unavailable (due to water restrictions or participation in ancillary markets) and locally balance the electricity supply and demand.

The operation of pumped hydro storage correlates with cross-border prices arbitrage. Economic benefits also occur for consumers through load shifts via batteries. However, regulatory reform and development of balanced approaches are essential to facilitate the penetration of behind the meter storage options. These insights could also be useful to countries with substantial hydro storage resources such as Norway, Sweden and Austria.

Power-to-gas technologies become commercial and generate clean fuels. Generation of hydrogen and methane from electrolysis with renewable electricity becomes commercial only in the Low Carbon scenario. About 28% of the variable renewable electricity generated in summer in 2050 enters into the power-to-gas pathway, and more than half of it is seasonally shifted (Fig. 5).

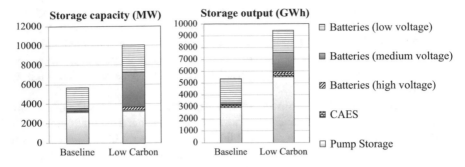

Fig. 4 Electricity storage capacity and production (stationary applications) in 2050

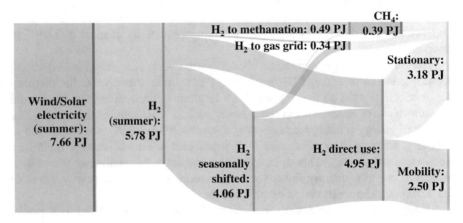

Fig. 5 Electricity production during summer that enters into the power-to-X pathway in 2050 (PJ/yr.) in the low carbon scenario

The power-to-gas technologies promise clean fuels for heating and mobility sectors but require a policy and regulatory framework. If supported by taxation incentives, requirements regarding the share of renewable fuels at the distribution levels, renewable fuels certification schemes, and subsidies, these technologies could be scaled up quickly once successfully demonstrated. For example, pilot projects in Belgium, Germany and Iceland have already reached the demonstration phase and can be scaled up with sufficient policy support (EC 2017).

3.2 Advancing the Low-Carbon Transition in End-Use Sectors

Electricity and energy efficiency play a predominant role in the decarbonisation. Electrification of the demand increases from 40% in the Baseline scenario to

60% in the Low Carbon scenario in 2050, where, oil and gas are phased-out, with residual uses of oil in transport and small-scale gas applications in industrial heating remaining by 2050. In contrast, heat produced from CHPs and heat produced from renewable sources account together for more than one-fourth of final energy consumption in 2050 (Fig. 6). Moreover, the final energy consumption in the Low Carbon scenario declines by 25% compared to the Baseline scenario in 2050. The deep decarbonisation pathway to 2050 comprises of rapid and aggressive deployment of highly efficient end-use technologies and rigorous application of energy codes and efficiency standards in all sectors. The insights from this subsection apply to many European countries.

Unlocking the energy savings potential in the buildings sector requires long-term and consistent price signals. The Low Carbon scenario requires a critical shift away from fossil fuels and moving towards close to zero emissions and efficient buildings. Heat pumps (60% in the total heat supply in buildings by 2050), renewables (21%), cogeneration and district heating (15%) and building renovation and insulation measures beyond the Baseline scenario are essential.

For this to happen, a comprehensive policy needs to be carefully designed, and unprecedented action needs to be taken to overcome economic barriers. Such barriers include unnecessary costs or early retirements of existing capital (e.g. the recent installations of gas-based heating equipment), and the high upfront capital costs of heat pumps, building conservation measures and solar PV. Since the insulation measures are not cost-efficient in the medium term, the concept of the efficient and close-to-zero emissions community could create economies of scale by implementing deep energy renovations across entire building blocks and therefore lower the costs and attract stakeholders. It could also be applied in managing and

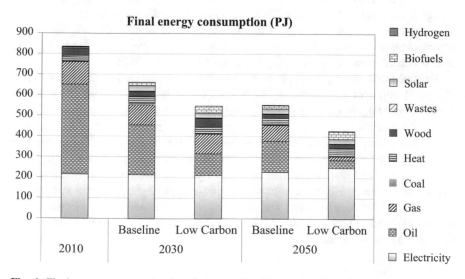

Fig. 6 Final energy consumption in industry, residential, commercial and transport

optimising the energy loads in buildings, allowing the shifting of electricity and heat loads relative to peak periods of demand.

Industry calls for energy and material efficiency strategies. The cumulative CO_2 emissions from industry over the period of 2020–2050 decline from 190 Mt in the Baseline to 140 Mt in the Low Carbon scenario. More than 70% of this reduction is attributable to the direct technology change, fuel and feedstock switching. The rest is attributable to additional energy conservation measures that reduce the energy service demands.

Energy efficiency and best available technologies (BAT) play a critical role in the decarbonisation of the industrial sectors. Implementation of energy management standards (e.g. ISO 50001) and process integration with the goal of minimising fuel consumption and emissions should be a priority. Public-private and cross-sectoral partnerships could efficiently be used to design and deploy integrated solutions. Moreover, material efficiency strategies should be further encouraged through price signals reflecting the energy and CO_2 footprint of production. Consumers should be made aware of this to avoid wasting materials, and the re-use of post-consumer scrap could be increased by implementing deposit refunding schemes upon product return. Improvement of the recycling rates and valorisation for electricity and heat production instead of landfill disposal could also help in achieving higher efficiency in materials usage.

Sustainable transport is a formidable task due to limited domestic clean fuels. The cumulative CO_2 emissions from the transport sector are about one-third less in the Low Carbon scenario compared to the Baseline over the period of 2020–2050. Much of this reduction is attributable to biofuels and electricity, which account for one-third each of the final energy consumption in transport by 2050. Because of environmental and food security concerns, the domestic biofuel production is limited (Steubing et al. 2010), and more than 95% of the biofuels in transport are imported (Prognos 2012). The decarbonisation of the transport sector also requires changing the nature and structure of transport demand, significant improvements in efficiency, policies and measures that increase the share of public transport modes and optimise freight transport (Prognos 2012).

Demand side management is a crucial pillar of a low carbon energy system. Temporal shifts of the electricity in electric-based heating systems on different hours of consumption and heat supply occur in both scenarios (up to 13% of total electricity in 2050). They are driven by congestion and arbitrage in heat supply costs. The shifts are enabled by water heaters and heat pumps (Fig. 7) and result in economic benefits for consumers due to the use of less energy in peak hours.

Such an active demand side management implies an awareness of the consumers to real-time prices, enabled through smart grids, monitoring and control systems, as well as aggregation of demand response and virtual power plants. However, the scalability of these new business models and the amount of flexibility that they can deliver remains uncertain.

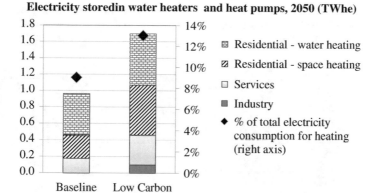

Electricity storedin water heaters and heat pumps, 2050 (TWhe)

Fig. 7 Amount of electricity shifted to different consumption hours via water heaters and heat pumps by 2050

3.3 The Cost for the Transition to a Low Carbon Energy System

A well-designed carbon price is part of the emission reduction strategy. The marginal cost per t CO_2 rises to 2150 CHF by 2050 (6 CHF/litre of light fuel oil), twice the carbon price projected in the NEP scenario of the Swiss Energy Strategy (Ecoplan 2012). It reflects the limited availability of low-carbon resources and the weak behavioural response of the high-income economic agents to a given price level. Carbon prices incentivise investment changes and can be introduced in different ways (e.g. carbon tax, cap-and-trade, credits). However, the carbon pricing mechanisms take time to develop, and policymakers need to establish an enabling environment and regulatory frameworks. Nonetheless, carbon prices are a source of government revenue, offer a good tax base that is difficult to evade, and can help offset economic burden via recycling (Stiglitz et al. 2017). International coordination can help avoid carbon leakages between countries with different carbon price levels and lower the overall cost of reducing emissions.

Investment lock-in creates stranded assets, and early action is necessary. In the Low Carbon scenario, about 1.6 GW of large gas power plants (90% of the 1.8 GW investment made in 2040) are retired before their technical lifetime over 2015–2050 (the remaining 200 MW contribute to secondary reserve). This is ten times more than in the Baseline. The early replacement of fossil-based heating supply in the Low Carbon scenario is twice the one in the Baseline.

Going beyond the NDC with only domestic emission reduction measures exponentially increases the per capita climate change policy costs by 2050. The average annual policy cost per capita (discounted at a 2.5% social rate), by excluding the transport sector, increases from 280 CHF/yr. in 2030 (less than 1% of GDP/capita) to 1780 CHF/yr. in 2050 (around 5% of GDP/capita) (Fig. 8). The scale of effort required indicates that the gap between the current Swiss

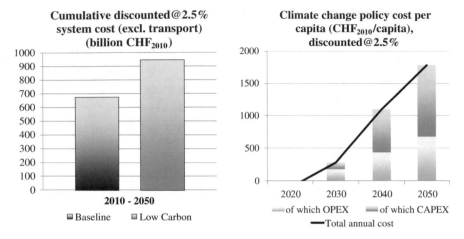

Fig. 8 Discounted system cost and policy costs per capita (electricity generation and heating sectors only, the transport sector is excluded)—1 CHF_{2010} = 0.96 USD_{2010} = 0.72 EUR_{2010}

commitments and the pathway to ambitious emissions reductions with domestic measures is immense. To this end, the current Swiss energy strategy is largely aiming to mitigate some of the CO_2 emissions outside Switzerland, including know-how and capital transfers to developing countries with lower mitigation costs. Such supporting mechanisms could also indirectly facilitate energy access in the developing countries and contribute to the broader global sustainable development goals (UN 2017a).

The daunting task is to mitigate emissions in heating supply. Total capital investments are almost tripled under ambitious climate change mitigation targets compared to Baseline. Capital expenditures in stationary sectors are twice those in

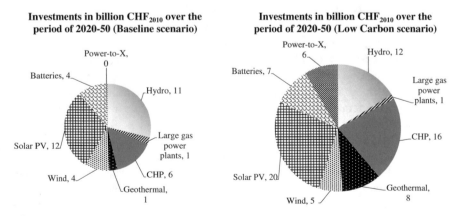

Fig. 9 Cumulative investment, discounted@2.5%, in key technologies for electricity supply (1 CHF_{2010} = 0.96 USD_{2010} = 0.72 EUR_{2010})

electricity supply, driven by the increase of investments in heat pumps in the buildings sector. Moreover, the emerged new business models shift capital from large-scale electricity generation technologies to storage and distributed generation (Fig. 9).

4 Conclusion

In this chapter, we assessed the technical feasibility of the Swiss climate change mitigation pledges, by taking into account the main pillars of the Swiss energy strategy: gradual nuclear phase-out, energy efficiency gains, and increased penetration of renewable energy. The shift to a below 2 °C pathway would require faster and deeper CO_2 emission reductions across both the energy supply and demand sectors than the current Swiss national determined contribution. Several of the insights and policy recommendations would be valid beyond the Swiss case.

The power sector undergoes a profound restructuring towards distributed renewable generation. Prosumers, microgrids, virtual power plants and storage are key components of the future low-carbon electricity supply and shift the transformation focus on integration rather than supply-side. Grid reinforcement and market reforming, not based on marginal cost pricing, are necessary and strong carbon pricing policies are needed backed-up with technology support measures to reduce investment risks.

Efficiency is crucial for achieving deep emission cuts in the end-use sectors. The energy saving potential in buildings is vast, but it requires a stronger involvement from the buildings-related stakeholders as in many cases owners may not reap the full benefits. Efficient and close-to-zero emissions communities could be an opportunity to create economies of scale in deep energy renovation across entire building blocks.

Since the Swiss NDC in the Baseline scenario already harvests the "low-hanging fruits" in the emission reductions in industry, promotion of best available technologies, fuel and feedstock switching, process integration, implementation of performance standards and material usage efficiency are required to achieve the additional emission cuts in a below 2 °C pathway. Price signals should be incorporated into consumer products related to environmental externalities of materials, to increase the collection, recycling and reuse of post-consumer scrap.

The additional abatement of the CO_2 emissions in the transport sector requires shifts in individual mobility behaviour and optimisation of freight transport. An important challenge for Switzerland is the limited domestic clean energy carriers, and the absence of a large research activity that explores ways in achieving the decarbonisation of the transport sector. Power-to-gas technologies can help in transforming excess renewable electricity into clean fuels for the heating and transport sector, but their commercialisation needs further policy support, and appropriate regulatory frameworks enabling these new business models.

Finally, while the Swiss energy strategy aims at mitigating some of the CO_2 emissions outside Switzerland, in developing countries with lower abatement costs, negative domestic emissions could be inevitable in the longer term. In this regard, policymakers should also work towards the development of appropriate legislation for carbon capture, transmission and storage.

Acknowledgements The research reported in this paper was partially funded by the Competence Centre Energy and Mobility (CCEM) through the project "Integration of Stochastic renewables in the Swiss Electricity Supply System (ISCHESS)", and by the Swiss Competence Centre for Energy Research through the project "Joint activity in Scenario and Modelling".

References

BAFU (2016a) Swiss greenhouse gas inventory. Swiss Federal Office For Environment—BAFU, Bern. https://www.bafu.admin.ch/dam/bafu/en/dokumente/klima/fachinfo-daten/kenngroessen_zurentwicklungdertreibhausgasemissioneninderschweiz.pdf.download.pdf/kenngroessen_zurentwicklungdertreibhausgasemissioneninderschweiz.pdf

BAFU (2016b) Switzerland's second biennial report under the UNFCCC. Federal Office for the Environment—BAFU, Bern. http://www.bafu.admin.ch/climatereporting

Bauer C, Hirschberg S (eds), Bäuerle Y, Biollaz S, Calbry-Muzyka A, Cox B, Heck T, Lehnert M, Meier A, Schenler W, Treyer K, Vogel F, Wieckert HC, Zhang X, Zimmermann M, Burg V, Bowman G, Erni M, Saar M, Tran MQ (2017) Potentials, costs and environmental assessment of electricity generation technologies. PSI, WSL, ETHZ, EPFL, Paul Scherrer Institut, Villigen. http://www.bfe.admin.ch/php/modules/publikationen/stream.php?extlang=en&name=en_8548 80113.pdf

BFE (2015) Schweizerische Elektrizitätsstatistik. Bundesamt für Energie. http://www.bfe.admin.ch/themen/00526/00541/00542/00630/index.html?lang=de&dossier_id=00765

BFE (2017) Energy strategy 2050 after the popular vote. Bundesamt für Energie (BFE). http://www.bfe.admin.ch/php/modules/publikationen/stream.php?extlang=en&name=en_210755710.pdf&endung=Energy%20Strategy%202050%20after%20the%20Popular%20Vote

Bretschger L, Zhang L (2017) Nuclear phase-out under stringent climate policies: a dynamic macroeconomic analysis. Energy J 38. https://doi.org/10.5547/01956574.38.1.lbre

Daly HE, Ramea K, Chiodi A, Yeh S, Gargiulo M, Ó Gallachóir (2015) Modal shift of passenger transport in a TIMES model: application to Ireland and California. In: Giannakidis G, Labriet M, Ó Gallachóir B, Tosato G (eds) Informing energy and climate policies using energy systems models: insights from scenario analysis increasing the evidence base. Springer International Publishing, Cham, pp 279–291. https://doi.org/10.1007/978-3-319-16540-0_16

EC (2017) Building up the future. Final report of special group on advanced biofuels to the sustainable transport forum. European Commission, Brussels

Ecoplan (2012) Energiestrategie 2050 - volkswirtschaftliche Auswirkungen. Bundesamt für Energie. https://www.newsd.admin.ch/newsd/message/attachments/35780.pdf

Fuchs A, Demiray T, Panos E, Kannan R, Kober T, Bauer C, Schenler W, Burgherr P, Hirschberg S (2017) ISCHESS—integration of stochastic renewables in the Swiss electricity supply system. ETH Zurich—Research Center for Energy Networks. PSI—Laboratory for energy systems analysis. https://www.psi.ch/lea/HomeEN/Final-Report-ISCHESS-Project.pdf

IEA (2016) Energy technology perspectives 2016. International Energy Agency, Paris. http://www.iea.org/Textbase/npsum/ETP2016SUM.pdf

IEA (2017) Energy technology perspectives 2017. International Energy Agency, Paris. https://www.iea.org/etp2017/

Kannan R, Turton H (2014) Switzerland energy transition scenarios—development and application of the Swiss TIMES energy system model (STEM)

Kannan R, Turton H (2016) Long term climate change mitigation goals under the nuclear phase out policy: the Swiss energy system transition. Energy Econ 55:211–222. https://doi.org/10.1016/j.eneco.2016.02.003

Kypreos S (1999) Assessment of CO_2 reduction policies for Switzerland. Int J Global Energy Issues 12:233–243. https://doi.org/10.1504/IJGEI.1999.000836

Lehtilä A, Giannakidis G (2013) TIMES grid modeling features. IEA—Energy Technology Systems Analysis Programme (ETSAP). http://iea-etsap.org/docs/TIMES-RLDC-Documentation.pdf

Lehtilä A, Noble K (2011) TIMES early retirement capacity. IEA—ETSAP. http://iea-etsap.org/docs/TIMES-Early-Retirement-of-Capacity.pdf

Lehtilä A, Giannakidis G, Tigas K (2014) Residual load curves in TIMES. IEA—Energy Technology Systems Analysis Programme (ETSAP). http://iea-etsap.org/docs/TIMES-RLDC-Documentation.pdf

Marcucci A, Turton H (2012) Swiss energy strategies under global climate change and nuclear policy uncertainty. Swiss J Econ Stat (SJES) 148:317–345. https://ideas.repec.org/a/ses/arsjes/2012-ii-8.html

Mathys N, Justen A (2016) Perspektiven des Schweizerischen Personen- und Güterverkehrs bis 2040. Hauptbericht Bundesamt für Raumentwicklung (ARE). https://www.are.admin.ch/dam/are/de/dokumente/verkehr/publikationen/Verkehrsperspektiven_2040_Hauptbericht.pdf.download.pdf/Verkehrsperspektiven_2040_Hauptbericht.pdf

Millar RJ, Fuglestvedt JS, Friedlingstein P, Rogelj J, Grubb MJ, Matthews HD, Skeie RB, Forster PM, Frame DJ, Allen MR (2017) Emission budgets and pathways consistent with limiting warming to 1.5 °C. Nat Geosci 10:741 https://doi.org/10.1038/ngeo3031. https://www.nature.com/articles/ngeo3031#supplementary-information

Osorio S, van Ackere A (2016) From nuclear phase-out to renewable energies in the Swiss electricity market. Energy Policy 93:8–22. https://doi.org/10.1016/j.enpol.2016.02.043

Panos E, Kannan R (2016) The role of domestic biomass in electricity, heat and grid balancing markets in Switzerland. Energy 112:1120–1138. https://doi.org/10.1016/j.energy.2016.06.107

Panos E, Lehtilä A (2016) Dispatching and unit commitment features in TIMES. International Energy Agency—Energy Technology Systems Analysis Programme (ETSAP). https://iea-etsap.org/docs/TIMES_Dispatching_Documentation.pdf

Pattupara R, Kannan R (2016) Alternative low-carbon electricity pathways in Switzerland and it's neighbouring countries under a nuclear phase-out scenario. Appl Energy 172:152–168. https://doi.org/10.1016/j.apenergy.2016.03.084

Prognos AG (2012) Die Energieperspektiven für die Schweiz bis 2050 (The energy perspectives for Switzerland until 2050). Bundesamt für Energie (BFE). http://www.bfe.admin.ch/php/modules/publikationen/stream.php?extlang=de&name=de_564869151.pdf

Schlecht I, Weigt H (2014) Swissmod: a model of the Swiss electricity market. FoNEW discussion paper 2014/01

Schulz TF, Kypreos S, Barreto L, Wokaun A (2008) Intermediate steps towards the 2000 W society in Switzerland: an energy–economic scenario analysis. Energy Policy 36:1303–1317. https://doi.org/10.1016/j.enpol.2007.12.006

Stauffacher M, Muggli N, Scolobig A, Moser C (2015) Framing deep geothermal energy in mass media: the case of Switzerland. Technol Forecast Soc Change 98:60–70. https://doi.org/10.1016/j.techfore.2015.05.018

Steubing B, Zah R, Waeger P, Ludwig C (2010) Bioenergy in Switzerland: assessing the domestic sustainable biomass potential. Renew Sustain Energy Rev 14:2256–2265. https://doi.org/10.1016/j.rser.2010.03.036

Stiglitz J, Stern N, Duan M, Edenhofer O, Gireaud G, Heal G, la Rovere E, Morris A, Moyer E, Pangestu M, Shukla P, Sokona Y, Winkler H (2017) Report of the high-level commission on carbon Prices. World Bank. https://static1.squarespace.com/static/54ff9c5ce4b0a53decccfb4c/t/59244eed17bffc0ac256cf16/1495551740633/CarbonPricing_Final_May29.pdf

Sutter D, Werner M, Zappone A, Mazzotti M (2013) Developing CCS into a realistic option in a country's energy strategy. Energy Procedia 37:6562–6570. https://doi.org/10.1016/j.egypro.2013.06.588

UN (2017a) Sustainable development goals. United Nations. http://www.un.org/sustainable-development/sustainable-development-goals/

UN (2017b) World population prospects: the 2017 revision. DVD edn. United Nations, Department of Economic and Social Affairs, Population Division. https://esa.un.org/unpd/wpp/Download/Standard/Population/

UNFCCC (2015) Switzerland's intended nationally determined contribution (INDC) and clarifying information. UNFCCC. http://www4.unfccc.int/submissions/indc/Submission%20Pages/submissions.aspx. Accessed 21.02.2015

von Kupsch B (2015) Bericht zum Strategischen Netz 2025 (Technical report on the "Strategic Grid 2025"). Swissgrid AG. https://www.swissgrid.ch/dam/swissgrid/company/publications/de/sn2025_technischer_bericht_de.pdf

Weidmann N, Kannan R, Turton H (2012) Swiss climate change and nuclear policy: a comparative analysis using an energy system approach and a sectoral electricity model. Swiss J Econ Stat (SJES) 148:275–316. https://ideas.repec.org/a/ses/arsjes/2012-ii-7.html

Welsch M, Howells M, Hesamzadeh MR, Ó Gallachóir B, Deane P, Strachan N, Bazilian M, Kammen DM, Jones L, Strbac G, Rogner H (2015) Supporting security and adequacy in future energy systems: the need to enhance long-term energy system models to better treat issues related to variability. Int J Energy Res 39:377–396. https://doi.org/10.1002/er.3250

Zuberi MJS, Patel MK (2017) Bottom-up analysis of energy efficiency improvement and CO_2 emission reduction potentials in the Swiss cement industry. J Clean Prod 142:4294–4309. https://doi.org/10.1016/j.jclepro.2016.11.178

France 2072: Lifestyles at the Core of Carbon Neutrality Challenges

Ariane Millot, Rémy Doudard, Thomas Le Gallic, François Briens, Edi Assoumou and Nadia Maïzi

Key messages

- Policymakers must find a compromise between pathways that lead to economic growth (GDP) but make it impossible to reach a carbon neutrality target, such as a digital society, and pathways with lower GDP that enable reaching a carbon neutrality target, such as a more collective society.
- Policymakers must ensure consistency between the intended technology deployment and evolving lifestyles.
- Policymakers should make use of their ability to set up support measures in order to shift our habits toward more sobriety, especially in the transport sector.
- The TIMES models are relevant to study the impacts of lifestyle changes as they have to satisfy energy services demands that are linked to lifestyles.

A. Millot (✉) · R. Doudard · T. Le Gallic · F. Briens · E. Assoumou · N. Maïzi
Centre for Applied Mathematics, MINES ParisTech, PSL Research University,
Sophia Antipolis, France
e-mail: ariane.millot@mines-paristech.fr

R. Doudard
e-mail: remy.doudard@mines-paristech.fr

T. Le Gallic
e-mail: thomas.le_gallic@mines-paristech.fr

F. Briens
e-mail: francois.briens@mines-paristech.fr

E. Assoumou
e-mail: edi.assoumou@mines-paristech.fr

N. Maïzi
e-mail: nadia.maizi@mines-paristech.fr

© Springer International Publishing AG, part of Springer Nature 2018   173
G. Giannakidis et al. (eds.), *Limiting Global Warming to Well Below 2 °C: Energy
System Modelling and Policy Development*, Lecture Notes in Energy 64,
https://doi.org/10.1007/978-3-319-74424-7_11

1   Introduction: Lifestyles as Levers in the Energy System Transition

1.1   Carbon Neutrality Challenge

Attempts to combat climate change have been the object of international negotiations for many years. The 2015 Paris Agreement set the specific target of: "holding the increase in the global average temperature to well below 2 °C above pre-industrial levels and pursuing efforts to limit the temperature increase to 1.5 °C". According to the Fifth IPCC's Assessment Report, to have a chance to respect this target, global emissions should be nil during the second half of the 21st century. Some countries have already committed to attaining carbon neutrality, such as Norway by 2030 (Neslen 2016) and Sweden by 2045 (Bairstow 2017). In France, the current commitments are to reduce greenhouse gas emissions by 40% in 2030 and by 75% in 2050 compared with 1990 levels, but the question of neutrality is being debated, and the current minister for ecology Nicolas Hulot is pushing to adopt a neutral carbon target by 2050 (Ministry of Ecological and Solidarity Transition 2017). This would not actually mean nil emissions by 2050 because compensation mechanisms would be employed (international carbon credits, carbon sinks, etc.).

In order to evaluate to what extent France could respect the Paris Agreement, we have made a prospective analysis of the French energy system by exploring options aimed at reaching carbon neutrality in 2072. This date allows us to go beyond the generally used but rapidly approaching 2050 horizon, to tackle the challenges impacting the second half of the century, as mentioned in the Paris Agreement. It also refers to the 1972 publication of the "Limits to Growth" report by the Club of Rome, which for the first time attempted to evaluate the long-term impacts of growth (on the environment, people, quality of life, etc.) and which provoked numerous debates on the sustainability of the development model of industrialized countries.

1.2   Lifestyles in the Framework of Prospective Analysis

The energy system responds to a demand for energy services (heat, mobility, etc.) whose analysis is often restricted to techno-economic drivers and merits being extended. For example, the demand for mobility does not only depend on the price of gas or the means of transport (technology) used, but also on the choice of location of residence, the balance between real and virtual mobility (e.g. tele-working from home, e-commerce) and citizens' propensity to travel far from home. In other words, it depends on a set of determinants that are characteristic of lifestyle. Lifestyles can therefore constitute an analysis framework of demand for real-life services (Schipper et al. 1989), which makes them important levers to tackle energy issues and climate change.

The role played by lifestyle was identified in 1987 by the authors of the Brundtland Report, who considered at the time that, "sustainable global development requires that those who are more affluent adopt life-styles within the planet's ecological means", stipulating, "in their use of energy, for example" (Brundtland et al. 1987). The question of lifestyle has thus gradually drawn the attention of the scientific community, from the first studies of their connection to energy evoked in the analyses that followed the oil crises (Maréchal 1977; Leonard-Barton 1981; Dillman et al. 1983) to their emphasis in the latest IPCC reports (IPCC 2007, 2014). The issue has thus motivated studies attempting to describe "sustainable" lifestyles (Druckman and Jackson 2010; Mont et al. 2014; Neuvonen et al. 2014), to identify ways to attain them (Tukker 2008), or to encourage citizens to adopt "low-carbon" lifestyles (Goodall 2010).

Although lifestyles have dramatically changed and diversified in the space of two or three generations, they are likely to be totally transformed during the 55 years that separate us from 2072, the target date of our study. Given the above, we have therefore decided to try to answer the following question: To what extent do lifestyles influence the energy system's capacity to achieve carbon neutrality? And amongst other, is a digital world compatible with the need to decarbonize the energy system, as usually thought?

1.3   Soft Linking of Three Models

To answer this, we employ modeling to evaluate the long-term effects of lifestyles on the energy system. Along with this ambition to represent a complex system, we aim to take a complex approach that involves understanding the state of mutual dependence of several elements coming from a broad disciplinary spectrum.

Firstly, the question of evolving lifestyles requires characterizing and quantifying current and past practices to make it easier to integrate them into a prospective energy approach. Secondly, from these lifestyles we determine a configuration of the economy and production on which changes in final energy service demand are closely dependent. Lastly, we need to capture the complexity of the energy system through a detailed understanding of the diversity of technological chains to identify the best possible decarbonization avenues.

Prospective energy exercises traditionally focus on changes in technologies and the economy. This involves using partial equilibrium energy system models (like TIMES), sometimes coupled with macroeconomic models (known as top-down). We consider that it is impossible to tackle the issue of carbon neutrality and the conditions for achieving it without integrating a human dimension, which requires also considering lifestyles in the prospective analysis. Therefore, capturing in the most pertinent way possible the connections between lifestyles, economics and energy systems is an essential modeling concern. The TIMES model is especially relevant to study the impacts of lifestyle changes as the model has to satisfy energy services demands that are linked to lifestyles.

We therefore chose for this study to combine (by soft linking) three models each of which is devoted to one of the three "links" in the chain. In the following section, we describe the models used and the parameters that they integrate.

2 Connecting Economics, Technique and Society: The Models

Each of the models employed represents one of the three dimensions by which we can apprehend the overall system (i.e. technology, economy, and lifestyle). The way they are connected is represented in Fig. 1, with the liaison being established through the different demands. The model representing lifestyles simulates future individual demands for mobility, housing, and goods and services. These demands are used in the input data of a macroeconomic input-output analysis model representing the relationships between the different production sectors of the French economy, which evaluates the evolution in the activities of these different branches. The energy system model, i.e. the TIMES-FR model, is ultimately fed with the usage demands (demand for mobility and residential demand) taken from the model representing lifestyles, and with the demands for industry, agriculture, services and mobility of goods taken from the input-output model. In the next subsections we look in detail at how each of these models operates.

2.1 Reconstructing Lifestyles

The notion of lifestyle relates to multiple dimensions that can be extrapolated from the social practices of individuals and households. Although lifestyle is hard to

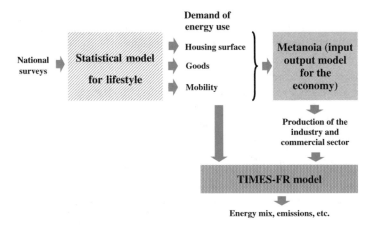

Fig. 1 Connections between the models used

define (Blok 2004), our use of the term refers to a certain set of key determinants, such as total housing surface area and total distance travelled per year. The simulation principle intuitively presented here is described in more detail in (Le Gallic et al. 2017). The proposed model employs the combined data from four national surveys relating to these practices, i.e. the population census, the housing survey, the national survey on transport and journeys, and the household budget survey.

The simulation process employed by the model involves building up an image of the population, its lifestyles and energy uses based on a set of variables resulting from these surveys. The variables used relate to people's attributes (e.g. age, gender), lifestyle practices (e.g. residence location practices, cohabitation practices) and energy usage indicators. More precisely, the latter include the surface area of housing, the distances covered by distance section, and the volumes of purchases of certain goods, because they are at the origin of an indirect consumption of energy (cf. Sect. 3.1).

Over a time period, the simulation involves enriching a simple demographic projection of the population (e.g. number of inhabitants by age) and progressively adding variables. The addition of variables thus highlights the correlations between variables existing in current or past surveys. These correlations take into account biological, economic, sociological and cultural phenomena. As an example, the surface area of housing per person currently depends on the location of the household (e.g. smaller dwellings in city centers because it's more expensive) or its size (due to pooling phenomena). Thus, if future households are less urban and more restricted because of changing lifestyles, the demand for housing surface area is likely to rise faster than the population, and this can be quantified by the simulation process.

The proposed representation ultimately relies on the identification of behavior patterns (e.g. mobility practices characteristic of people of a given age in a given situation; surface area of housing characteristic of households with a defined size and location). By considering these patterns as the structure of future behavior, we adopt the principle of *mimicking* behaviors. Underlying this is a hypothesis of reproduction of all or part of the social structures, psychological mechanisms, economic approaches, and the value systems in force.

2.2 From Lifestyle to Economic Impact: METANOIA Macroeconomic Model

Assessing the environmental impacts of different lifestyles and social arrangements requires taking into account the specific configuration of the economy—especially the productive structure and processes—that they imply. For this purpose, we use METANOIA (Macro-Economic Tool for the Assessment of Narratives using Output-Input Analysis), a dynamic simulation macroeconomic model of the French economy, designed to run medium- to long-term scenarios. An in-depth description of this model can be found in (Briens 2015). This model has been built using public data only—essentially from the national statistics bodies, INSEE and EUROSTAT. It features a sectorial disaggregation of the economy into 37 branches (e.g. agriculture, health, transport, construction, etc.), and various specific modules: demography (cohort model), demand for goods and services, residential sector, transport, agriculture, capital stocks (different types of productive assets are considered: dwellings, machinery, transport equipment, intangible fixed assets, etc.), foreign exchanges, employment, fiscal apparatus and the public administration budget (Fig. 2). For the sake of simplicity, there is no explicit monetary sector in our model.

In a nutshell, sectorial final demand stems mostly from exogenous hypotheses aiming at reflecting changes in consumption patterns, lifestyles, and the social organization, and expressed in terms of evolution of household demand for different goods and services, their mobility and housing choices, etc. This final demand

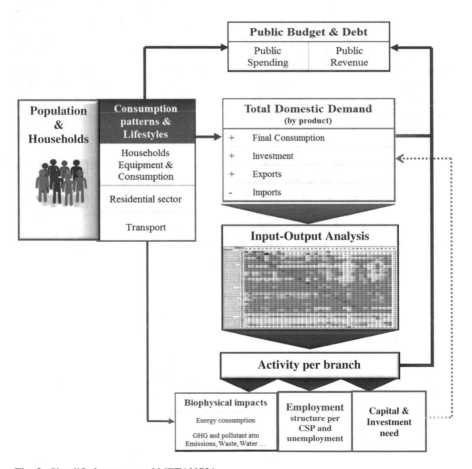

Fig. 2 Simplified structure of METANOIA

"drives" the evolution of the production of the different branches, which is determined using input-output analysis, so as to take into account the complex direct and indirect relationships and interdependencies between the various branches of the economy (Briens 2015). Using input-output analysis makes it possible to define the production of the different branches of the economy that corresponds to given levels of final demand, based on a matrix of technical coefficients. In order to reflect technical changes in the production processes, the technical coefficients of the input-output table may evolve according to exogenous hypotheses. For each branch, the production level in turn determines the need for investment or the amount of labor required on the basis of assumptions regarding the evolution of labor, capital, energy and emissions intensities in that branch. Socio-economic outcomes, including employment and unemployment, public budget balance and debt, depend on policy choices (these can include working time, public expenditure, fiscal and redistributive policies).

2.3 Integrating Economics and Lifestyle into the French Energy System: TIMES-FR Model

The TIMES-FR model initiated in (Assoumou 2006) represents the French energy system. This bottom-up techno-economic model is based on an optimality paradigm: by minimizing the total discounted cost of the energy system over the considered horizon (we applied a social discount rate of 4%), it selects technologies that can satisfy energy service demands. Each sector of the energy system features a detailed representation of the technologies available and the associated energy values, along with their technical and economic characteristics, availability date for new technologies, and deployment potential (e.g. for wind or solar). Energy service demands include demands from the residential and tertiary sectors (heating, cooking, hot water and specific electricity), disaggregated demands for mobility of goods and passengers depending on the transport mode (road, rail, air or river), and lastly demands from agriculture and industry. As part of the recalibration of the model's reference year (now 2014), a breakdown of demands from industry and the tertiary sector was introduced to improve the association with the METANOIA model: 21 sectors in industry and 8 branches in the tertiary sector. The agriculture sector was also subdivided, including a demand for tractors and another for buildings. This more detailed disaggregation than in standard models means that we can introduce means to decarbonize this sector, for example oil substitutes used for tractors (gas, biodiesel, electricity).

3 What Are the Levers for Carbon Neutrality?

In this section, and in order to illustrate how much lifestyle weighs in the overall balance, we have chosen to explore the consequences of two contrasting lifestyle scenarios on the economy and the energy system thanks to the breakdown of the models and approaches described above. The starting point of the exercise therefore involves outlining two lifestyle evolution scenarios based on a set of coherent hypotheses.

The first, entitled "digital society" reflects this trend: "In this more individualistic and technological society, people are motivated by a desire for personal achievement and long life."

The second is called "collective society" and describes alternative aspirations: "In this society organized around social connections and cooperation, people are motivated by a desire to be—and do—with others."

3.1 Impact of Lifestyles on Energy Uses

3.1.1 Hypotheses for the Lifestyles Model

The hypotheses attributed to different dimensions of lifestyle are briefly described in Table 1.

Table 1 Overview of hypotheses from both scenarios by lifestyle dimension

Dimensions	Digital society	Collective society
Demography	+18% Higher life expectancy, lower birth rate	+18% Continuation of current trends (inc. aging of the population)
Cohabitation practices	More single households	Development of shared forms of housing
Relationship with technologies	Higher equipment level	Maintenance of current practices
Mobility practices and relationship to space	More virtual activities (e.g. teleworking)	Contraction of activity area
Work attitude	Greater place of work in society	Reduced place of work in time organization
Location of dwelling	Preference for urban centers in cities	Reduction in semi-urban areas to the benefit of urban areas
Living standard, income and distribution	Higher living standard	Drop in the number of households with high and low incomes
Tourism and leisure travel practices	Development of long-distance destinations	Development of local destinations

3.1.2   Results of the Lifestyles Model

The simulations of the two scenarios developed provide a quantified glimpse of the influence of future lifestyles on energy uses in 2072 and very contrasting results depending on the options selected.

Regarding the uses of the residential sector, the extent of the challenge of a transition towards a low-carbon housing is very different in the two scenarios. Thus, the housing surface area to be heated and cooled slightly increases in the "collective society" scenario (+10% compared to currently) and increases much more in the "digital society" scenario (+34%). The variations are similar for specific electricity and hot water uses. These simulations show to what extent future cohabitation practices and future location choices will impact the "quantity" of residential energy uses to satisfy. The increase in the relative proportion of shared housing solutions on the one hand, and the proportion of collective housing solutions on the other, can help reduce the housing surface to be heated and cooled, and to be renovated and built. Thus the "cost" of achieving energy performance for the housing stock could be lower.

Simulations of the demand for goods in 2072 also showed significant differences between scenarios, which varied depending on the type of good considered: between +4 and +134% for the "digital society" scenario compared to the "collective society" scenario (Fig. 3). They also shed light on certain effects connected to pooling facilities and in particular the changing relationship to technologies.

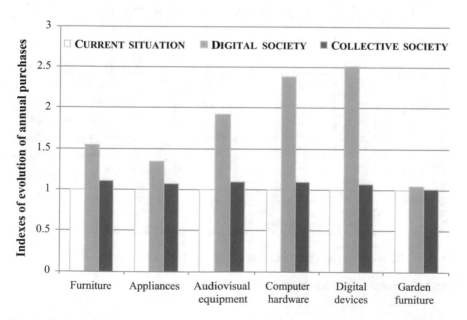

Fig. 3 Consumption of six types of durable goods for the two scenarios and the current situation (indexes of evolution of annual purchases)

Lastly, the simulations of uses connected to mobility produced marked contrasts, both for local mobility and long-distance mobility. The "collective society" scenario led to a 35% drop in distances covered annually for daily journeys due to a contraction of the living area and a reorganization of activities (Fig. 4). However, the number of journeys per person remains practically identical to the current situation in this scenario. An 11% drop in distances covered is also simulated for the "digital society" scenario, this time resulting mainly from a decrease in the number of journeys made. In fact, part of the actual mobility is substituted by virtual mobility (e.g. teleworking from home). Nevertheless, the model's capacity to take into account the systemic impacts of this kind of hypothesis remains limited (e.g. effects of the transfer of personal mobility connected to purchases towards delivery of goods, rebound effects of teleworking, impact on domestic energy consumption of an increased presence in the home).

For long-distance mobility, changes of a cultural nature were explored and produced a broad range of results (from −34 to +115% compared to the current situation). These simulations of long-distance mobility indicate on the one hand the decisive impact of lifestyle choices on mobility demand, and on the other hand the antagonisms that can emerge between local and long-distance mobility. These antagonistic evolutions introduced into the "digital society" scenario show that a significant reduction in personal local mobility can be more than compensated by more intensive long-distance mobility practices.

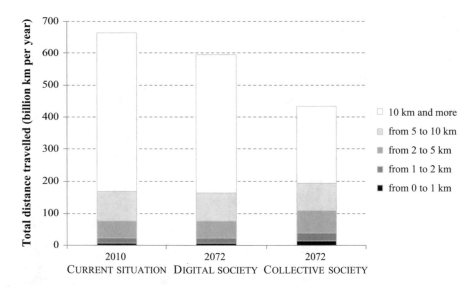

Fig. 4 Short-distance mobility demand under the three scenarios in 2072

3.2 Impacts of Lifestyles on the Economic System

3.2.1 Hypotheses for the Input-Output Analysis

Concerning the macroeconomic modeling, in both of the scenarios studied, the hypotheses concerning the households' changing consumption of goods, housing surface areas and mobility were defined from the results of the model's lifestyle simulations. Other hypotheses concerning the evolution of the population's activity rate, hourly productivity, the speed of obsolescence of capital (life expectancy and scrappage functions for assets), the development of external commerce, and the evolution of technical coefficients on the input-output table were chosen exogenously in order to coherently reflect the social transition and the lifestyles considered. Table 2 gives a succinct overview of the different hypotheses employed.

3.2.2 Results of the METANOIA Model

The simulation of these two scenarios in the METANOIA model gives contrasting results, in particular in terms of total final demand split between each of the 37 branches of the economy. This goes up by 62% for the "digital society" scenario compared to 8% for the "collective society" scenario from 2014 to 2072 (Fig. 5). These evolutions are particularly marked in the "digital society" scenario for the branches connected to computing (manufacture of equipment or telecommunications), and branches connected to health, while for the "collective society" scenario, the demand goes down in the branches of computer equipment manufacture, clothing, and branches involving vehicle construction.

These changes in demand logically translate into production indicators (Fig. 5): the "digital society" scenario, which is devised to reflect a more individualistic society that consumes high quantities of goods and services, leads to a 73% increase in GDP from 2014 to 2072 (split between the industry and tertiary branches, in particular transport, where activity doubles), compared to 9% for the "collective society" scenario, which corresponds to more sober lifestyles and more collective consumption patterns. In this latter scenario, most branches of the economy undergo moderate growth in their activity and we see a drop in the activity of branches connected to transport (manufacture of equipment and transport of goods).

Looking at the evolution of indicators connected to employment, we observe that in the "digital society" scenario, despite an increase in the total number of jobs, unemployment gets worse, due to the greater duration of work and active life (postponement of exit from the labor market), and improved productivity (Fig. 6). In the "collective society" scenario, the increase in total employment is moderate, while unemployment reaches very low levels, attaining 3% in 2072 (Fig. 6).

Table 2 Hypotheses for the METANOIA model

Dimensions		Digital society	Collective society
Employment	Labor force	Extension by 10 years of the retirement age	Activity rate unchanged
	Annual productivity gains	Divided by 3 by 2070	Divided by 5 by 2070
	Working time policies	No change	Annual work duration: −20% on average
Capital stocks		Lifespan of capital reduced in all sectors: −20% for computing, software and machines and equipment by 2072 (obsolescence through rapid innovation), and −10% for the remainder	Lifespan of capital increased in some sectors: +20% for computing, software and machines and equipment, and vehicles by 2072, and +5% for non-residential buildings
Foreign trade	Imports	Slight increase in imports: from +5% to +20% for heavy industry (extractive industry, metallurgy, etc.)	Relocation: Drop in imports of 30% for agriculture and 10% for most industrial products
	Exports	Evolution of exports in proportion to imports	Evolution of exports in proportion to imports
Transport	Modal share of passenger transportation	No change	Increase in walking and cycling for short distances, collective transport (public transport and trains) for middle and long distances
	Freight	No change	Transfer of 10% of road transport to rail
Household consumption		Increased consumption in some branches (housing, catering, health, leisure activities, social housing) and shorter lifespan for equipment	Drop in consumption or moderate increase in most branches and longer lifespan for equipment

3.3   Impacts of Lifestyles on the Energy System

In this section, we discuss the energy system's capacity to satisfy the demand for services that results from both lifestyle scenarios described above under the constraint to reach CO_2 neutrality by 2072 (we consider only CO_2 emissions in this study and no other GHGs).

3.3.1   Influence of Lifestyles on the Marginal Cost of CO_2

In order to evaluate the efforts required to reach the carbon-neutral objective, we employ as an indicator the marginal cost of the carbon constraint, which translates the

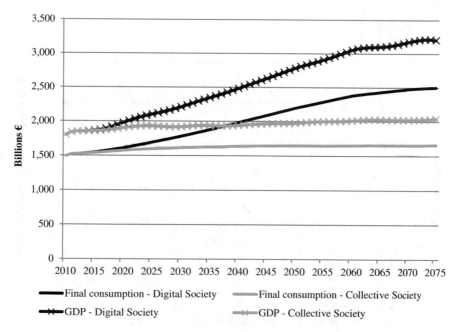

Fig. 5 Evolution of GDP and final consumption

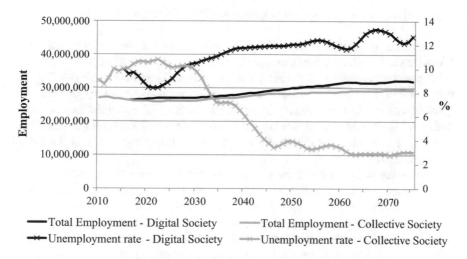

Fig. 6 Total employment and unemployment rate

impact of an additional one-ton reduction in CO_2 on the total discounted cost of the system (thus in some ways translating the state of tension in which our system finds itself under constraint). The analysis of this indicator highlights (1) a non-linearity of

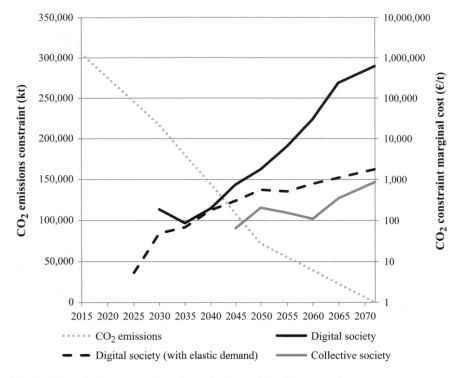

Fig. 7 CO$_2$ emissions constraint and marginal cost of the CO$_2$ constraint

the effort made to decarbonize in both scenarios and (2) the essential weight of lifestyles vis-à-vis the system's propensity to reach carbon neutrality.

The evolution patterns of the marginal cost are similar in both scenarios (Fig. 7); they translate a non-linearity between the decarbonization trajectory and the marginal cost of the CO$_2$ constraint. Up to 2035, the state of constraint is low, while emissions have already gone down by 40% compared to 1990. After this date, there is a strong upward effect due to the zero-emission target especially in the digital society scenario.

The values of the marginal cost shown by both scenarios are on the other hand very different, and this difference increases as it approaches the 2072 horizon (Fig. 7). At this date, the marginal cost for the "digital society" scenario reaches a value 700 times higher than for the "collective society" scenario. In fact, in the "digital society" scenario, the system finds itself in an "over-constrained" state. To achieve carbon neutrality, materials and energy are imported by the system at a very high price. Only the introduction of a hypothesis of elastic demand, which would illustrate consumers' propensity to adjust their demand for energy services in line with the marginal cost of energy (compared to a reference case with no carbon constraint) brings about a relaxation in tensions in the modeled system, and the return to a lower CO$_2$ marginal cost, but therefore meaning changes in consumption behaviors … and thus lifestyles.

3.3.2 Influence of Lifestyles on Final Energy Consumption

A shift towards more sober lifestyles, for example in the spirit of a "collective society" scenario, thus appears to be crucial for the transition towards an energy system that has less of an impact on the environment. Beyond the significant reduction of final energy consumption in the collective scenario (−29% between 2014 and 2072), the results show that the largest drop occurs in the transport sector (Fig. 8). Moreover, final energy consumption in this sector is reduced by 40%

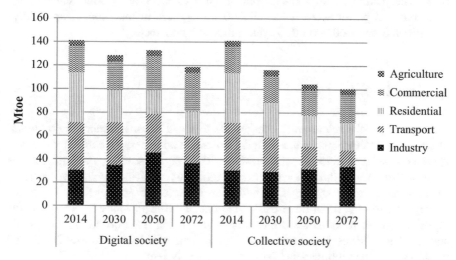

Fig. 8 Final energy consumption by sector

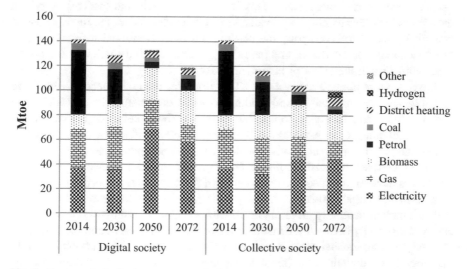

Fig. 9 Energy mix in final energy consumption

compared to the digital society scenario. Lifestyle changes can thus significantly influence the transport sector.

Moreover, the results show that the optimal configuration of the energy system associated with this scenario does not favor a single vector, but rather a multi-energy configuration in which we observe a varied range of technological solutions solicited to decarbonize the system (Fig. 9).

A significant reduction in emissions thus inevitably involves considerable technological substitutions: this raises the questions of technical feasibility and social acceptability, like for example the use of CCS on a wide scale in the industrial and power sector up to 50,000 $ktCO_2$/year in the "collective society" scenario and up to 60,000 $ktCO_2$/year in the "digital society".

4 Conclusion

Numerous appeals are currently being made to move towards a more digital world in which the interactions between digital transition and energy transition will make it easier to achieve a decarbonized society. The "digital society" scenario that we have described was devised with this in mind, whereas the "collective society" scenario fits in more with grassroots/bottom-up attitudes of degrowth. While the energy transition is often described as expensive, difficult and even unlikely, the digital transition receives a more enthusiastic reaction from economic and public stakeholders, who see in it a means to guarantee the road to growth. But is a digital world compatible with decarbonizing the energy system?

Our study shows that a digital society-type scenario makes the attainment of carbon neutrality unrealistic, even with the deployment of new technologies and deep changes in the energy mix. This result is subject to the availability of the decarbonization means currently available in our model and to the deployment of incentive policies to accompany the phasing-out of conventional technologies. In brief, technological evolutions and the move towards a digital society would not be sufficient to reach the target of carbon neutrality.

Based on this observation, we can either hope for the emergence of disruptive technologies (e.g. nuclear fusion) leading to rapid decarbonization, or reflect on social choices and lifestyles compatible with a carbon-neutral objective. In our exercise, attaining carbon neutrality is only possible if lifestyles change significantly compared to present trends, coupled with technological change.

As a digital society leads to higher GDP growth than a collective society, policymakers have to find a compromise between these pathways in order to enable a carbon-neutral target. But moving away from a collective scenario pathway only makes a carbon neutrality target more difficult to achieve. Policymakers must also ensure consistency between the intended technology deployment and the evolution of lifestyles. The societal organization stemming from a digital society will amplify the impacts of digitalization (more long-distance trips, more individual houses, more datacenters, etc.), unlike a collective society. Finally, the transport sector appears to be one of the key sectors for decarbonisation and can particularly be influenced by lifestyle changes. Since existing urban plans and transport infrastructures are closely linked to mobility habits, significant state support is vital in order to counter the inertia of this system.

To conclude, let's remind that the target to reach neutrality in France by 2072 may not be sufficient for the 1.5 °C scenario as industrialized countries should preferably attain this level by 2050. Thus the pathways introduced in this chapter would need to be even more ambitious to reach a target well below 2 °C. However, they highlight the main hurdles to reaching neutrality that would be exacerbated by an earlier target date.

As far back as 1972, the authors of the "Limits to Growth" report launched an alert on the non-sustainability of our system, and called for a deep-seated change in our development mode and thus society. As we declare the need to limit climate change, in order to achieve carbon neutrality it seems necessary to act not just on the deployment of decarbonized technologies, but on our lifestyles, which we should urgently question, and collectively redirect. This could constitute a brake as much as a lever in the energy transition.

References

Assoumou E (2006) Modélisation MARKAL pour la planification énergétique long terme dans le contexte français. École Nationale Supérieure des Mines de Paris

Bairstow J (2017) Sweden says it will go 'carbon neutral by 2045.' In: Energy live news—energy made easy. http://www.energylivenews.com/2017/06/20/sweden-says-it-will-go-carbon-neutral-by-2045/. Accessed 17 Aug 2017

Blok K (2004) Lifestyles and energy. In: Cleveland CJ (ed) Encyclopedia of energy. Elsevier, New York, pp 655–662

Briens F (2015) La Décroissance au prisme de la modélisation prospective : Exploration macroéconomique d'une alternative paradigmatique. Phdthesis, Ecole Nationale Supérieure des Mines de Paris

Brundtland G, Khalid M, Agnelli S, Al-Athel S, Chidzero B, Fadika L, Hauff V, Lang I, Shijun M, Morino de Botero M, Singh M, Okita S et al (1987) Our common future ('Brundtland report')

Dillman DA, Rosa EA, Dillman JJ (1983) Lifestyle and home energy conservation in the United States: the poor accept lifestyle cutbacks while the wealthy invest in conservation. J Econ Psychol 3:299–315. https://doi.org/10.1016/0167-4870(83)90008-9

Druckman A, Jackson T (2010) The bare necessities: how much household carbon do we really need? Ecol Econ 69:1794–1804. https://doi.org/10.1016/j.ecolecon.2010.04.018

Goodall C (2010) How to live a low-carbon life: the individual's guide to stopping climate change, 2nd edn. fully updated. Earthscan, London

IPCC (2007) Climate Change 2007: synthesis report. Contribution of working groups I, II and III to the fourth assessment report of the Intergovernmental Panel on Climate Change, IPCC. Intergovernmental Panel on Climate Change, Geneva, Switzerland

IPCC (2014) Climate Change 2014: synthesis report. Contribution of working groups I, II and III to the fifth assessment report of the Intergovernmental Panel on Climate Change, IPCC. Intergovernmental Panel on Climate Change, Geneva, Switzerland

Le Gallic T, Assoumou E, Maïzi N (2017) Future demand for energy services through a quantitative approach of lifestyles. Energy. https://doi.org/10.1016/j.energy.2017.07.065

Leonard-Barton D (1981) voluntary simplicity lifestyles and energy conservation. J Consum Res 8:243. https://doi.org/10.1086/208861

Maréchal P (1977) Crise de l'énergie et évolution des modes de vie, CREDOC

Meadows DH, Meadows DL, Randers J, Behrens WW III (1972) The limits to growth: a report for the Club of Rome's project on the predicament of mankind, 2nd edn. Universe Books, New York

Ministère de la transition écologique et solidaire (2017) Plan Climat, France. Available at https://www.ecologique-solidaire.gouv.fr/sites/default/files/2017.07.06%20-%20Plan%20Climat.pdf Accessed 25 August 2017

Mont O, Neuvonen A, Lähteenoja S (2014) Sustainable lifestyles 2050: stakeholder visions, emerging practices and future research. J Clean Prod 63:24–32. https://doi.org/10.1016/j.jclepro.2013.09.007

Neslen A (2016) Norway pledges to become climate neutral by 2030. The Guardian. Available at http://www.theguardian.com/environment/2016/jun/15/norway-pledges-to-become-climate-neutral-by-2030 Accessed 17 Aug 2017

Neuvonen A, Kaskinen T, Leppänen J, Lähteenoja S, Mokka R, Ritola M (2014) Low-carbon futures and sustainable lifestyles: a backcasting scenario approach. Futures 58:66–76. https://doi.org/10.1016/j.futures.2014.01.004

Schipper L, Bartlett S, Hawk D, Vine E (1989) Linking life-styles and energy use: a matter of time? Annu Rev Energy 14:273–320. https://doi.org/10.1146/annurev.eg.14.110189.001421

Tukker A (ed) (2008) Perspectives on radical changes to sustainable consumption and production. Greenleaf, Sheffield

From 2 °C to 1.5 °C: How Ambitious Can Ireland Be?

Xiufeng Yue, Fionn Rogan, James Glynn and Brian Ó Gallachóir

Key messages

- The most stringent technicaly feasible carbon budget for Ireland to contribute is 360MtCO2 from 2015 to 2070. This target is 1.5 °C compatible but extremely challenging.
- Cost effective decarbonisation rates are non-linear and any strategy compatible with a well below 2 °C can only be achieved through much stronger near-term mitigation efforts than suggested by the current policy.
- Strong mitigation efforts in the near term suffer economic losses from stranded assets but are required to avoid carbon "lock-in" and consequences of delayed actions.
- Flexibility of the TIMES model framework enables us to develop a sensitivity analysis with large number of scenarios and explore incremental carbon budget constraints. This enables us to gain insights on energy system dynamics between incremental scenarios instead of traditional scenario schemes based on a few scenarios which bounds the solution space.

X. Yue (✉) · F. Rogan · J. Glynn · B. Ó Gallachóir
MaREI Centre, Environmental Research Institute, University College Cork,
Cork, Ireland
e-mail: xiufeng.yue@ucc.ie

F. Rogan
e-mail: f.rogan@ucc.ie

J. Glynn
e-mail: james.glynn@ucc.ie

B. Ó Gallachóir
e-mail: b.ogallachoir@ucc.ie

© Springer International Publishing AG, part of Springer Nature 2018
G. Giannakidis et al. (eds.), *Limiting Global Warming to Well Below 2 °C: Energy System Modelling and Policy Development*, Lecture Notes in Energy 64,
https://doi.org/10.1007/978-3-319-74424-7_12

191

1   Introduction

The Paris agreement adopted in 2015 (UNFCCC 2015) aims to keep the global temperature rise well below 2 °C and contains the ambition to pursue efforts to limit the temperature increase even further to 1.5 °C. As a member state of the European Union, Ireland had to define a national climate target linked to the 2 °C target, which is considered to translate to an 80–95% reduction in greenhouse gas emissions relative to 1990 levels by 2050 (European Council 2009a), with an intermediate reduction target of 20% by 2020 (European Council 2009b), and at least 40% reduction by 2030 as outlined by the EU NDC (European Union 2015). These targets are not sufficient to reach a 1.5 °C and each country will have to propose more ambitious reduction, as agreed in the Paris Agreement on Climate Change. Ireland is currently not on track to meet the 2020 target. The total greenhouse gas emissions (GHGs) rose by 3.5% in 2016, and the cumulative emissions are projected exceed obligations by 11.5–13.7Mt of CO_2 equivalent over the period 2013–2020 (EPA 2017). This makes achieving the long-term goals and further reduction even more difficult.

The global mean surface warming has a near linear relationship with the cumulative amount of CO_2 in the atmosphere (Allen et al. 2009; Matthews et al. 2009). Pursuing efforts to remain below 1.5 °C warming requires further reductions in cumulative emissions over the century. However, it is difficult to determine the national carbon budget due to scientific and political uncertainties. The global carbon budget is usually expressed with a likely chance of remaining below a temperature threshold, where uncertainty arises from interactions with non-CO_2 GHGs and climate sensitivity uncertainty. In addition, the allocation of the global carbon budget for each country has usually been considered according to certain equity principles (Clarke et al. 2014), such as responsibility based on historical emissions, capability based on GDP, equality based on population, or a combination of multiple approaches. However, the results from different approaches have significant variations, and there is no consensus on which method is most equitable.

In this analysis, instead of attempting to share out a portion of the remaining global CO_2 budget for Ireland, we explore how far Ireland can reduce the carbon budget level and pursue efforts to remain below 1.5 °C. A high granularity analysis on the carbon budget over the period 2015–2070 was carried out. Starting from a carbon budget level with no emission constraint, the carbon budget is gradually reduced with small step changes in each subsequent scenario until the model becomes infeasible, resulting in over 100 scenarios. The feasibility of reaching different levels of mitigation target (Gambhir et al. 2017) is assessed based on the degree of challenge measured by model solvability, rate of decarbonization, carbon price and energy system cost. The challenge levels of the carbon budget scenarios are then compared with the current 80% reduction target to determine feasible carbon budgets. An alternative set of carbon scenarios are generated with different assumptions to explore the role of bioenergy imports and carbon capture and sequestration (CCS) technologies.

2 Methodology

2.1 Irish TIMES Model

TIMES (The Integrated MARKAL-EFOM System) (Loulou et al. 2016) is an economic model generator for local, national, multi-regional, or global energy systems, which provides a technology-rich basis for representing energy dynamics over a multi-period time horizon. TIMES assumes that each agent has perfect foresight on the market's parameters, and computes the inter-temporal dynamic equilibrium by maximizing the total surplus over the entire time horizon with decisions made on equipment investment and operation, primary energy supply and energy trade for each region. TIMES is thus a vertically integrated model of the entire extended energy system.

The original Irish TIMES dataset was extracted from the Pan European TIMES (PET36) project by the Energy Policy and Modelling Group (EPMG) at University College Cork (UCC), and was updated with local detailed data, calibrated to the national energy system with macroeconomic projections from the Economic and Social Research Institute (ESRI). The inputs to the Irish TIMES include estimates of end-use energy service demands derived from the macroeconomic model HERMES (Bergin et al. 2013), estimates of existing stocks of energy related equipment, characteristics of available future technologies, as well as present and future sources of primary energy supply and their potential.

The Irish TIMES model has been used to provide detailed insights on the possible pathways in achieving the challenging emission reduction targets for Ireland, and inform the development of national legislation on climate change and energy policy. Scenario analysis has been carried out to explore energy system trajectories to meet emission reduction targets (Chiodi et al. 2013a, b), the security of supply dimensions of a decarbonized energy system (Glynn et al. 2017a), and bioenergy availability (Chiodi et al. 2015). To facilitate scenario generation for sensitivity analysis, we developed a scenario ensemble tool on top of the existing TIMES model data handling system, in order to allow users to perform parametric sensitivity analysis by batch generating, queueing, and executing a large number of scenarios (Yue 2016).

2.2 A Series of Diverse Scenarios

In this chapter, two single scenarios and five carbon budget (CB) scenario sets have been developed. The Reference (REF) scenario and CO_2_80 scenario represents the pathways with and without climate policies, and provide a benchmark of comparison with carbon budget scenairos. Each of the carbon budget scenario set, including CB, CB15, NOCCS, NOBECCS, NOBIO, consists of large number of scenarios with different levels of carbon budget.

- **REF scenario**: The Reference Energy System scenario is the least cost optimal pathway that delivers energy service demands without any climate policies. Cumulative CO_2 emissions from the energy system from 2015 to 2070 are $2100MtCO_2$. This scenario includes the current macro-economic outlook, but does not explicitly include any carbon constraint policies on the energy system.
- **CO_2_80 scenario**: The CO2_80 scenario sets a linearly declining annual emission constraint according to to the EU climate & energy package and EU low-carbon economy roadmap from 2020 to 2050, achieving an 80% reduction in all energy related CO_2 emissions by 2050 on 1990 levels. ETS sectors are included in the total energy system CO_2 annual cap constraint. Beyond 2050 no further policy instrument is assumed and the emission level remains constant. This scenario has cumulative emissions of $950MtCO_2$.
- **CB**: The carbon budget scenarios apply decreasing levels of carbon budget constraints from 2015 to 2070. The model is free to choose the optimal emission pathways that reach the given carbon budgets. The carbon budget level starts from the REF scenario of $2100MtCO_2$. Each subsequent scenario reduces the carbon budget by 10Mt until the model is no longer feasible. This approach, rather than the opposite one (start with a very small budget and increase it) was followed in order to be able to analyse the evolution of the energy system with increasing ambition until infeasibility. Based on the historical emission trends and future projections of Ireland, significant mitigation actions are unlikely to be taken in the near future. Therefore, it is assumed that mitigation efforts will not start until 2020, and the emission pathways of CB scenarios are assumed to adhere with the pathway of REF scenario until 2020.

Besides the CB scenarios, alternative carbon budget scenarios including NOBECCS, NOCCS and NOBIO scenarios are generated using the same approach to quantify the impacts if certain technology options on the carbon budget feasibility, usually considered as key in emission mitigation, are no longer available.

- **NOBECCS**: The NOBECCS carbon budget scenario set disables bioenergy carbon capture and sequestration for power generation.
- **NOCCS**: This set of carbon budget scenarios disables all carbon capture and sequestration technologies for power generation. It should be noted that only CCS in the power sector is disabled, and CCS usage in the industry sector is still allowed.
- **NOBIO**: This set of carbon budget scenarios disables all bioenergy imports, and only allows bioenergy from indigenous sources.
- **CB15**: This set of carbon budget scenarios assume that significant mitigation efforts have been taken from 2015 instead of 2020.

3 Model Results

Even though there is uncertainty and lack of consensus on equity principles when determining a share of the remaining carbon budget, global estimates on carbon budget (Rogelj et al. 2016; Millar et al. 2017) and national level analysis (Pye et al. 2017) suggest that 1.5 °C consistent carbon budget is in general at least 50% lower relative to the 2 °C target. The carbon budgets for limiting temperature rise by 1.5 °C and 2 °C with 66% certainty are 400GtCO$_2$ and 1000GtCO$_2$ respectively from 2011 onward to net-zero emissions (IPCC 2014). Based on the equity standard by population, the carbon share of Ireland is 766MtCO$_2$ for 2 °C target and 223MtCO$_2$ for 1.5 °C target (Glynn et al. 2017b), which is approximately 20 and 75% less than the cumulative emission of 950MtCO$_2$ under the current climate policy. The analysis on the feasibility of the carbon budget scenarios indicate that reducing cumulative emissions to 360MtCO$_2$ is technically feasible (compatible with a 1.5 °C carbon budget), while 530MtCO$_2$ is a more realistic carbon budget level.

3.1 Feasible Carbon Budget Levels for Ireland

To determine feasible carbon budget targets, the carbon budget scenarios are assessed based on multiple criteria that measure level of challenge, which include the model feasibility, decarbonization rate from 2020 to 2030, timing of carbon neutrality, marginal abatement costs in 2050 and energy system cost.

Whether or not the model can produce a solution is a key indicator of low carbon pathway feasibility under the techno-economic assumptions reflected in the model. As increasingly more stringent emission constraints are applied, the capacity of available mitigation options is gradually exhausted, and the model needs to deploy very expensive backstop options for further mitigation. The results from the carbon budget scenarios show that the minimum feasible carbon budget for Ireland is 360MtCO$_2$. However, this budget level is extremely challenging. Reaching the 360MtCO$_2$ carbon budget requires 80% annual emission reduction to be delivered by 2030, and the carbon cost is €1600/tCO$_2$ in 2030 and €5200/tCO$_2$ in 2050.

Since it is assumed that no significant amount of carbon reduction can be achieved before 2020, the 2020 emission level is fixed (Fig. 1). The gap between the milestone year trendlines measures the required reductions in annual emissions over each 10-year period, and indicates that achieving a more stringent carbon budget mainly depends on stronger near-term mitigation efforts between 2020 and 2030. Compared to the constant 3.8% per annum decarbonization rate under the current policy framework, the annual decarbonization rate rises significantly to 10% in CB_530Mt scenario (Fig. 2). High decarbonization rates indicate significant challenge as phasing out fossil fuels and early retirement of existing technologies may raise concerns in terms of social acceptance, stranded assets and economic losses.

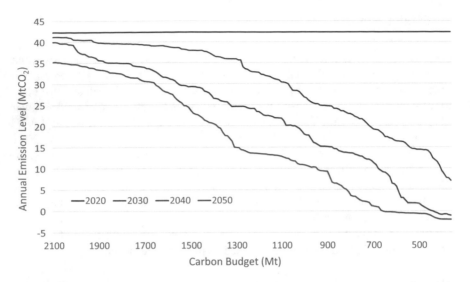

Fig. 1 Annual emission levels for milestone years for 175 carbon budget scenarios from REF scenario to CB_360Mt

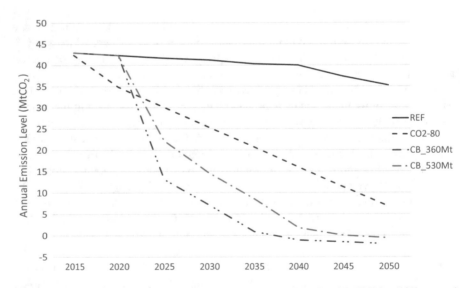

Fig. 2 Optimal annual emission pathways for REF, CO₂-80, CB_530Mt (10% annual decarbonization rate from 2020 to 2030), and CB_360Mt (minimum feasible carbon budget) scenario

Stabilizing temperature essentially requires net-zero anthropogenic carbon emissions (Matthews and Caldeira 2008). The timing of carbon neutrality is important and 1.5 °C consistent targets need to reach net-zero carbon emissions by mid-century (Rogelj et al. 2016). The milestone years emission levels (Fig. 1) show that reaching 660MtCO$_2$ carbon budget requires negative emissions by 2050. Further reduction in carbon budget requires earlier carbon neutrality. The CB_440Mt scenario has net-zero emission by 2040, and CB_510Mt scenario reaches carbon neutrality by 2045.

Deep decarbonization involves drastic energy system transformation and poses significant long-term challenges. The level of difficulty can be assessed with the CO$_2$ mitigation prices, which measures the marginal effort in mitigating an additional tonne of CO$_2$. A very high CO$_2$ price indicates a lack of available mitigation options in delivering the required mitigation level. The CO$_2$ price in the CO2-80 scenario is €114/tCO$_2$ in 2030, rising to €484/tCO$_2$ in 2050 (2016 prices). Due to stronger early actions, with the same CO$_2$ price in 2050 the total emissions can be reduced from 950MtCO$_2$ to 820MtCO$_2$ as a result of an optimised mitigation pathway rather than a linearly declining emissions cap. At a carbon cost of €1000/tCO$_2$, the total emissions can be cut to 530MtCO$_2$. The carbon cost has a nonlinear relationship with the carbon budget level (Fig. 3). Further reductions beyond 530MtCO$_2$ requires significantly higher marginal abatement costs, which increases to €2000/tCO$_2$ in the CB_430Mt scenario. Comparing carbon costs across scenarios shows that achieving the same level of annual emissions reduction in 2030 requires much higher carbon cost than in 2050. For example, the CB_360Mt scenario has 80% annual emissions reduction by 2030 with a carbon cost of €1600/tCO$_2$, tripling the 2050 CO$_2$ price of the CO$_2$-80 scenario.

Another key indicator of the carbon budget feasibility is the annual energy system cost (Fig. 4). The system cost accounts for 6.5% of GDP in 2030 and 8.0%

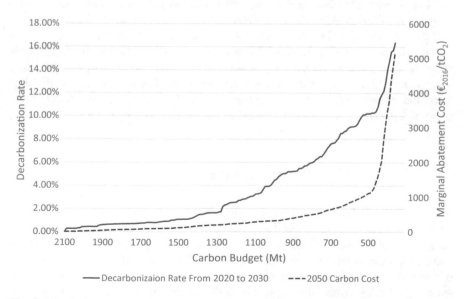

Fig. 3 Marginal abatement cost in 2050 and decarbonization rate over 2020 to 2030 for 175 carbon budget scenarios from REF scenario to CB_360Mt

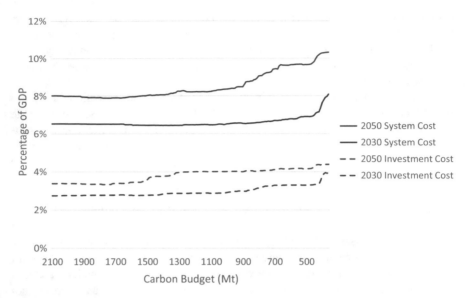

Fig. 4 Energy system cost and the investment portion as a percentage of GDP in 2030 and 2050 for 175 carbon budget scenarios from REF scenario to CB_360Mt

in 2050 for the REF scenario, increasing to 6.9% in 2030 and 9.7% in 2050 for the 530MtCO$_2$ scenario, indicating that reducing total emissions from 2100MtCO$_2$ to 530MtCO$_2$ requires an additional 1.7% GDP total costs. The energy system cost includes investment costs, fuel costs, and operation and maintenance costs. The proportion of investment costs in energy system cost remains relatively stable across scenarios. The investment cost as a percentage of GDP rises from 2.8% (2030) and 3.4% (2050) for the REF scenario, to 3.3–4.2% for the CB_530Mt scenario. Compared to the marginal abatement cost, the rise of energy system cost and the investment cost with emission constraint level is not as drastic. Pushing from the current policy to an ambitious mitigation target of 530MtCO$_2$ only requires 27% additional investment cost to the energy sector in 2050.

3.2   Energy System Transformation

In 2015, the total final energy consumption (TFC) in Ireland was 11.3 Mtoe, with 77% fossil fuels consumption and an electrification level of 19% (SEAI 2016). The energy demand of the CO2-80 scenario is characterized by significant reductions in fossil fuel consumption that decreases to 20% by 2050, as well as an increased level of electrification to 24%. The remaining energy demand is provided by renewable sources primarily from bioenergy. The TFC results of more amibitious carbon

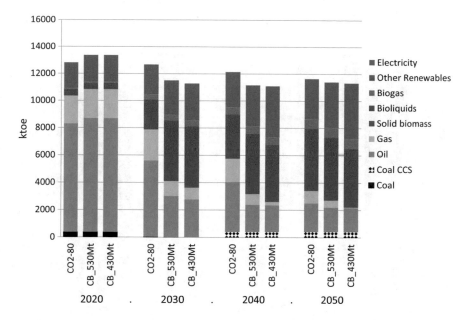

Fig. 5 Final energy consumption by fuel for CO_2-80, CB_530Mt and CB_430Mt scenarios

budget scenarios with 2050 carbon prices of €1000/tCO_2 (CB_530Mt) and €2000/tCO_2 (CB_430Mt) indicate that transition towards more ambitious emission targets requires higher levels of electrification and bioenergy consumption that replace fossil fuels (Fig. 5). Overall fossil fuel consumption declines with mitigation stringency as a result of efficiency gains from electrifying private cars and reduced demands due to higher carbon costs.

The choice of mitigation options in the carbon budget scenarios are similar to the CO2_80 scenario and the sectoral proportions in TFC remain relatively stable (Fig. 6). Cost effective mitigation options include switching to electric vehicles and plug-in hybrids for private cars; bioenergy for navigation and freight transportation; replacing gas with biomass for industrial boilers; switching to coal CCS for cement and lime production; increased penetration of biogas and biomass for the residential sector. Transition towards more stringent carbon budget constraints consistent with a 1.5C goal mainly requires replacing gas by electrifying the residential and services sectors, and large deployment of BECCS technology (Fig. 7).

Negative emission technologies are critical for carbon neutrality due to emissions from unmitigable sources, such as cement and lime production, passenger trains, hybrid vehicles. In the Irish TIMES model, BECCS is able to capture as much as 3.5MtCO_2, which is essential to bring the minimum achievable annual emissions level from 1.7MtCO_2 (96% reduction) to a level of negative emissions of −2MtCO_2. More stringent carbon budget levels require earlier deployment of BECCS technology, which starts to penetrate in 2035 for CB_520Mt scenario and in 2040 for CB_640Mt scenario.

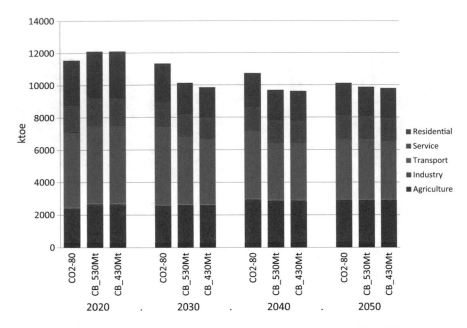

Fig. 6 Final energy consumption by sector for CO_2-80, CB_530Mt and CB_430Mt scenarios

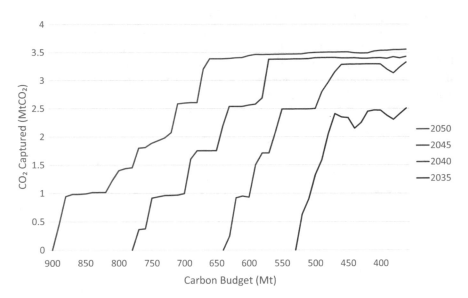

Fig. 7 Total amount of CO_2 captured by BECCS technology for 54 carbon budget scenarios from CB_890Mt scenario to CB_360Mt scenario

Due to limited mitigation options, reduction of annual emission beyond $-2MtCO_2$ is extremely costly. Attaining ambitous carbon budget levels therefore relies on strong efforts in the near term. As shown in Figs. 8 and 9, at ambitious carbon budget levels, the TFC trend in 2030 decreases at a fast rate, while the 2050 trend stays relatively stable after reaching net-zero emission. The timing of technology deployment is closely related to the level of annual emission reduction. For example, the 360MtCO2 budget scenario has a similar energy system configuration in 2030 as the system of CO2_80 scenario in 2050. Results in the levels of technology penetration also show that early transition into a low carbon economy requires phasing out of fossil fuel based technologies before the end of their

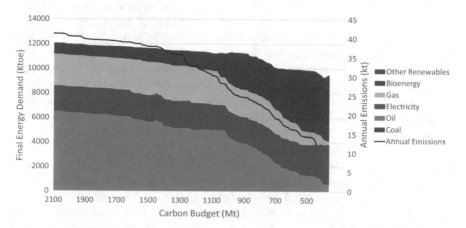

Fig. 8 Final energy demand in 2030 for 175 carbon budget scenarios from REF scenario to CB_360Mt

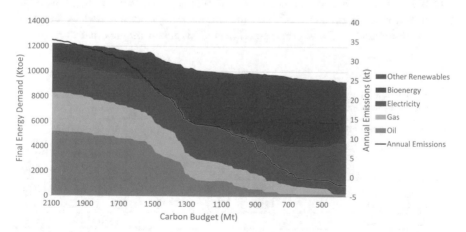

Fig. 9 Final energy demand in 2050 for 175 carbon budget scenarios from REF scenario to CB_360Mt

lifetime, creating stranded assets. Technologies with heavy investments and long lifetime have higher risks of being stranded, and such technologies include gas-fired power plants, space and water heating in residential and commercial sectors.

3.3 Implications from Variant Scenarios

Table 1 compares the carbon budgets for different CO_2 mitigation costs and associated rates of decarbonisation under alternative assumptions. The importance of key mitigation options can be quantified by the difference in carbon budget level attainable at the same degree of challenge (i.e. CO_2 mitigation cost and rate of decarbonisation). For example, if the availability of BECCS technology is disabled, the minimum feasible level of carbon budget is increased from $360MtCO_2$ to $480MtCO_2$, and the carbon budget that can be achieved at the same level of difficulty is increased by approximately $100MtCO_2$. Compared to the BECCS technologies, CCS power plants have less impacts on the carbon budget feasibility. Without CCS technologies, the energy system can deploy more biomass power plants for deep decarbonization, and the change in carbon budget level attainable at the same degree of challenge is within $20MtCO_2$ from NOBECCS scenarios.

In the CO_2-80 scenario, import accounts for 100% of bio liquids and 32% of the biomass, rising to 100% and 59% in the CB_530Mt scenario. The minimum solvable carbon budget of the NOBIO scenarios is $660MtCO_2$, over $200MtCO_2$ higher budget levels compared to the original carbon budget scenarios under the same degree of challenge, indicating that bioenergy imports have a more critical role than CCS technologies. With no bio liquids supply, the navigation and road freight transportation can no longer be fully decarbonized and has significantly higher penetration of hydrogen and biogas. BECCS has no penetration in any of the NOBIO scenarios due to limited supply of biomass. The decarbonization of the power sector relies on gas CCS, and the biomass is primarily used by industrial boilers, residential heat and commercial heat. This result implies that even though

Table 1 Carbon budget level ($MtCO_2$) for carbon budget scenarios under different level of challenge measurement

Scenario	Minimum solvable	2050 Carbon price (€$_{2016}$/tCO$_2$)			Annual average decarbonization rate from 2020 to 2030		
		€500	€1000	€2000	5%	7%	10%
CB	360	800	530	430	960	730	540
NOBECCS	480	880	670	570	990	800	650
NOCCS	490	870	660	560	1020	820	670
NOBIO	660	1010	810	710	1000	810	740
CB15	260	750	450	350	640	450	300

BECCS is critical for negative emissions, it is comparatively more cost intensive and should only be deployed when sufficient biomass supply can be secured.

The mitigation feasibility is also strongly impacted by the timing of mitigation efforts. The results of the CB15 scenarios (Table 1) show that with strong early actions starting from 2015, $100MtCO_2$ emissions can be further mitigated at the same degree of challenge. The minimum level of carbon budget with a model solution for the CB15 scenarios is $260MtCO_2$. The 2050 carbon price is €1000/tCO_2 for the CB15_450Mt scenario and €2000/tCO_2 for the CB15_350Mt scenario.

4 Conclusion

The aim of the chapter was to explore how ambitious Ireland can be in constraining its carbon budget towards a 1.5 °C consistent level, and to provide insights in the degree of challenge and required actions at ambitious reduction targets. The sensitivity analysis on the carbon budget constraints shows that the cumulative emissions can be cut significantly, and constraining the carbon budget from the current $950MtCO_2$ climate target down to $360MtCO_2$ is technically feasible. Even though $360MtCO_2$ can be considered as a well below 2 °C carbon budget level, it requires over 80% emission reduction to be delivered by 2030 and €5000/tCO_2 carbon cost by 2050. R&D that provide cost effective new mitigation options, particularly in carbon intensive industries, can be important in reducing the economic impacts and making the roadmap to 1.5 °C carbon budget more economically feasible.

Analysis on the level of challenge for carbon budget scenarios suggests that the $530MtCO_2$ carbon budget, which lies approximately midway between 1.5 and 2 °C target, is a more economically feasible carbon budget target with current technologies. However, compared to the current climate policy scenario CO_2-80, the $530MtCO_2$ scenario is still much more challenging, and requires a 10% p.a. emissions rate decline to 2030, resulting in a 57% emission reduction below 1990 levels by 2030, which is significantly higher than the 40% reduction target set by the EU NDC. The decarbonization rate from 2020 to 2030 is tripling the rate of the scenario corresponding to the current policies, and the carbon cost in 2050 is twice as much as that of CO_2-80 scenario at €1000/tCO_2.

With the current technological assumptions, reducing the net annual emission level of Ireland below $-2MtCO_2$ is extremely costly. Pushing towards more ambitious long-term climate target therefore relies on much stronger mitigation efforts by 2030 than the reduction target suggested by the NDC. However, strong mitigation efforts in the near term suffer economic losses from stranded assets such as gas-fired power plants and space and water heating in residential and commercial sectors. Decisions as to whether Ireland should accept such losses and pursue a 1.5 °C compatible carbon budget target should be made immediately. Investment in energy system are often capital intensive. Adhering to the existing near-term emission target may raise risks of "lock-in" to an energy system configuration that meets the near-term target but is unsuitable for a long-term 1.5 °C roadmap.

References

Allen MR, Frame DJ, Huntingford C et al (2009) Warming caused by cumulative carbon emissions towards the trillionth tonne. Nature 458:1163–1166. https://doi.org/10.1038/nature08019

Bergin A, Conefrey T, Fitzgerald J, Kearney I (2013) Working Paper No. 460 July 2013 The HERMES-13 macroeconomic model of the Irish economy

Chiodi A, Deane P, Gargiulo M, Ó Gallachóir B (2015) The role of bioenergy in ireland's low carbon future—is it sustainable? J Sustain Dev Energy, Water Environ Syst 3:196–216. https://doi.org/10.13044/j.sdewes.2015.03.0016

Chiodi A, Gargiulo M, Deane JP et al (2013a) Modelling the impacts of challenging 2020 non-ETS GHG emissions reduction targets on Ireland's energy system. Energy Policy 62:1438–1452. https://doi.org/10.1016/j.enpol.2013.07.129

Chiodi A, Gargiulo M, Rogan F et al (2013b) Modelling the impacts of challenging 2050 European climate mitigation targets on Ireland's energy system. Energy Policy 53:169–189. https://doi.org/10.1016/j.enpol.2012.10.045

Clarke LE, Jiang K, Akimoto K, et al (2014) Assessing transformation pathways. Clim Chang 2014 Mitig Clim Chang Contrib Work Gr III to Fifth Assess Rep Intergov Panel Clim Chang 413–510

Edenhofer O, Pichs-Madruga R, Sokona Y, Farahani E, Kadner S, Seyboth K, Adler A, Baum I, Brunner S, Eickemeier P, Kriemann B (2014) IPCC, 2014: summary for policymakers. Clim Chang

European Union (2015) Intended nationally determined contribution of the EU and its member states 1–7. https://doi.org/10.1613/jair.301

Gambhir A, Drouet L, McCollum D et al (2017) Assessing the feasibility of global Long-Term mitigation scenarios. Energies 10:89. https://doi.org/10.3390/en10010089

Glynn J, Chiodi A, Ó Gallachóir B (2017a) Energy security assessment methods: quantifying the security co-benefits of decarbonising the irish energy system. Energy Strateg Rev 15:72–88

Glynn J, Gargiulo M, Chiodi A, et al (2017b) Ratcheting national mitigation ambition using equitable carbon budgets to achieve the Paris Agreement goals. Clim Policy

Loulou R, Remme U, Kanudia A, et al (2016) Documentation for the TIMES Model Part II. IEA Energy Technol Syst Anal Program 1–78

Matthews HD, Caldeira K (2008) Stabilizing climate requires near-zero emissions. Geophys Res Lett 35:1–5. https://doi.org/10.1029/2007GL032388

Matthews HD, Gillett NP, Stott PA, Zickfeld K (2009) The proportionality of global warming to cumulative carbon emissions. Nature 459:829–832. https://doi.org/10.1038/nature08047

Millar RJ, Fuglestvedt JS, Friedlingstein P et al (2017) Emission budgets and pathways consistent with limiting warming to 1.5 °C. Nat Geosci 10:741–747. https://doi.org/10.1038/NGEO3031

Pye S, Li FGN, Price J, Fais B (2017) Achieving net-zero emissions through the reframing of UK national targets in the post-Paris agreement era. Nat Energy 2:17024. https://doi.org/10.1038/nenergy.2017.24

Rogelj J, Schaeffer M, Friedlingstein P et al (2016) Differences between carbon budget estimates unravelled. Nat Clim Chang 6:245–252. https://doi.org/10.1038/nclimate2868

SEAI (2016) Energy in Ireland 1990–2015. 86. http://www.seai.ie/Publications/Statistics_Publications/Energy_in_Ireland/Energy-in-Ireland-1990–2015

UNFCCC (2015) Adoption of the Paris Agreement. Report No. FCCC/CP/2015/L.9/Rev.1

Yue X (2016) Techniques for running large numbers of scenarios in TIMES. IEA-ETSAP Work Cork, Ireland

European Council (2009a) Brussels European Council 29/30 OCTOBER 2009-Presidency Conclusions. Council of the European Union.

European Council (2009b) 406/2009/EC of the European Parliament and of the Council of 23 April 2009 on the effort of Member States to reduce their greenhouse gas emissions to meet the Community's greenhouse gas emission reduction commitments up to 2020. Official Journal of the European Union, 5

The Pivotal Role of Electricity in the Deep Decarbonization of Energy Systems: Cost-Effective Options for Portugal

Júlia Seixas, Sofia G. Simoes, Patrícia Fortes and João Pedro Gouveia

Key messages

- The electrification targets are found to be useful but not sufficient to reach deep decarbonization such as nearly zero emissions or carbon neutrality for energy production and consumption.
- Further electrification of end-use sectors is necessary, especially the massive deployment of electric vehicles, electric dryers and the electrification of heat pumps in buildings. Currently, these technologies are not addressed in any Portuguese energy and climate policy initiative.
- Further analysis of carbon capture is also required.
- Portuguese energy and climate decision-makers should shift their current policy focus from broad national emission targets and incentives for the decarbonization of the power sector, and address the role of specific end-use low carbon technologies that will be required to move towards carbon neutrality.
- In this context, TIMES models are robust tools to support technology oriented policy-making.

J. Seixas (✉) · S. G. Simoes · P. Fortes · J. P. Gouveia
CENSE—Center for Environmental and Sustainability Research,
NOVA School of Science and Technology, NOVA University Lisbon, Caparica, Portugal
e-mail: mjs@fct.unl.pt

S. G. Simoes
e-mail: sgcs@fct.unl.pt

P. Fortes
e-mail: p.fs@fct.unl.pt

J. P. Gouveia
e-mail: jplg@fct.unl.pt

© Springer International Publishing AG, part of Springer Nature 2018
G. Giannakidis et al. (eds.), *Limiting Global Warming to Well Below 2 °C: Energy System Modelling and Policy Development*, Lecture Notes in Energy 64,
https://doi.org/10.1007/978-3-319-74424-7_13

207

1 Introduction

The Paris Agreement requires the deep decarbonization of all developed economies to achieve a global carbon neutral balance by mid-century. Energy efficiency and clean electricity are often touted as the most cost-effective strategies to ensure deep cuts in greenhouse gas (GHG) emissions from economic activities (Berst 2008; Williams et al. 2012; EURELECTRIC 2017). Williams et al. (2014) showed that it is technically feasible to achieve an 80% reduction in GHG emissions below 1990 levels by 2050 in the United States, and that multiple alternative pathways exist to achieve these reductions using existing commercial or near-commercial technologies. High levels of energy efficiency, decarbonization of electricity generation, electrification of most end uses, and switching the remaining end uses to lower carbon fuels, were the strategies to achieve such deep decarbonization. Moreover, the cost of achieving these reductions does not appear prohibitive, with an incremental cost to the energy system equivalent to less than 1% of gross domestic product (GDP) in the base case, without including potential non-energy benefits, for example, avoided human and infrastructure costs of climate change and air pollution. Also, the European Roadmap for a Low Carbon Economy in 2050 (EC 2011) showed the feasibility to reduce GHG by 80% in 2050 in comparison with 1990, mostly based on the deep decarbonization (around 90%) of the power, building and industrial sectors. An additional investment of around 1.5% of European Union GDP per annum was identified, which probably overstates the cost given recent reductions in the cost of low emission technologies.

The cost-effectiveness of such emissions' cuts depends on the energy system characteristics and how they may evolve. However, it is becoming apparent worldwide that the increase of electricity use in all end-uses across economic sectors, coupled with the increase of renewables share in the power sector and ambitious efficiency levels, is a robust deep decarbonization option. A technical assessment of the Portuguese energy system decarbonization up to 2050 was conducted in 2011 (Seixas et al. 2012), showing the feasibility of a cost-effective GHG emission reduction of 70% at acceptable costs, mainly if the impact on external primary energy trade and air quality were considered. However, achieving a neutral emission balance for the Portuguese economy, as politically stated at COP22 in Marrakesh, requires a deeper decarbonization, close to zero emissions.

The goal of this study is to assess the role of the electricity in the deep decarbonization of the Portuguese energy system, including scenarios of significant electrification of energy end-uses out to 2050. We are interested in finding how the power system, already characterized by a high share of renewable energy sources (RES), could evolve towards such an ambitious mitigation goal, as well as the impacts it implies, namely on emissions reduction, major energy shifts in the different economic sectors, additional investments, and costs of electricity.

The Portuguese energy system is an excellent case-study to analyse significant end-use electrification as a strategy towards a deep decarbonization due to three main reasons: (i) the power system features 61% of RES installed capacity in 2015 and an increased penetration of renewable electricity may be very challenging; (ii) the electricity consumption per capita is very low (around 15.9 GJ/capita compared to other EU countries (e.g. 22.6 GJ/capita in Germany or 18 GJ/capita in Spain in 2015, PORDATA 2016) meaning that there is scope to expand its use in all economic sectors (electricity represented 26% of the final energy consumption in 2014), and, (iii) there is political consensus among political parties for a carbon neutral economy by 2050. Therefore, this study is timely to assess such a deep decarbonization pathway.

2 Methodology for Modelling Energy Systems for Deep Decarbonization Assessment

2.1 Overview of the Model

We have used the TIMES_PT model, a peer-reviewed model for the Portuguese energy system, in use for more than 15 years (Simões et al. 2008; Gouveia et al. 2012a, b; Fortes et al. 2014; Simões et al. 2014, 2015; Fortes et al. 2015), including to support public policies (Seixas et al. 2010, 2012, 2014; APA 2012).

TIMES (The Integrated MARKAL-EFOM system) is a linear optimization energy model generator developed by ETSAP-IEA. The ultimate objective of the model is the satisfaction of the energy services demand at the minimum total system cost (i.e., net surplus maximization), subject to technological, physical and policy constraints. To this end, the model makes simultaneous decisions regarding technology investment, primary energy supply and energy trade (more information about TIMES, including model equations in Goldstein et al. (2016) and Loulou et al. (2016a, b). TIMES_PT considers the full range of the Portuguese energy system processes including energy supply (production, imports and exports), transformation (power and heat production), distribution and end-use energy demand in industry, residential, services, agriculture and transport (Fortes et al. 2015) in 5-year time steps from 2005 to 2050.

The model is sustained by a highly detailed technology database, containing more than two thousand supply and demand technologies, both current and emergent (see Fortes et al. (2015) for an example of the economic and technical data of selected power generation technologies of TIMES_PT database). TIMES_PT also considers economic and physical information of the energy resources available to satisfy the demand, including imports, and Portugal's RES potentials, which are estimated from national studies and validated by national experts (Seixas et al. 2012, 2014). The emission factors considered are from APA (2012).

For the purpose of this study, the following improvements were made to TIMES_PT: (i) a thorough update of the energy technologies database (technical and economic parameters), (ii) updated RES potentials for solar photovoltaics (PV) and concentrated solar power (CSP) (now 13.4 GW for solar PV roof size and a combined potential of 12.0 GW for both CSP and solar PV plant size); (iii) the modelling of the electricity transmission and distribution networks costs as fixed costs instead of variable costs, and (iv) the detailed analysis of the electricity sector per energy service.

The socio-economic development and the respective demand projections are the driving forces of the whole energy system modelled in TIMES_PT. In this study all the modelled scenarios adopt the same socioeconomic growth rate (average annual GDP growth of around 1.5% from 2020 onwards and a population decrease of 7.7% until 2050), which represent an average of the two socio-economic scenarios considered in the National Program on Climate Change 2020 (NPCC) (APA 2015; EC 2014). We consider the import prices for oil, gas and coal consistent with the New Policies Scenario in the World Energy Outlook 2016 (IEA 2016).

Moreover, unless explicitly mentioned, all scenarios include the following common exogenous assumptions: (i) average annual hydrological conditions (with seasonal variations) for all periods from 2015 onwards; (ii) consideration of the following policy instruments—oil products tax (ISP), road transport tax (IUC), carbon tax of 5€/tCO$_2$ applied to EU-ETS sectors; (iii) zero net electricity imports after 2015; (iv) no nuclear energy, and (v) no new 'conventional' coal power plants installed after the decommissioning of the existing coal power plants (Sines and Pego) which are assumed to occur in 2035–2040, respectively. Unless otherwise mentioned, the current 2030 and 2035 policies and targets on GHG emissions (APA 2015) and RES, RES subsidies or feed-in tariffs and energy efficiency targets are not explicitly modelled. These assumptions allow evaluation of the cost-effectiveness of energy technology options and endogenous resources to support deep decarbonization targets.

2.2   Decarbonization and Electrification Scenarios

Several groups of scenarios were assessed in TIMES_PT to study how, when and how much electricity will likely contribute to the decarbonization of the Portuguese economy:

(1) *Reference scenario (REF)*: the 2020 energy and climate policy goals are extended out to 2050, with the objective to identify the role of electricity assuming the continuation of current policy settings;

(2) *Decarbonization scenarios*: the tax of 5€/tCO$_2$ applied to EU-ETS sectors in REF was substituted by an overarching cap on GHG energy- and industrial processes-related emissions reduction of 50%, 60%, 75% and 85% in 2050 compared with 1990 values (named respectively, CO$_2$-50%, CO$_2$-60%, CO$_2$-75% and CO$_2$-85%). This allows assessing how the electricity sector interplays with other final energy carriers in response to stringent climate mitigation targets;

(3) *Electrification scenarios*: successive shares of electricity in final consumption of 40%, 50% and 70% (named respectively, ELC40, ELC50 and ELC70) are imposed from 2030 until 2050, to assess the technological feasibility of expanding low-emission electricity generation and the induced emissions reduction. In 2015 the share of electricity in total final energy consumption was 26%. The electrification targets were chosen considering the electrification shares obtained in 2050 in the REF scenario of 33% and in the CO_2-85% scenario of 51% electricity consumption.

Although Portuguese GHG emissions are rather small when compared to other countries, the considered carbon mitigation caps assume that, to achieve a below 2° global warming level, even countries like Portugal will have to lower the emissions of their energy system well below 1990 levels, or even achieve carbon neutrality.

3 Results Towards Deep Decarbonization Pathways

In this section, the results are compared across the scenarios: the REF scenario; the decarbonization scenarios CO_2-50 to CO_2-85 (reduction of 50–85% of GHG emissions in 2050/1990); and, the electrification scenarios ELC40, ELC50 and ELC70 (40, 50 and 70% of electricity in the final energy balance). We focus mostly on assessing the impacts on GHG emissions reduction, electrification trends in the different economic sectors, changes in the power generation mix, additional investment costs, and the wholesale cost of electricity.

3.1 What Are the Features of the Deep Decarbonization Pathways Selected?

Deep decarbonization pathways do not impact the electricity share in final consumption by 2030, but lead to the electrification of total final energy consumption approximately through a linear relationship by 2050 (Fig. 1). This means that, no matter what the mitigation target is, we observe an increase in the use of electricity from the current share of 28% (2016) to around 30% in 2030. However, the electrification of final consumption in the next 20 years out to 2050 appears very dependent on the mitigation target: 34% (REF), 36% (CO_2-60%), 44% (CO_2-75%) and 51% (CO_2-85%).

It is worthy to note the REF scenario shows a substantial decarbonization in 2050, with only a very modest CO_2 tax (5€/t CO_2). GHG emissions of the energy and industrial processes in 2050 are 38% lower than 1990 emissions, while in 2015 they were 17% above 1990. This is a very substantial decarbonization, which is mostly achieved due to the increased uptake of cost-effective energy efficiency technologies and renewable electricity. It should be underlined that the TIMES_PT

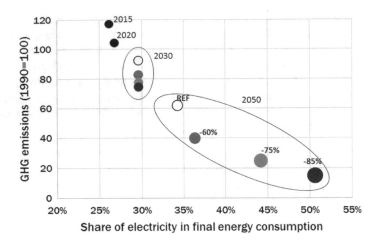

Fig. 1 Share of electricity in total final energy consumption and corresponding GHG emission mitigation for decarbonised scenarios in 2015, 2020, 2030 and 2050

feature of optimization, based on the cost-efficacy criterion, assumes that all actors in the energy system are perfectly rational in decisions for technology substitution.

On the other hand, electrification targets as such, without explicit decarbonization targets, do not necessarily lead to substantial emissions reduction than what was already achieved in the REF scenario (Fig. 2). Renewable electricity and energy efficiency are cost-effective up to an electrification of 40% of total final energy consumption (ELC 40%). For higher shares of electricity, and in the absence of mitigation targets, it is more cost-effective to generate electricity from natural gas plants.

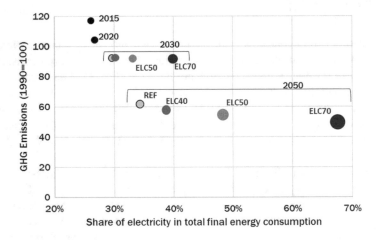

Fig. 2 Share of electricity in total final energy consumption and corresponding GHG emission mitigation for the electrification scenarios in 2015, 2020, 2030 and 2050

The modelled scenarios translated different GHG mitigation trajectories (Fig. 3). The slope of the trajectory between 2020 and 2050 has a significant impact on the results, particularly regarding CO_2 marginal abatement costs. With the current mitigation trajectories, the 2050 marginal abatement costs vary between 183 and 2930 €$_{2011}$/tCO$_2$e (Table 1). The upper range value corresponds to the CO_2-85% scenario and indicates that, with the currently considered technologic portfolio, although technologically feasible, it is likely economically infeasible to comply with such deep decarbonization. Other technology options need to be considered, such as electric heavy trucks, electric industrial kilns and carbon capture and utilization (CCU) in the cement industry to name a few.

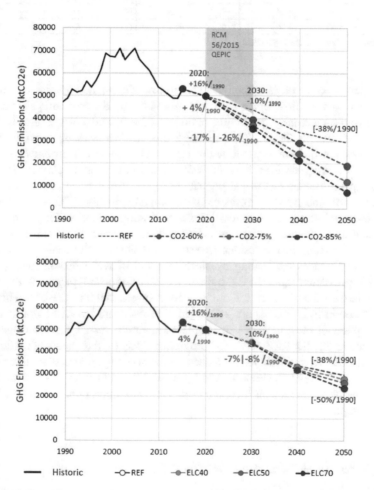

Fig. 3 Evolution of energy related GHG emissions in the REF and decarbonised scenarios (top) and in the REF and electrification scenarios (bottom). GHG emission values from 1990 till 2015 are historic. The 2020 and 2030 GHG national mitigation targets (QEPIC 2015) are included for comparison purposes

Table 1 Selected indicators from the modelled scenarios in 2030 and 2050

Scenario/year	% RES in final energy consumption[a]		% GHG energy emissions reduction compared to 1990		% RES electricity		CO_2 marginal abatement cost ($€_{2011}/tCO_2$)	
	2030 (%)	2050 (%)	2030 (%)	2050 (%)	2030 (%)	2050 (%)	2030	2050
REF	37	55	−8	−38	68	91	5	5
CO_2-50%	39	65	−17	−50	74	91	17	119
CO_2-60%	40	68	−17	−60	76	92	33	183
CO_2-75%	40	88	−23	−75	79	94	37	411
CO_2-85%	41	95	−26	−85	82	98	36	2930
ELC40	37	59	−7	−42	68	89	–	–
ELC60	39	64	−8	−45	67	86	–	–
ELC70	40	76	−8	−50	66	86	–	–

[a]Calculated as in the directive 28/2009/EC

The power sector plays a dominant role in lowering GHG emissions in the REF and CO_2-50% scenarios, which is mainly motivated by the cost competitiveness of the renewable electricity generation options (RES-e), and not necessarily by the mitigation target. In fact, in 2050 the RES-e share increases up to 91% in REF, which does not include any mitigation target. This share corresponds to a GHG emissions intensity of electricity in 2050 of around 52 gCO_2e/kWh. The other sectors lag the power sector and contribute to decarbonization depending on the stringency of the GHG emissions cap: already in the CO_2-60% scenario, the commercial sector replaces fossil fuels for heating with electricity; whereas the residential and transport sectors only start mitigating their direct emissions in the CO_2-75% scenario. In the residential sector, we observe the replacement of gas boilers and furnaces for heating by electricity and of water heating with gas and liquefied petroleum gas (LPG) boilers by solar thermal panels with electric backup. In the transport sector, there is likely a substantial increase of the biodiesel consumption in freight transport (10-fold in 2050 compared to REF), noting there is no the option of electric trucks. It should be mentioned that electric vehicles are cost-effective in the REF scenario, providing 74% of the passenger kilometres (pkm) travelled by car in 2050. With decarbonization targets, this share increases up to 84% in the CO_2-60% scenario, and up to 90% in both the CO_2-75% and CO_2-85% scenarios, in 2050. This level represents the maximum upper bound on electric vehicles deployment.

3.2   What Is the Increase in Electrification?

Due to the adoption of cost-effective energy efficient technologies, final energy consumption (FEC) is slightly reduced by 2050 in all scenarios from 2015 levels of between −2.8% (in REF) and −7.7% (in CO_2-85%). Nonetheless, in all modelled scenarios, the electricity consumption increases both in absolute terms (46 TWh in 2015, to 59 TWh in REF in 2050 or 62–82 TWh in 2050 in the decarbonization scenarios, and 64–100 TWh in 2050 in the electrification scenarios), and in relative share for total FEC (Fig. 4). Therefore, we also observe an increase in electricity consumption per capita from 4.4 MWh/capita in 2015 to 6.5–8.6 MWh/capita in the decarbonization scenarios, or up to 6.1-10.5 MWh/capita in the electrification scenarios (Fig. 5).

The modelling exercise shows an increase in the share of electricity in FEC in all scenarios from the observed 26% in 2015: 35% in REF; 35–41% in the decarbonization scenarios; and 39–67% in the electrification scenarios by 2050. All end use sectors increase electricity consumption due to the cost-effectiveness of electric technologies and efficiency gains from fuel switching to electric technologies. The sector with the highest electricity share is the commercial (services) sector, followed by residential buildings, industry and transport. The extreme electrification scenario ELC70 illustrates the maximum possible deployment of electric end-use technologies (almost 70% overall, and almost 100% in the services sector).

The electric intensity of GDP tends to decrease in all scenarios, since the main driver for decarbonization is energy efficiency, including electrification. However, for GHG mitigation targets higher than 60% below 1990 levels, electric intensity of GDP increases after 2040, as deep decarbonization is facilitated by carbon free electricity.

3.3 What Are the Most Cost-Effective Processes in Each Sector for Electrification?

Based on the consolidated modelling results, the following processes are identified as cost-effective in all scenarios:

(1) *Electric vehicles that supply 75% of the pkm travelled in private cars in the REF scenario in 2050.* In the CO_2-85% scenario, this value reaches 96% of pkm travelled and 93% in the ELC70 scenario. The corresponding value for the year 2015 was approximately 0.04% of pkm driven in private cars. The pace of deployment of EVs is restrained in the model as described in Sect. 2, replicating the inertia in adopting new more efficient technologies. Without such inertia, i.e. considering adoption solely based on cost-effective criteria, it was found in a set of alternative model runs (not presented here for the sake of brevity) that EVs would be deployed at a much faster pace and would entirely replace the existing car fleet by 2030;

(2) *Electric heat pumps in residential and commercial buildings supplying space heating, space cooling and, to a lesser extent, water heating.* Regarding residential buildings, heat pumps supply the same share of 35% of the space heating needs and 83% of the space cooling needs in 2050 in both the REF and CO_2-85% scenarios. In the ELC70 scenario, this value is also 35% of the space

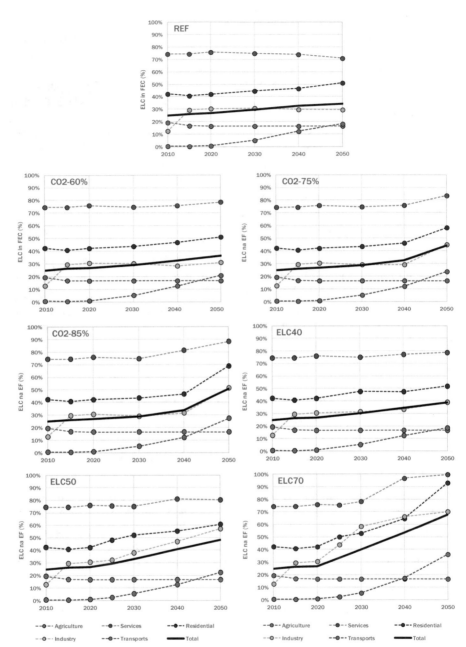

Fig. 4 Share of eletricity in total final energy consumption in the reference (REF), decarbonization (CO_2) and electrification (ELC) scenarios

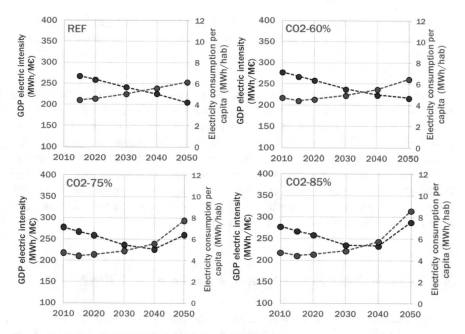

Fig. 5 Evolution of GDP eletric intensity and electricity consumption per capita in the modelled decarbonised scenarios

heating needs in 2050, but heat pumps are more expensive to operate for space cooling needs in 2050, which are supplied with cheaper fans and air conditioners (deliver all the space cooling needs in 2050). The corresponding value for the year 2015 was approximately 8% of residential space heating being delivered by heat pumps and 7% of the residential space cooling needs. Regarding commercial buildings, heat pumps supply only 15% of the space heating needs and no space cooling needs in 2050 in the REF scenario. In the CO_2-85% scenario, heat pumps supply 52% of the space heating needs and no space cooling needs in 2050 (as in REF). In the ELC70 scenario, this value reaches 71% of the space heating needs in 2050. Regarding space cooling needs, in all scenarios for the services sector, heat pumps are more expensive to operate for space cooling needs in 2050, which are supplied with cheaper fans and air conditioners (deliver all space cooling needs in 2050). The corresponding value for the year 2015 was approximately 10% of services space heating being delivered by heat pumps and practically none of the space cooling needs of commercial buildings. Similar to EVs, the deployment of electric heat pumps for space heating is driven by their cost-effectiveness and not by the GHG emissions or electrification targets. In the REF scenario, these technologies achieve what is considered to be their maximum feasible deployment.

(3) *Residential building insulation options* (including building envelop retrofitting
    and double-glazed windows) that allow lowering the space heating demand by
    31% in the REF scenario in 2050, 36% in the CO_2-85% scenario, and 33% in
    the ELC70 scenario.

Although not cost-effective in all modelled scenarios it is worth mentioning that
steam production in industrial processes, especially in the chemical industry, is
based on electricity ranging in 2050 between 47 and 100%, respectively for
CO_2-85% and ELC70. The electrification of steam production in "other industry
processes" only ranges from 22–100% steam generated from electricity, respectively for ELC50 and ELC70. Also, the generation of process heat (kilns and
furnaces) from electricity is cost-effective for "other industry processes" from 19%
estimated as generated from electricity in 2015, up to 26% in the REF scenario,
81% in CO_2-85% scenario and 100% in ELC70 scenario.

3.4 How Will the Power System Transition Look like?

Renewables play a dominant role in electricity generation (Fig. 6), even in the
absence of a GHG mitigation policy beyond the continuation of the current policies
in place up to 2020, from 60% penetration of renewable power in 2020, to 68%
penetration in 2030 up to 91% penetration by 2050. More stringent decarbonization
targets accelerate renewable penetration to 98% by 2050. Hydropower, onshore
wind and solar PV are the most cost-effective technology options, with the first two
reaching their maximum technical potential. Offshore wind and CSP emerges as a
cost-effective option in 2050 under stringent GHG mitigation targets (i.e., −75 and
−85%).

The increase in electricity generation and the changes in the power mix lead to
an increase in power sector costs (Fig. 7). Note that we do not include the
investment costs of existing technologies up to 2015. Investment costs are the cost
component with the highest growth in the share of total costs over the modelled
period, from around 36% of total costs in 2015 to around 41–47% in 2030 and
51–59% in 2050 for the REF and CO_2-85% scenarios, respectively. This is due to
the substantial increase of RES-e capacity that requires significant investment. On
the contrary, and as expected, fuel costs decline dramatically in some scenarios,
particularly the decarbonization scenarios. If grid costs are excluded, this leads to an
increment of the power sector's share of fixed costs in total costs (without grid
costs), which range from 78% of total costs in 2015 to more than 87% in all
scenarios by 2050.

Finally, the wholesale electricity prices generated by TIMES_PT in the
CO_2-85% scenario in 2050 are estimated to be around 42% higher than in the REF
scenario. On the contrary, in the ELC70 scenario, wholesale electricity prices are
16% lower than in REF, mainly because there is a much lower marginal CO_2 price

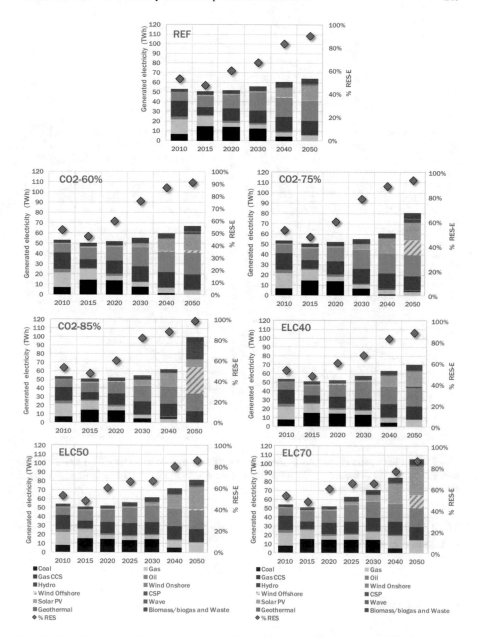

Fig. 6 Electricity production per technology in the reference (REF), decarbonization (CO_2) and electrification (ELC) scenarios

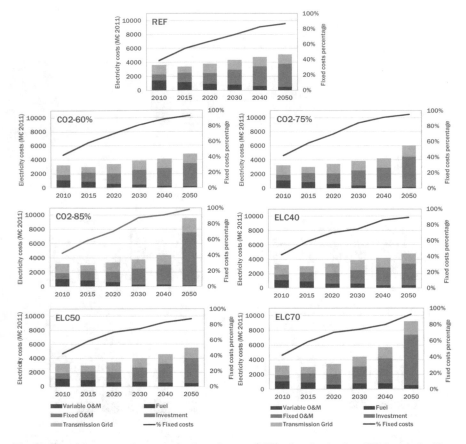

Fig. 7 Electricity generation costs in the reference (REF), decarbonization (CO₂) and eletrification (ELC) modelled scenarios. Share of fixed costs over variable costs (excluding grid costs) represented on the tight hand axis

in ELC70 and thus natural gas is still cost-effective despite the 5€/tCO₂ price considered in this scenario.

4 Conclusion

In this work we have studied the role of electricity in the deep decarbonization of the Portuguese economy. We have modelled three groups of scenarios using the TIMES_PT bottom-up partial equilibrium model: a reference case (REF) with current energy-climate targets extended until 2050; four similar scenarios with alternative GHG emissions reduction of 50, 60, 75 and 85% by 2050 when compared to 1990 levels; and finally, a group of three electrification scenarios, without explicit overall GHG mitigation targets, but imposing a 40, 50 and 70% share of electricity in total final energy consumption.

The electrification targets are found to be useful but not sufficient to reach deep decarbonization. In the highest electrification scenario, GHG emissions in 2050 are only 69% lower than 1990 levels, due to the deployment of gas to generate electricity in 2050. We conclude that although fuel switching to electricity promotes deep decarbonization, this is not a linear relationship. On the contrary, in the CO_2-85 scenario, practically all electricity is generated from RES sources and other sectors, particularly freight transport and industry, make contributions to achieve a reduction of 85% less GHG emissions from 1990 levels.

The technologies that contribute the most to the electrification of the Portuguese economy are electric vehicles, heat pumps (both in residential and commercial buildings), and dryers in some industry sectors. The electrification is ensured by significant deployment of RES power plants, mainly solar PV and wind offshore, as well as CSP to a smaller extent. In all scenarios, onshore wind and hydro play a key role, not necessarily driven by electrification or decarbonization, but instead by their cost-effectiveness.

The electrification scenarios are not found to be necessarily more energy efficient than the deep decarbonization scenarios. The wholesale electricity generation costs per MWh increase 47–62% in 2050 from 2015 levels for the decarbonization scenarios up to −75% reduction and increase 209% in the −85% scenario. This is due to the increased deployment of higher cost CSP and offshore wind power.

Since significant electricity consumption may not necessarily lead to a desired GHG emissions reduction target, we conclude that additional policy instruments may be needed, such as a CO_2 price. Another identified necessity regarding energy and climate policy making in Portugal is to shift the current policy focus from broad national emission targets and from incentives to the decarbonization of the power sector, to address the role of specific end-use low carbon technologies, which will be required to move towards carbon neutrality. It should be noted however, that significant technologies not currently included our analysis were identified as potentially important, namely long-distance electric trucks, carbon capture and utilization (CCU) in the cement industry, and electric kilns for the ceramic and glass industries.

The results presented herein should be interpreted with caution due to a number of limitations of the methodology and model used. The TIMES_PT model uses perfect foresight and does not consider any budget constraint; thus, technology replacement occurs more rapidly than observed in reality. Moreover, the increase in electricity prices will most probably lead to a decreased demand, not considered in this work (inelastic demand). In addition, further work should investigate realistic deployment rates of electric vehicles and heat pumps, via sensitivity analysis. We are also currently conducting a stakeholder elicitation process with Portuguese industry stakeholders to better assess the feasibility of the electrification of industrial processes. In these cases, using a TIMES model proves to be a good approach to communicate on technology shifts required to move towards a 2° world.

Another issue to be considered is what could happen to existing gas infrastructure in buildings if electrification dominates in the future. Additionally, as we observe in the results, the techno-economic potential for the deployment of RES technologies significantly constrains the model results, and thus should be reviewed and subjected to sensitivity analysis, particularly for solar PV, considering the significant changes in that technology in recent years. Moreover, TIMES_PT assumes a limited temporal resolution (12 times slices), which represents a limitation for load analysis in electricity systems with high penetration of renewable electricity generation.

Acknowledgements This work is part of a project funded by EDP—Energias de Portugal, S.A. The content of this work, however, is of the sole responsibility of its authors and it should in no way be viewed as reflecting any views or opinions on the part from EDP-Energias de Portugal. The authors want to thank the collaboration and fruitful discussions with Ana Quelhas and Andreia Severiano from EDP.

References

APA (2014) Portugal national inventory submission 2014. Portuguese Environment Agency
APA (2012) RNBC Roteiro Nacional de Baixo Carbono 2050 - Opções de Transição para uma Economia de Baixo Carbono Competitiva em 2050. RNBC National Low Carbon Roadmap 2050- Transition Options for a Low Carbon Economy. Agência Portuguesa do Ambiente. Amadora. Lisboa, Portugal. Available at: https://www.apambiente.pt/_zdata/DESTAQUES/2012/RNBC_COMPLETO_2050_V04.pdf
APA (2015) *PNAC - Programa Nacional para as Alterações Climáticas 2020/2030*. Agência Portuguesa do Ambiente. Amadora. Lisboa, Portugal. National climate change plan. Available at: http://sniamb.apambiente.pt/infos/geoportaldocs/Consulta_Publica/DOCS_QEPIC/150515_PNAC_Consulta_Publica.pdf
Berst J (2008) The electricity economy—new opportunities from the transformation of the electric power sector. Global Environment Fund and Global Smart Energy, USA
EC, COM (2014) 15 final. Communication from the commission to the council, the European parliament, the council, the European economic and social committee and the committee of the regions. A policy framework for climate and energy in the period from 2020–2030, European Commission, Brussels, 2014.—European Commission. http://eur-lex.europa.eu/legal-content/EN/TXT/?uri=CELEX:52014DC0015
EC (2011) Communication from the commission to the European parliament, the council, the European economic and social committee and the committee of the regions. A roadmap for moving to a competitive low carbon economy in 2050 (COM(2011) 112 final). European Commission, Brussels
EURELETRIC (2017) A bright future for Europe-The value of electricity in decarbonizing the European Union. pp 47. Brussels. Available at: http://www.eurelectric.org/media/318404/electrification_report_-_a_bright_future_for_europe-2017-030-0291-01-e.pdf
Fortes P, Pereira R et al (2014) Integrated technological-economic modeling platform for energy and climate policy analysis. Energy 73:716–730
Fortes P, Alvarenga A, Seixas J, Rodrigues S (2015) Long-term energy scenarios: Bridging the gap between socio-economic storylines and energy modeling. Technol Forecast Soc Change 91:161–178
Goldstein G, Kanudia A, Lettila A, Remme U, Wright E (2016) Documentation for the TIMES model: PART III. Energy Technology Systems Analysis Programme, Paris

Gouveia JP, Fortes P, Seixas J (2012a) Projections of energy services demand for residential buildings: insights from a bottom-up methodology. Energy 47:430–442

Gouveia JP, Dias L, Fortes P, Seixas J (2012b) TIMES_PT: Integrated energy system modeling. proceedings of the first international workshop on information technology for energy applications. In: Carreira P and Amaral V (eds) CEUR workshop proceedings, vol. 923. ISBN 978-989-8152-07-7

IEA (2016) World energy outlook 2016. International Energy Agency, Paris

Loulou R, Goldstein G, Kanudia A, Lettila A, Remme U (2016a) Documentation for the TIMES model: PART I. Energy Technology Systems Analysis Programme, Paris

Loulou R, Lettila A, Kanudia A, Remme U, Goldstein G (2016b) Documentation for the TIMES model: PART II. Energy Technology Systems Analysis Programme, Paris

Pordata (2016) Europe—Environment, Energy and Territory. PORDATA. Available at: (www.pordata.pt)

Simões S, Cleto J et al (2008) Cost of energy and environmental policy in Portuguese CO_2 abatement—scenario analysis to 2020. Energy Policy 36(9):3598–3611

Simoes S, Fortes P, Seixas J, Huppes G (2015) Effects of exogenous assumptions in GHG emissions models—A 2020 scenario study for Portugal using the Times model. Technol Forecast Soc Change 94:221–235

Simoes S, Seixas J, Fortes P, Huppes G (2014) The Savings of Energy Saving: Interactions between energy supply and demand side policies—quantification for Portugal. Energy Effi J 7 (2):179–201

Seixas J, Simões S, Fortes P, Dias L, Gouveia J, Alves B, Maurício B (2010) Novas Tecnologias Energéticas: RoadMap Portugal 2050 [New Energy Technologies: RoadMap Portugal 2050]. Lisbon

Seixas J, Fortes P, Dias L, Dinis R, Alves B, Gouveia J, Simões S (2012) Roteiro Nacional de Baixo Carbono: Portugal 2050 - Modelação de gases com efeito estufa, Energia e Resíduos [Low Carbon RoadMap: Portugal 2050 - Energy and Waste Greenhouse emissions]. Lisbon. Available at: http://www.apambiente.pt/_zdata/RNCB/EnergiaResiduos_10_07.pdf

Seixas J, Fortes P, Dias L et al (2014) Cenários de emissões de GEE e opções tecnológicas de descarbonização para Portugal em 2020 e 2030. GHG Emission Scenarios and carbon mitigation technological options. Faculdade de Ciências e Tecnologia- Universidade Nova de Lisboa. Estudo Técnico de Suporte ao PNAC 2020. Technical studies supporting the National Climate Change Plan. Agência Portuguesa do Ambiente, Lisboa

Williams JH, Haley B, Kahrl F et al (2014) Pathways to deep decarbonization in the United States. The U.S. report of the Deep Decarbonization Pathways Project of the Sustainable Development Solutions Network and the Institute for Sustainable Development and International Relations

Williams H, DeBenedictis A et al (2012) The technology path to deep greenhouse gas emissions cuts by 2050: the pivotal role of electricity. Sci 335(53):53–59

Part III
The Decarbonisation Pathways Outside Europe

The Canadian Contribution to Limiting Global Warming Below 2 °C: An Analysis of Technological Options and Regional Cooperation

Kathleen Vaillancourt, Olivier Bahn and Oskar Sigvaldason

Key messages

- The Canadian contribution to limiting global warming below 2 °C involves significant improvements in energy efficiency, the rapid decarbonization of electricity production, and large investments for replacing burning of fossil fuels with massive electrification complemented with bioenergy.
- Canada would benefit from achieving greater cooperation between jurisdictions because of the large diversity in the composition of regional energy systems.
- Using a detailed and regional optimization energy model brings value to Canadian decision-makers by analyzing complex energy systems with multiple GHG reduction options in various sectors, and in a dynamic time framework.

1 Canadian Perspectives on Climate Mitigation

As part of the Paris Agreement, Canada committed to reduce its greenhouse gas (GHG) emissions by 30% below 2005 levels by 2030. In its Mid-Century Long-Term Low-Greenhouse Gas Development Strategy (Environment and Climate Change Canada 2016a), Canada further defined its overall strategy for

K. Vaillancourt (✉)
ESMIA Consultants, Blainville, Québec, Canada
e-mail: kathleen@esmia.ca

O. Bahn
GERAD and Department of Decision Sciences, HEC Montréal,
Montréal, Québec, Canada
e-mail: olivier.bahn@hec.ca

O. Sigvaldason
SCMS Global, Toronto, Ontario, Canada
e-mail: sigvaldason.oskar@gmail.com

© Springer International Publishing AG, part of Springer Nature 2018   227
G. Giannakidis et al. (eds.), *Limiting Global Warming to Well Below 2 °C: Energy
System Modelling and Policy Development*, Lecture Notes in Energy 64,
https://doi.org/10.1007/978-3-319-74424-7_14

reducing climate change impacts, by committing to also achieving 70–90% reductions in GHG emissions by 2050 relative to 2005.

Canada is not a major contributor to the world's GHG emissions (only about 1.6% of the total in 2013). However, at 20.1 tons of CO_2 equivalent (tCO_2-eq) per capita (2013), it is ranked as one of the highest per capita emitters in the world (World Resources Institute 2017). In 2015, total GHG emissions were 722 million tons of CO_2 equivalent) (Mt CO_2-eq), with the energy sector accounting for 81% of this total (Environment Canada 2017). Emissions vary greatly between jurisdictions, with the highest per capita emissions being from jurisdictions with economies having major dependence on production and delivery of fossil fuels. With its federal structure and based on constitutional allocation of responsibilities, each of the provinces and territories have primary responsibility for planning and development of resources within its jurisdiction, including energy. The Pan Canadian Framework on Clean Growth and Climate Change, released in 2016 (Environment and Climate Change Canada 2016b), builds on existing commitments and actions by the Provinces and Territories; it includes meeting or exceeding Canada's international emissions targets, and transitioning Canada to a low carbon economy. Some of the provinces, including Quebec, Manitoba and Ontario, have even committed to GHG mitigation targets within their respective jurisdictions, greater than Canada's 30% reduction commitment by 2030 (Boothe and Boudreault 2016).

These commitments require special consideration of the energy sector in Canada. The country is endowed with abundant energy resources. It is one of the world's main energy producers, with enormous proven reserves. In 2015, it ranked second in production of uranium (fourth for proven reserves), fourth for production of crude oil (third for proven reserves), fifth for production of natural gas (seventeenth for proven reserves), and twelfth for production of coal (fifteenth for proven reserves) (Natural Resources Canada 2017). However, these resources are not evenly distributed across the country. For example, most of the oil reserves are in Alberta, and the uranium reserves are in Saskatchewan. Because of long-distance transport, it is more cost efficient to import oil from the Middle East into Eastern Canada, while oil is being exported from Western Canada to the United States and the Far East. As a consequence, there is not only extensive trade between Canada and the rest of the world for energy commodities (both export and import); there are also major trade flows between the respective jurisdictions in Canada.

The main objective of this chapter is to present results of cost-efficient strategies for Canada for achieving major reductions in GHG emissions from the energy sector, consistent with its global obligations. More precisely, we identify several decarbonization pathways for the Canadian energy sector associated with different GHG reduction targets and defining premises. This is carried out with a TIMES (The Integrated MARKAL-EFOM Model) model for Canada, which is part of the NATEM (North American TIMES Energy Model) platform (Vaillancourt et al. 2017). This application illustrates that TIMES models are powerful tools to map the energy transition needed to achieve GHG reduction targets compatible with the well below 2 °C) temperature goal and provide useful insights for decision markers at a national level.

This work builds on a previous large scale study exploring deep decarbonization pathways for Canada and identifies priority actions for the Canadian energy sector using both an optimization and a simulation model to derive optimal solutions for achieving progressive reductions in GHG emissions, from 30 to 70% by 2050 compared with 1990 levels (TEFP 2016). The novelty of this chapter includes a comparative assessment of official reduction target scenarios with more ambitious reduction scenarios, scenarios with reduction levels up to 75 and 80% by 2050, and a regional analysis of energy supply and trade, including the contribution of high voltage interconnections as an option for purchasing dependable capacity from their neighboring jurisdictions.

2 The North American TIMES Energy Model

Our approach is based on the application of the NATEM-Canada optimization model. NATEM belongs to the TIMES family of models developed within the Energy Technology Systems Analysis Program (ETSAP) of the International Energy Agency (Loulou et al. 2016). As such, NATEM relies on the concept of a Reference Energy System (RES) that describes energy value chains from primary energy to useful energy.

For the energy supply side, two sectors are distinguished: electricity and heat production (including combined heat and power); and supply of all other energy forms (in particular, primary fuels and derivatives from fossil and biomass sources). For energy demand, five sectors are considered: agriculture, commercial, industrial, residential, and transportation. At each point of the value chains, technologies are competing, not only existing ones but also new or improved ones. The model accounts for direct GHG emissions from fuel combustion as well as fugitive emissions from the energy sector. This enables NATEM to capture substitution of energy forms (e.g., switching to biofuels) and energy technologies (e.g., use of renewable power plants instead of fossil-fired facilities) to comply with GHG reduction targets. NATEM is a dynamic linear programming model. The objective is to minimize the net present worth cost of the energy system. The main outputs, in addition to the overall minimum present worth cost, include the selected decision variables and the magnitude of such variables for contributing to overall GHG mitigation for Canada. These variables also include transfers of energy commodities between jurisdictions, as well as international imports and exports.

NATEM-Canada models the RES of each of the 13 Canadian provinces and territories, as well as flows of energy and material commodities between these jurisdictions. For reporting purposes, Canada is sometimes divided into two broad regions: Western Canada, including the three prairie provinces, British Columbia and the three Territories; and Eastern Canada including Quebec, Ontario and the Atlantic provinces. The model includes representation of more than 4500 specific technologies and 475 different commodities. NATEM-Canada is calibrated to a

2011 base year, and represents the 40-year period from 2011 to 2050. Costs are presented in 2011 Canadian dollars (CAD$, at parity with US$ in 2011), with future costs being discounted at 5% per annum.

3   Scenarios with Increasing Levels of Mitigation Efforts

The results reported herein include four separate GHG mitigation scenarios, with comparison to the results of a reference scenario (REF) representing a business-as-usual case. In the REF scenario, demands for energy services are projected from the 2011 base year through 2050 using a coherent set of socio-economic and sector-specific drivers provided by the Canadian Energy System Simulator (CanESS) model (TEFP 2016): (i) a doubling of the gross domestic product (GDP) between 2010 and 2050, (ii) a slowing population growth (0.98% over 2010–2035 and 0.63% beyond 2035), (iii) a slowing GDP per capita growth (1.86% over 2010–2035 and 1.31% beyond 2035). The REF scenario is also based on government policies already in place, such as Corporate Average Fuel Economy (CAFE) standards in transport (NHTSA 2011), requiring a maximum consumption of litres per 100 kilometres for cars and light trucks; federal and provincial regulations on minimum content of biofuels in gasoline (5%) and diesel (2%) (Government of Canada 2010); coal-fired electricity standards (Environment Canada 2013); and existing carbon market between Quebec and California through a floor price ranging from 9.9$/tonne in 2012 to 24.9$/tonne in 2030 (in CAD$, at parity with US$ in 2011).

For comparison with this REF scenario, we analyze four separate GHG mitigation scenarios with increasing levels of mitigation efforts for Canada (Table 1). *The first scenario (GHG13)* represents the intended national contribution of Canada with a 30% reduction by 2030 from 2005 levels (UNFCCC 2016). It is equivalent to 13% reduction, relative to 1990. This target is then assumed to be held constant to 2050. *The second scenario (GHG50)* has a reduction target of 50% in 2050, compared with 1990 levels. This is a target derived from MERGE (Model for Evaluating the Regional and Global Effects of GHG Reduction Policies, proposed by Manne et al. 1995; and modified in Bahn et al. 2011) where all regions are (optimistically) assumed to fully cooperate to limit the temperature increase to 1.5 °C from 2020 on, and reductions are made where it is cheaper to abate GHG

Table 1   GHG emission reduction targets in Canada

Target	% reduction in 2050 from 1990 (%)	Cap in 2050 (Mt)	Absolute reduction in 2050 from the reference scenario (Mt)
GHG13	−13	378	−371
GHG50	−50	219	−530
GHG75	−75	108	−641
GHG80	−80	87	−662

emissions. *The third scenario (GHG75)* has a target reduction of 80% by 2050 from 2005 levels as proposed in the Canadian Mid-Century Strategy as a target consistent with the below 2 °C temperature goal (Environment and Climate Change Canada 2016a). Compared with 1990 levels, this represents a 75% reduction. *The fourth scenario (GHG80)* has a more ambitious reduction target of 80% by 2050 relative to 1990. Although this is a further reduction of only 5%, it serves to demonstrate several limits of what can be achieved with current assumptions.

In summary, the first mitigation scenario corresponds to an existing but limited government policy, the second represents an optimistic view of full cooperation at the World level, the third corresponds to the Canadian vision of its responsibility in reaching the below 2 °C temperature goal, and the forth is an attempt to reach an even more ambitious GHG reduction target.

These limits are applied to GHG emissions from fuel combustion in the energy sector. These emissions represent 71% of total emissions in Canada in 1990 (Environment Canada 2016).

Through a sensitivity analysis, we assessed also the potential for increasing trade of electrical energy and dependable capacity between neighboring jurisdictions; scenarios and results are summarized in Sect. 5.

4 Main Findings for the Energy Transition

4.1 Reducing GHG Emissions in All Sectors

For the REF scenario, combustion related GHG emissions increase by 48% between 2013 and 2050. The breakdown of emissions by sector (Fig. 1) shows that the transportation sector accounts for one third of total emissions, caused by a continued reliance on petroleum-based fuels. In 2050, emissions from energy supply (production of fossil fuels) and industries represent, respectively, 22 and 17% of total

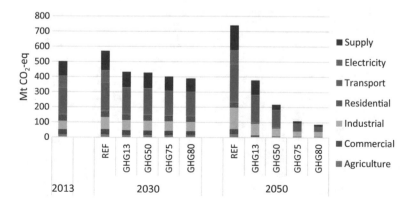

Fig. 1 GHG emissions by sector in Canada

emissions. GHG emissions increase in all sectors. However, the carbon intensity of the economy is reduced thanks to cost-efficient energy efficiency and energy conservation measures, including CAFE standards in the transportation sector.

Results for the four mitigation scenarios show dramatic changes from the REF scenario. First, the results suggest a fully decarbonized electricity supply by 2030 and a progressive decarbonization of the other sectors to 2050. The residential and commercial sectors are fully decarbonized by 2050 for the three most stringent mitigation scenarios. The dominant remaining challengers in 2050 are in the transportation, industrial and fossil fuel supply sectors. The challenges for the transportation sector are associated with non-fossil fuel options at higher costs for heavy freight and long-distance transport, limits on availability of biofuels, and limits on electricity storage systems. For the industrial sector, there are limits of availability of proven conversion options for switching from use of fossil fuels. For fossil fuel supply, the limitations are associated with the high implicit GHG mitigation cost for production and delivery of fossil fuels, relative to its value in the global market.

4.2   Achieving Major Transformations in Final Energy Demand

For the REF scenario, final energy demand is projected to increase by 45% between 2013 and 2050 (Fig. 2), with half of the additional demand coming from the industrial sector. In 2050, final energy is mainly consumed by industries (35%) and the transport sector (34%). The energy mix continues to be dominated by oil products and electricity in the long term, representing 47 and 26% of total consumption in 2050, respectively.

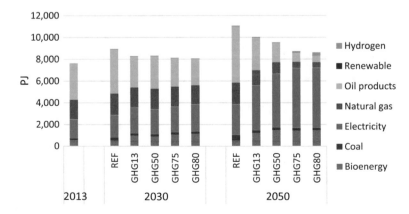

Fig. 2   Final energy consumption by type in Canada

However, for all mitigation scenarios, there are major transformations in the energy system. These include: (i) reductions in demands due to impacts of price elasticities, especially for heavy freight, as well as air and marine transport (up to 18% compared with their projected initial value for 2050); (ii) significant energy efficiency improvements through technological substitutions; (iii) greater penetration of electricity in all sectors (between 41 and 64% of total consumption in 2050), arising dominantly from reduced use of oil, from 47% in the REF scenario, to as little as 7% in the GHG80 scenario; (iv) reduced use of natural gas, from 18% in the REF scenario, to 6% in the GHG80 scenario; (iv) increased use of bioenergy in 2050, from 5% in the REF scenario to 17% in the GHG80 scenario. These transformation trends are generally similar across jurisdictions.

Significant transformations occur especially in the transport sector (Fig. 3). Electrification represents a central option for decarbonizing passenger road (cars, school and city buses and light truck), rail transportation (urban and interurban), and light-duty freight vehicles. As for medium and heavy freight, a portion of conventional diesel fuel is replaced with biofuels, compressed natural gas and renewable natural gas. Liquefied natural gas (LNG) is primarily used for marine transportation in replacement of heavy fuel oil, also resulting in significant reductions in GHG emissions. Hydrogen is part of the optimal energy mix in 2050 under the most ambitious scenarios only, given the high cost associated with the development of its supply chain. Hydrogen from electrolysis of water provides a role as a reduction option for heavy freight when the amount of biomass feedstock available for biofuel production or other uses is constrained by limitations on overall resource availability.

While electricity already accounts for a large proportion of the energy mix by 2050 in the residential, commercial and agricultural sectors in the REF scenario (54%), this proportion reaches 86–87% in scenarios GHG50–GHG80 as a replacement for oil products and natural gas (Fig. 4). In addition, decentralized

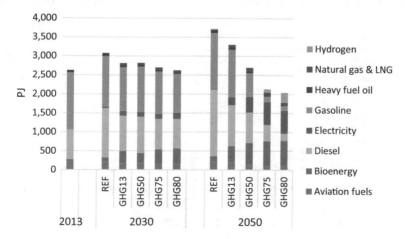

Fig. 3 Final energy consumption in the transport sector in Canada

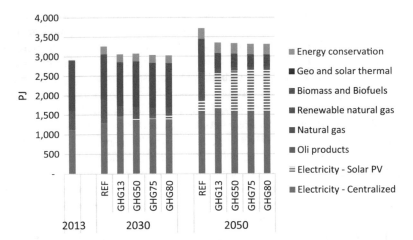

Fig. 4 Final energy consumption in the residential, commercial and agriculture sectors in Canada

electricity generation through photovoltaic solar panels replaces a significant portion of the electricity from the grid: from 35 (GHG13) to 40% (GHG50–GHG80) of total electricity used in 2050. Decentralized production occurs initially in the commercial sector where costs are slightly lower than in the residential sector. The results follow the optimistic cost reductions expected in the future for solar panels and used as assumption in the model (EIA 2014; NREL 2013). The maximum potential of energy conservation measures is reached in all scenarios including the REF scenario. Energy conservation is therefore not limited by the costs of measures, but by the maximum potential savings that it is possible to achieve.

Electrification also represents a significant reduction option for the industrial sector (Fig. 5). In addition, a significant portion of natural gas is replaced with renewable natural gas (biogas) that represents up to 7% of the total consumption in 2050.

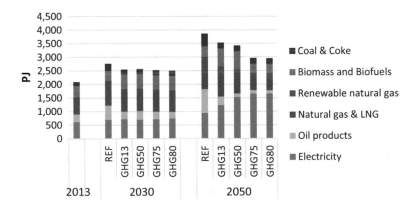

Fig. 5 Final energy consumption in the industrial sector in Canada

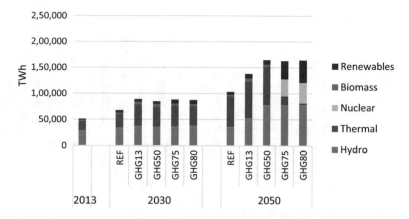

Fig. 6 Electricity generation in Western Canada

This proportion of natural gas also includes 2% of LNG, mainly used in the mining sector in remote areas.

The role of biomass varies across time. There is a decrease of biomass consumption in the pulp and paper industry due to projected decline in demand. This reduction is offset by the use of biomass as an alternative to coal in the cement and iron & steel industries in the least stringent scenarios (GHG13 and GHG50). However, by 2050, in the most stringent scenarios (GHG75 and GHG80), the direct use of biomass in the industrial sector is limited, due to competition for other uses, including massive production of second-generation biofuels.

There are significant energy efficiency gains in the most energy-intensive industries, including aluminum, chemicals, and iron and steel. For example, efficiency gains for the aluminum sector are as high as 2.0% per annum over the period 2013–2050. With the recent Canadian regulation on coal-related emissions, coal could either be replaced by a less polluting source or combined with carbon capture and storage (which is not included as a technical option in this chapter). And lastly, a portion of coke used in the iron and steel industries remains in the mix as no option exists in the existing database to replace it.

4.3   Decarbonizing Electricity Supply

Massive electrification of end uses represents a consistent observation across sectors for the GHG mitigation scenarios, with the amount of electrification increasing with increasingly stringent GHG reduction targets. Major investments are required to increase the supply of carbon free electricity generation, in both Western Canada (Fig. 6) and Eastern Canada (Fig. 7).

While electricity generation already increases significantly in the REF scenario, it increases further for GHG mitigation and reaches a 2.4—fold increase between

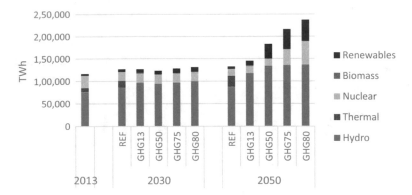

Fig. 7 Electricity generation in Eastern Canada

2013 and 2050 in the GHG80 scenario. Due to a very uneven distribution of primary resources in Canada (as described in Sect. 4.4), the electricity generation mix varies between jurisdictions.

Electricity generation in Western Canada, currently, comes primarily from hydro power plants located in British Columbia and Manitoba, and thermal power plants. For the REF scenario, the increasing demand for electricity by 2050 is dominantly supplied by an additional 19 GW of gas-fired power generation, especially in Alberta, for steam generation in the oil sands industry. Thermal generation from existing coal-fired plants is decreasing, from 8 to 1 GW, due to planned phasing out (Environment Canada 2013).

In the GHG mitigation scenarios, thermal generation is replaced with addition of hydro developments, primarily in Manitoba and the Northwest Territories (there is a moratorium on new large-scale hydro projects in British Columbia). The remaining portion of thermal generation includes the existing gas-fired power plants used to provide dependable electricity and refurbished coal-fired power plants with carbon capture and storage facilities (6 GW). Nuclear power plants are built in Alberta (19 GW) to achieve deep reduction levels by 2050 (GHG75 and GHG80) and provide dependable capacity. The electricity mix is completed with important additional investment in intermittent renewables from wind energy (46 GW supported with 18 GW of pumped storage) and development of the small geothermal potential in Alberta and British Columbia (2 GW).

Electricity generation in Eastern Canada is already close to being carbon free, with 89% of electricity generation from non-emitting sources. The installed capacity for electricity generation currently includes 55 GW of hydro plants located in Quebec, Ontario and Newfoundland and Labrador. In the REF scenario, another 7 GW of hydro capacity is expected by 2050 to satisfy the projected growth in demand for electricity. Projected growth in electricity demand is also satisfied with greater use of gas-fired power plants, primarily in Ontario. Wind farms are projected to slightly increase. The installed capacity from nuclear power is projected to decrease with the already scheduled refurbishment program for only a portion of the

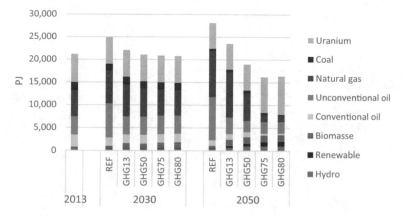

Fig. 8 Primary energy production in Western Canada

existing reactors in Ontario. Finally, coal-fired plants are phased out as scheduled, and not refurbished with carbon capture and storage facilities. Renewable electricity also includes generation from small scale solar plants in Ontario and one tidal plant in Nova Scotia.

Additional electricity supply is needed in the GHG mitigation scenarios compared with the REF scenario, up to 38 GW of additional hydro capacity, 18 GW of nuclear capacity (in Ontario, where little hydro potential remains) and 49 GW of wind capacity, with support from 26 GW of pumped storage capacity. While thermal generation already plays a marginal role in the REF scenario, it is almost non-existent in the GHG mitigation scenarios. The technical potential for hydro-electric generation capacity is estimated at 125 GW in Eastern Canada, with 82 GW in Quebec alone, i.e. more than twice the installed capacity in 2013. The hydraulic resource represents a major asset in meeting the growing demand for electricity for the more ambitious GHG mitigation scenarios.

In both regions, electricity generation from biomass remains small as the various types of biomass feedstocks are used primarily for liquid biofuel production, renewable natural gas production, space heating, and as a replacement of coal in some industries.

4.4 Shifting Away from Fossil Fuel Supply

The primary resources are distributed very unevenly across the country. Primary energy production for both domestic consumption and exports is illustrated separately for Western Canada (Fig. 8) and Eastern Canada (Fig. 9).

In Western Canada, where most of conventional and unconventional fossil fuel reserves are located, the REF scenario shows the primary mix being dominated by

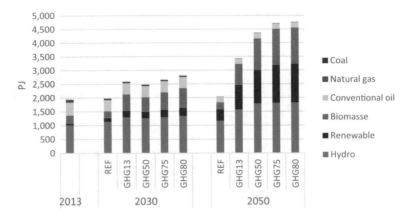

Fig. 9 Primary energy production in Eastern Canada

an increase in oil and gas production by 68% from 2013 to 2050. In 2050, fossil fuels (oil, gas and coal) represent 76% of the primary energy production mix.

Major reductions (up to 66%) in oil and gas production occur in the GHG mitigation scenarios from 2013 to 2050. Domestic needs are dramatically reduced with the electrification of end uses. Moreover, there are major reductions in international exports, especially oil, as the system-wide costs for in situ bitumen extraction in Alberta are greater than revenues from exports for the most severe scenarios.

By comparison, the primary energy production mix in Eastern Canada is strongly dominated by the exploitation of hydraulic, other renewables (existing or scheduled wind plants, and planned decentralized solar generation) as well as biomass feedstock resources (Fig. 9).

4.5   Shifting Away from Fossil Fuel Delivery

Obviously, with fossil fuel and uranium reserves that greatly exceed domestic needs, Western Canada is a net exporter of energy to the rest of Canada, the United States and the rest of the World (Fig. 10). While most of fossil fuel exports are currently destined to international markets (89% for oil, 78% for gas, 98% for coal), these shares are expected to decrease over the long term with increasing exports to Eastern Canada. For the REF scenario, international exports are projected to decline to 68% for oil, 42% for gas, and 75%, for coal.

Major reductions in production and delivery of fossil fuels occur in the GHG mitigation scenarios driven by reduced domestic demand, as well as the need to reduce emissions for supply and delivery of fossil fuels, which are emissions intensive with high implicit costs, resulting in uneconomic production. Moreover, it is assumed that the international community will reduce its demands for fossil fuels

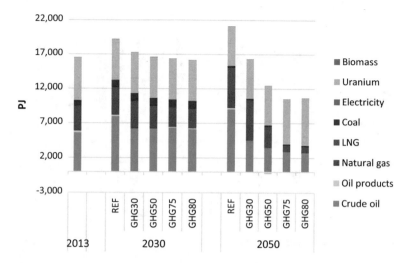

Fig. 10 Net exports in Western Canada

at the same ratio as Canada (the model is flexible to represent such assumed relationships). Such a drastic reduction will certainly impact the Canadian GDP as well as any major structural change in the energy system. However, studying the macroeconomic effects of such climate policies are beyond the scope of this chapter. Production of uranium remains relatively constant across scenarios and time, dominated by exports. There is a slight decrease in exports by 2050 for the most severe scenarios, in response to increased projected domestic demand, primarily for additional nuclear reactors in Alberta and Ontario.

In contrast, Eastern Canada is a net importer of fossil fuels and uranium (Fig. 11). While most of the crude oil imports come from international markets in 2013, these are progressively replaced with synthetic oil from Alberta in the REF scenario given the lower price for oil sands (although the Energy East pipeline project is on hold, transport by rail and heavy trucking remains).

For the GHG mitigation scenarios, all imports are decreasing, with associated reductions in refining activities. While natural gas comes from Alberta through existing pipelines, an increasing amount could come from shale gas deposits in the United States. Uranium is imported exclusively from Saskatchewan. Eastern Canada is a net exporter of electricity to the United States through existing interconnections from Quebec (4.3–6.3 GW from 2020 with scheduled projects), Ontario (2.1 GW) and New Brunswick (1.1 GW) to New England, New York and the Midwest.

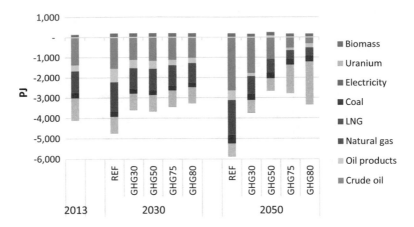

Fig. 11 Net exports in Eastern Canada

4.6 How Much Should Carbon Prices Be?

Marginal costs for GHG mitigation increase significantly over time, and reach high values by 2050, from 343$/tCO$_2$-eq to 2130$//tCO$_2$-eq (in CAD$), depending on mitigation targets (Fig. 12). The less stringent scenarios (GHG13 and GHG50) would require a doubling of the cost of gasoline to reflect such marginal costs.

The analyses show that limiting GHG emissions below 200 MtCO$_2$-eq by 2050 represents a sensitive point, with abatement costs increasing exponentially due to the lack of reduction options, especially for the transportation, industrial and fossil fuel supply sectors. Additional reduction options are needed to explore more aggressive transformation pathways. These include carbon capture and storage facilities in various industries, third-generation biofuels, CO$_2$ trade between juris-dictions to allow a greater role for bioenergy with carbon capture and storage (which is not limited to jurisdictions with both biomass feedstock and sequestration potential), and additional options for long-distance and heavy freight transport. Moreover, it is possible that our price elasticity assumptions underestimate the reduction in service demands in such ambitious reduction contexts. Finally, the analysis does not include the costs of not acting (adaptation costs and damages), and consequently, does not calculate the net costs and benefits ratio of achieving ambitious targets.

5 Greater Cooperation Between Jurisdictions

In Canada, each of the 13 jurisdictions is responsible for electricity self-sufficiency. This includes both guaranteed electrical energy and dependable generating capacity for meeting energy demand at all time, especially during peak periods. There are

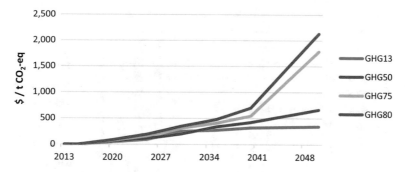

Fig. 12 Marginal cost of abatement

three exceptions, resulting in reliance on neighboring jurisdictions: the Churchill Falls interconnection with Newfoundland and Labrador supplying Quebec through a contract to 2041; a contract between Newfoundland & Labrador and Nova Scotia for 20% of Muskrat Falls production for 35 years; and a contract between New Brunswick and Prince Edward Island for transfer of 5% of the nuclear energy generated at Point Lepreau. Other interconnections exist, but electricity imports cannot be used to account for dependable capacity. We assessed the potential for increasing trade of electrical energy and dependable capacity between neighboring jurisdictions. This includes additional investment in high-voltage interconnections. An interesting low-cost option for dependable capacity is also to add generating capacity at existing and future hydro sites, referred to as the "incremental hydro" option. This option applies for both inter-jurisdictional trade and for complementing electricity generation from intermittent renewable generating sources. This option is included in the following alternate mitigation scenarios, respectively identified with an "a": GHG75a and GHG80a.

Since energy resources are not distributed evenly across the country, important trade movements and modifications in trade patterns occur in order to satisfy a growing demand both in a business-as-usual situation (REF scenario) and in GHG reduction scenarios. The role of incremental hydro capacity at existing hydro sites along with the option of purchasing dependable capacity from other jurisdictions reduces the need for investments in more expensive dependable capacity. These effects are illustrated (Fig. 13) by comparing the installed capacity by type in the two most ambitious scenarios (GHG75 and GHG80) in 2030 and 2050 in Western Canada. Decarbonization of the electricity sector is indeed more difficult in this region due to the lack of clean energy source that can be used as dependable capacity. With these additional options, Alberta for instance, benefits from capacity imports from the Northwest Territories and the development potential of the hydro resources of the Mackenzie River. Globally for Western Canada, these options allow reductions in investments in thermal based capacity by 13% and in storage capacity by 43% to achieve an 80% reduction by 2050 from 1990 levels (GHG80 scenario). For the slightly less stringent scenario (GHG75), this also contributes to

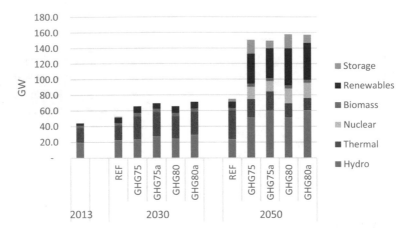

Fig. 13 Installed capacity for electricity generation in Western Canada

reducing nuclear based capacity by 16% in 2050. The total energy system cost is reduced by 5% in both scenarios compared with their formal version. These results suggest that Canada would benefit from achieving greater cooperation between neighboring jurisdictions. While our assumptions for these alternate scenarios were very conservative, namely in terms of potential, it would be interesting to further explore the role of incremental hydro as a cost-effective option to reduce GHG emissions and especially in nuclear phase-out scenarios.

6 Conclusion

In the context of the Paris Agreement, we have explored alternative strategies for Canada to meet its portion of globally agreed GHG mitigation targets to 2050 and beyond using the 13-region NATEM model. This includes implementing changes in the composition of energy systems in the various jurisdictions and achieving greater cooperation between neighboring jurisdictions, including greater regional integration. The results demonstrate the importance of energy conservation and efficiency as the primary lowest cost options for reducing GHG emissions. Not surprising, massive electrification dominates the many transformations required to meet mitigation targets. Interestingly, decentralized electricity generation plays an increasing role under climate mitigation. There are also major increases in the use of bioenergy. The most immediate transformation is the rapid decarbonization of electricity production, followed by rapid electrification in the end-use sectors, and electrification of the fossil fuel supply chain. When removing the electricity self-sufficiency constraint, the total energy system cost is reduced by 5%.

This analysis reflects the tremendous value of using an optimization model, such as the NATEM model, for deriving optimal solutions and associated marginal abatement costs for different mitigation scenarios. This becomes especially valuable when analyzing complex energy systems with multiple technology options in multiple jurisdictions, with multiple sectors, and analyzed in a dynamic time framework. This very rigorous approach has the additional benefit of demonstrating accurately trade-offs between results of different scenarios, for providing valued perspectives on decisions which are either sensitive to input parameters, and for demonstrating required actions, both short term or long term, which are essential for satisfying progressively increasing GHG mitigation demands.

Defining pathways for achieving ambitious GHG reductions is a complex task involving many uncertainties. This analysis did not consider all the factors that could impact the optimal solutions. In addition, some limitations are inherent to the current version of the model and do not provide a full perspective on the problem such as GHGs from other sectors and non-GHG pollutants.

Finally, it is increasingly argued that the carbon neutrality of the global energy sector should be achieved in 2050 to maintain the temperature below 1.5 °C by 2100. However, additional modeling efforts are required to reach even lower reduction levels and study the carbon neutrality goal, namely: the integration of more reduction options in the industrial sector and of net negative GHG options, as well as the optimization of transport modal shifts and other behavioral changes. Despite these limitations, however, the analysis of ambitious reduction scenarios illustrates the fact that optimization energy models such as NATEM are powerful tools for identifying robust trends and options.

References

Bahn O, Edwards N, Knutti R, Stocker TF (2011) Energy policies avoiding a tipping point in the climate system. Energy Policy 39:334–348

Boothe P, Boudreault FA (2016) By the numbers: Canadian GHG emissions. Ivey Business School at Western University, Lawrence National Centre for Policy and Management, p 15

EIA—Energy Information administration (2014) Annual energy outlook 2014—energy outlook database. U.S. Department of Energy [Online]. https://www.eia.gov/outlooks/aeo/data/browser/

Environment and Climate Change Canada (2016a) Canada's mid-century long term low greenhouse gas development strategy. Canada's Submission to the United Nations Framework Convention on Climate Change (UNFCCC), 87 p

Environment and Climate Change Canada (2016b) Pan-Canadian framework on clean growth and climate change: Canada's plan to address climate change and grow of the economy. 78 p

Environment Canada (2013) Reduction of carbon dioxide emissions from coal-fired generation of electricity regulations. Can Gaz 145(3) and 146(19)

Environment Canada (2016) National inventory report: greenhouse gas sources and sinks in Canada 1990–2014—executive summary. Canada's Submission to the United Nations Framework Convention on Climate Change, 15 p

Environment Canada (2017) National inventory report 1990–2015: GHG sources and sinks in Canada—executive summary [Online]. http://www.ec.gc.ca/ges-ghg/default.asp?lang=En&n=662F9C56-1

Government of Canada (2010) Renewable fuels regulation. Can Gaz 144(18)

Loulou R, Goldstein G, Kanudia A, Lehtila A, Remme U (2016) Documentation for the TIMES model, energy technology systems analysis program (ETSAP) of the international energy agency (IEA) [Online]. http://iea-etsap.org/docs/Documentation_for_the_TIMES_Model-Part-II_July-2016.pdf

Manne A, Mendelsohn R, Richels R (1995) MERGE: a model for evaluating regional and global effects of GHG reduction policies. Energy Policy 23:17–34

Natural Resources Canada (2017). Energy fact book 2016–2017 [Online]. https://www.nrcan.gc.ca/sites/www.nrcan.gc.ca/files/energy/pdf/EnergyFactBook_2016_17_En.pdf

NHTSA—National Highway Traffic Safety Administration (2011) Summary of fuel economy performance. US department of transportation

NREL—National Renewable Energy Laboratory (2013) Non-hardware ("soft") cost-reduction roadmap for residential and small commercial solar photovoltaics, 2013–2020 [Online]. https://www.nrel.gov/docs/fy13osti/59155.pdf

TEFP—Trottier Energy Futures Project (2016) Canada's challenge and opportunity—transformations for major reductions in GHG emissions [Online]. http://iet.polymtl.ca/en/tefp

UNFCCC—United Nations Framework Convention on Climate Change (2016) Canada's INDC submission to the UNFCCC [Online]. http://www4.unfccc.int/Submissions/INDC/Published%20Documents/Canada/1/INDC%20-%20Canada%20-%20English.pdf

Vaillancourt K, Bahn O, Frenette E, Sigvaldason O (2017) Exploring deep decarbonization pathways to 2050 for Canada using an optimization energy model framework. Appl Energy 195:774–785

World Resources Institute (2017) Climate analysis indicators tool 2017 [Online]. http://www.ec.gc.ca/ges-ghg/default.asp?lang=En&n=662F9C56-1. Accessed Oct 2017)

Modeling the Impacts of Deep Decarbonization in California and the Western US: Focus on the Transportation and Electricity Sectors

Saleh Zakerinia, Christopher Yang and Sonia Yeh

Key messages

- Significant adoption of wind and solar technologies in all Western States is crucial for achieving California's climate target as well as the carbon neutrality target in 2050. However, carbon neutrality in 2050 in all Western States is not feasible without carbon capture and storage technology.
- In order to take advantage of the spatial and temporal differences of the Western US, it is very important to completely deregulate the Western Electricity Coordinating Council grid.
- Alternative fuel vehicles contribute to more than 50% of total vehicle miles traveled in California in 2050 for achieving California's climate target. This penetration is smaller when carbon capture and storage is allowed.
- Although the CA-TIMES model does not have a very high temporal resolution, its capability in modeling the entire energy system of California as well as the entire electricity sector of the rest of the Western United States make it a perfect tool for exploring various policies in different sectors and their interactions with each other.

S. Zakerinia (✉) · C. Yang
Institute of Transportation, University of California, Davis, CA, USA
e-mail: mzakerinia@ucdavis.edu

C. Yang
e-mail: ccyang@ucdavis.edu

S. Yeh
Department of Space, Earth and Environment, Chalmers University
of Technology, Göteborg, Sweden
e-mail: sonia.yeh@chalmers.se

© Springer International Publishing AG, part of Springer Nature 2018   245
G. Giannakidis et al. (eds.), *Limiting Global Warming to Well Below 2 °C: Energy System Modelling and Policy Development*, Lecture Notes in Energy 64,
https://doi.org/10.1007/978-3-319-74424-7_15

1 Introduction

California is the leading state in the United States in reducing greenhouse gas emissions (GHG) to combat climate change. The state set a binding target to bring back its GHG emissions down to the 1990 level by 2020. It also legislated a long-term GHG emissions goal of reducing GHG emissions to 80% below 1990 levels by 2050 as well as a medium-term target of reducing emissions to 40% below 1990 levels by 2030. California is on the path to reach its 2020 GHG reduction goal through multiple policy mechanisms such as cap-and-trade; however, the possible pathways and their cost for deep decarbonization for reaching 2030 and 2050 targets are greatly uncertain.

There is a large body of studies and policy recommendations on how to achieve California's medium and long term goals (Williams et al. 2012; Wei and Nelson 2013; Yang et al. 2015; Greenblatt 2015). All of these studies show that decarbonization scenarios for California and other regions to 2030 and 2050 have a number of relatively robust trends, including significant adoption of plug-in electric vehicles and investments in large quantities of renewable wind and solar generation.

The first strand of literature focuses on possible paths for electricity decarbonization using grid models. Several papers such as Fripp (2012), Williams et al. (2012), Nelson et al. (2012) and Mileva et al. (2013) use different electricity planning models to study how California can decarbonize its electricity sector and increase the penetration of renewable sources of electricity. These studies also examine different policies such as the Renewable Portfolio Standard (RPS) and their impact on decarbonizing the electricity sector in California and the Western United States. The results of these papers shows there is no maximum possible penetration of wind and solar power—these resources could potentially be used to reduce emissions to 90% or more below 1990 levels without reducing reliability or severely raising the cost of electricity. However, the results emphasize the role that shifting electricity demand to times when renewable electricity is most abundant and storage technologies has in reducing the costs and increasing the reliability of grids with high renewables penetration.

A second strand of literature upon which we build is the impact of adopting electric vehicles on emissions reduction as well as the electricity grid. Hadley and Tsvetkova (2009) show plug-in hybrid electric vehicle (PHEV) penetration of the vehicle market is likely to create substantial changes for the electrical grid. Jansen et al. (2010) apply a dispatch model, their results reveal that the hourly emissions intensity of the grid depends upon the PHEV fleet charging scenario, and the hourly model resolution of changes in grid emissions intensity can be used to decide on preferred fleet-wide charging profiles. Axsen et al. (2011) indicate that compared to conventional vehicles, consumer-designed PHEVs cuts marginal (incremental) GHG emissions by more than one-third in current Californian energy scenarios and by one-quarter in future energy scenarios—reductions similar to those simulated for all-electric PHEV designs. The authors also show that long-term GHG reductions depend on reducing the carbon intensity of the grid.

These two developments in disparate sectors (electricity and transportation) are linked via the use of electricity for the transportation sector. As explained above, there has been a reasonable amount of work already looking at the emissions impacts of electric-vehicle charging. Much less work exists that fully investigates the system-wide benefits of electric vehicles on the electricity grid, including how flexible vehicle charging and load management can be used for improving grid stability and operation, especially in the grids with high-renewable penetration that are expected in the 2030–2050 timeframe.

In a grid with a high penetration of low-carbon generation (such as renewables), the use of electricity as a transportation fuel helps to reduce GHG emissions and criteria pollutants from the transportation sector. The benefits for the electricity grid are perhaps as important as this reduction in pollutant emissions. Using vehicle charging as flexible loads, along with energy storage and demand response to help shape overall load and ramp up when renewable generation exceeds other demands can reduce overall grid capacity requirements, reduce the use of peaking plants, reduce emissions and lower overall system costs.

Detailed electricity dispatch models typically have very high temporal and spatial resolution to capture important dynamics of the electricity system (such as renewable generation profiles, demand variability and energy storage systems). They are typically used on a short-timeframe and are not typically used to optimize the power generation capacity of a system. However, long-term energy system models like CA-TIMES do not have the same level of temporal resolution that dispatch models do (CA-TIMES only has 48 timeslices per year). But their strength is their long modeling time scales (e.g. multiple decades) and the integrated representation of many energy sectors and endogenous investment and operational optimization of end-use fuel demand and supply. To the best of our knowledge, there is no study that fully investigate the system-wide benefits of alternative fuel vehicles on the electricity grid especially grids with high renewables penetration in the long-term framework in the Western United States. In this study, we develop a related tool for analyzing the interactions between grids with high renewables penetration and charging of electric transportation. We improve an existing long-term energy system model, CA-TIMES, to encompass the larger electricity system and improve the spatial resolution (by expanding the electricity sector and including all of the Western United States). This modeling platform will enable the analysis of system integration and interactions between the convergent evolution of the electricity and transportation sectors. Given California is a leading state in the United States to combat climate change, the proposed framework in this study can help us understand how various climate strategies in California impact the technological development in other Western states. Moreover, we are able to fully investigate the system-wide benefits of alternative fuel vehicles on the electricity grid especially those with high renewable penetration.

2   Model Description

The CA-TIMES model is a bottom-up, technologically-rich, integrated economic optimization model that is based upon the MARKAL/TIMES framework (Loulou et al. 2005). It seeks to build an energy system by investing in technologies and processes that meet projections for future energy system demands by minimizing the costs and using linear optimization modeling. The model represents all of the main sectors of the energy system in the state including energy supply (energy resources, energy imports/exports, fuel production, conversion and delivery, and electricity production) and energy demands (residential, commercial, transportation, industrial and agricultural end-use sectors).

Energy service demands in each of these sectors must be satisfied by a flow of energy commodities that are generated and transformed by various upstream technologies, such as vehicles, appliances, transmission and distribution systems, power plants, fuel production facilities and resource extraction. The model chooses the mix of technologies that has the least discounted system cost to meet the energy service demands and satisfy various constraints such as limits on resource availability, growth/share constraints and policy constraints.

The structure of the CA-TIMES model is laid out across numerous energy sectors. Projected service demands to 2050 are described exogenously as input assumptions for the residential, commercial and transport sectors, and projected fuel demands are specified for the industrial and agricultural sectors, i.e. fuel consumption in the agricultural and industrial sectors are defined exogenously and we do not endogenously satisfy demands in these two sectors, however, it is assumed that efficiency improves exogenously in some scenarios. The specification of energy service demands, e.g. heating, cooling, lighting, water heating, etc. in the residential and commercial sectors and vehicle miles traveled (VMT) across different transportation sectors (light-duty, heavy-duty, aviation, marine, etc.) means that the model has the flexibility to meet these demands from a set of technologies (appliances and vehicles) that can differ in the fuel type, energy efficiency and capital and operating costs. As a result, the model has the flexibility to trade-off the higher costs for these technologies for greater emissions reductions and choose the appropriate mix of these technologies. In the industry and agriculture sectors, end-use demands are not specified and an assumption about the fuel demanded by the entire sector is used. Therefore, the model can only reduce emissions in these two sectors from changes in carbon intensity of the input fuels (i.e. electricity, natural gas and liquid fuels). In CA-TIMES we account for important greenhouse gases including CO_2, CH_4 and N_2O. Yang et al. (2015) fully described the details of CA-TIMES.

The previous version of CA-TIMES only had one region (CA) and it was assumed that electricity that is not generated in CA is imported from out of state. This work is focused on increasing the spatial resolution of the CA-TIMES model in order to include a representation of the Western US electricity grid into the long-term energy system model of California. We used the SWITCH model (Wei et al. 2012) data to construct our regional boundaries. In the SWITCH model, they divided the Western Electricity Coordinating Council (WECC) into 50 regions, we aggregate these regions into 11 regions which are roughly equivalent to state borders (Fig. 1). We model transmission between two zones as the sum of individual transmission lines between those two regions, and the capacity of the line is equal to the sum of thermal limits of all of those lines. Using this method we are able to correctly account for all of the transmission capacity between our defined regions, however, we are not able to model each existing transmission line individually. We aggregate existing transmission lines from the SWITCH model which are taken from Federal Energy Regulatory Commission (FERC) data on the thermal limits of individual power lines (FERC 2012). We also assume that 1% of transmitted power in each transmission line is lost for each 100 miles that is transmitted. This value is taken from the SWITCH model and it represents the typical loss factors for high voltage and long distance transmission. Figure 1 shows the region boundaries and existing transmission lines between regions in CA-TIMES model. CA-TIMES regions include CA (California), OR (Oregon), WA (Washington), NV (Nevada), AZ (Arizona), NM (New Mexico), CO (Colorado), UT (Utah), ID (Idaho), WY (Wyoming) and MT (Montana).

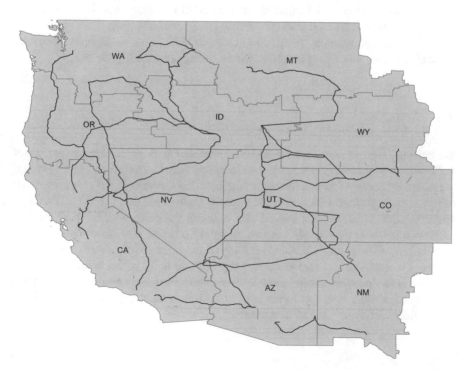

Fig. 1 Regions and existing transmission lines that are included in CA-TIMES, bold blue lines show transmission lines and gray lines represent region boundaries

In the same way as the SWITCH model, we consider generators with the primary fuel of coal, natural gas, fuel oil, nuclear, water (hydroelectric, including pumped storage), geothermal, biomass, biogas and wind from EIA (2010) database. Existing biomass co-firing units on existing coal plants are not included in this model. The excluded generators contribute to the small fraction of total capacity (less than 5%) and their exclusion does not impact our results. Moreover, we consider the main input fuel with which the existing power plants generated most of their electricity. The heat-rate of each generator type in each region is calculated by dividing generators fuel consumption by their total electricity output in 2010.

For the cost and efficiency of the new power plants, we used data from USEPA9r model (Lenox et al. 2013). We also assume the existing transmission lines can be expanded and the average cost of a high voltage transmission line is $1130/MW/km. We also use a cost multiplier to account for terrain types such as mountainous or urban terrain.

In this version of the model, we assume the price of coal, oil and natural gas is exogenous and it is equal in all regions. The availability of solar and wind generators is also taken from the USEPA9r model. In the current version of the model, we assume the electricity demand in all regions except California is exogenously defined; the exogenous demand for the ten regions other than California are taken from the projected demands in the SWITCH model.

The temporal resolution of the current version of CA-TIMES is 48 timeslices (6 seasons of two months each season is divided in 8 slices and each slice represents 3 actual hours). The temporal resolution of CA-TIMES is not as high as electricity dispatch models, but it is appropriate for modeling long-term policies.

2.1 Senarios

We can vary different assumptions and parameters such as technology, resource and policy-related inputs to analyze the results given these assumptions. In this study, we primarily focus on two scenarios. In the baseline scenario (BAU) we assume all of the existing policies that are currently implemented to 2020 (Table 1). The GHG scenario adds an emissions cap to the existing policies reflecting California's emissions target to reduce GHG emissions by 40% below 1990 levels in 2030 and 80% below 1990 levels in 2050. On the top of these scenarios, we also run a more stringent scenario with a zero emissions cap in 2050 in all of the Western US; this scenario is in line with a well below 2 °C scenario. It is important to note that the cap for states other than California is only on the electricity sector since we only model their electricity sectors in our model.

Table 1 Scenarios description

Baseline scenario (BAU)	Deep GHG reduction scenario (GHG)	Zero emissions scenario
• 2020 GHG cap (return GHG emissions back to 1990 level by 2020) • The fuel economy standards (CAFE) • Renewable portfolio standard (RPS) • Zero emission vehicle (ZEV) mandate • Low carbon fuel standard (LCFS) • Taxes and credits/ subsidies for biofuels, renewable electricity and electric vehicles	• All of policies in the baseline scenarios • Reduce GHG emissions by 40% below 1990 levels in 2030 and 80% below 1990 levels in 2050 in CA only	• All of policies in the baseline scenarios • Zero emission in 2050 in all of Western US states (electricity sector in non-CA states)

3 Results

3.1 BAU Scenario

In the baseline scenario, overall electricity demand remains roughly at the 2010 level in 2050 in California (Fig. 2). The import of renewable electricity from other states to California decreases over time and electricity generation from renewable sources, mainly solar and wind, increases over time in order to fulfill RPS policy, moreover, technology costs of renewable power plants (especially solar PVs) decrease over time and 59% of California's in-state generation comes from solar and wind power in 2050. The majority of in-state generation comes from natural gas in 2010, however, generation from renewable power plants ramps up after 2015 to fulfill RPS policy. According to RPS policy in California 50% of electricity generation (including imports) in 2030 should come from renewable sources including solar, wind and geothermal.

The light duty vehicles (LDV) sector will be dominated by combustion engine vehicles to 2050 (Fig. 2). However, the model adopts a modest number of battery electric vehicles and plugin hybrids up to 2025 to fulfill Zero Emission Vehicle (ZEV) mandate and then it adopts a slight number of Fuel Cell Vehicles (FCV) up to 2050.

There is a slight electricity demand increase in all of these states up to 2050 due to population growth (Fig. 3). California is a net importer of electricity in all periods, on the other hand, Montana, Nevada and New Mexico export electricity in all periods. It is important to note that the model makes these decisions solely based on cost variables, availability and constraints including transmission and policy constraints. In the first periods most of California's import comes from Arizona, Wyoming and Montana. In order to model RPS policy accurately, we keep track of

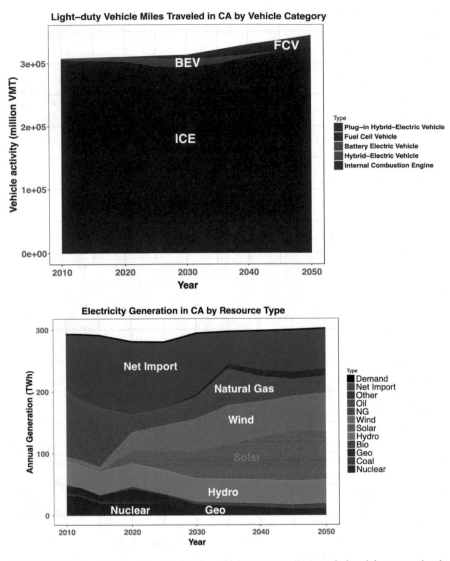

Fig. 2 Light-duty vehicle miles traveled by vehicle category (top), and electricity generation by resource type in California (bottom) in the BAU scenario

renewable and fossil-fuel based electricity transmission separately. All of the electricity exports from Arizona are fossil-fuel based and they mainly come from natural gas, coal and nuclear power plants. However, after 2025 most of the imports to California are renewable-based electricity (except imports from Arizona) to meet RPS policy. Arizona, Colorado, Oregon, Washington, Montana, New Mexico and Nevada also have their own RPS policy which is 15% (from 2025), 20% (from 2030), 25% (from 2025), 15% (from 2020), 15% (from 2020),15% (from 2020) and

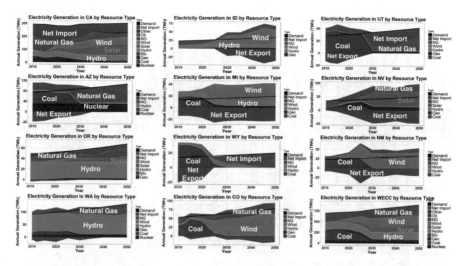

Fig. 3 Electricity generation by resource type in the Western US states in the BAU scenario. Negative import means export

25% (from 2025), respectively. Stringent RPS policy in California as well as RPS in other states (without any carbon cap) leads to generation of about 43% of electricity in the Western US from solar and wind.

The main resource of electricity generation in the Western US is natural gas (Fig. 3). In 2010, about 30% of electricity generation in the western US comes from natural gas, 30% from coal and 23% from hydro resources. In 2050, we have 27% of electricity generation from natural gas, 27% from wind, 16% from solar and the remaining from hydro, geothermal and biomass.

It can be seen in this scenario there is no significant adoption of alternative fuel vehicles in California and the slight adoption of ZEVs is only due to ZEV mandate policy, moreover, RPS plays a very important role in decarbonization of electricity sector under the BAU policy.

3.2 GHG Scenario

In the GHG scenario, we impose a cap on California's emissions target reflecting Californian commitments but not on the rest of states (Table 1). Furthermore, we assume carbon capture and storage (CCS) is not available and no new nuclear power plant is allowed based on current policies.

In this scenario, we use an iterative algorithm to account for the carbon intensity of imported electricity in California, in other words, how much emissions is imported into California through imported electricity, since California is the only region that has carbon cap and the exact source of imported electricity is not known.

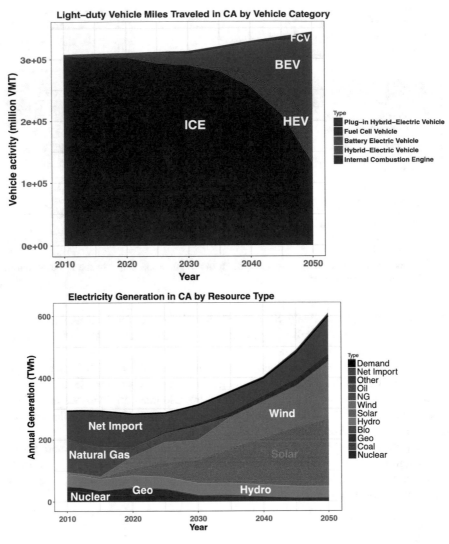

Fig. 4 Light-duty vehicle miles traveled by vehicle category (top), and electricity generation by resource type in California (bottom) in the GHG scenario

First, the model is run to calculate the carbon intensity of all states other than California for each year/timeslice; second, we assign the calculated carbon intensity to the imported electricity into California and the model is run again; if it is cheaper for the model to import electricity and its related emissions is low enough in the second run, more electricity will be imported and its carbon intensity will change; we recalculate the carbon intensity of imported electricity into California; we repeat this stage until the calculated carbon intensity for imported electricity converges and we use the obtained carbon intensity for imported electricity into California.

Electricity consumption doubles in 2050 compared with 2010 due to aggressive electrification in all sectors. The existing natural gas power plants should be expired by 2030 and California should invest in wind and solar energy heavily in order to meet its emissions goals (Fig. 4). California's electricity generation also becomes carbon free in 2050. Moreover, by 2040, all of the imported electricity to California should be from renewable resources. Carbon free electricity as well as aggressive electrification across different sectors will pave the road to reach aggressive emissions target by 2050. The adoption of ZEVs and partial ZEVs (hybrid electric vehicles) up to 2030 is very similar to the BAU scenario, however, ZEVs, especially battery electric vehicles, will take up to 60% of the LDV sector in 2050 (Fig. 4). Adopting BEVs along with carbon free electricity can help the LDV sector to significantly reduce its carbon emissions. Moreover, we assume up to 30% of total BEVs can do smart charging which can help the renewable-intensive grid to balance supply and demand in various timeslices.

In this scenario, we assume the electricity demand in the rest of the Western states does not change compared with the BAU scenario. We have seen that California is the net importer in the GHG scenario and it imports most of the renewable based electricity from other states (Fig. 5). In this scenario, Arizona is the major exporter to California in the first few periods (up to 2025), then there will not be that much renewable plants investment in Arizona, except some solar plants that are mainly used for in-state consumption. Most of other states will invest in various types of renewable plants in the second half of the time horizon (after 2025), export them to California and use them to fulfill their own RPS policy (they do not have a carbon cap under this scenario). There is a significant investment in wind generation in the Rocky mountain region where wind availability is abundant

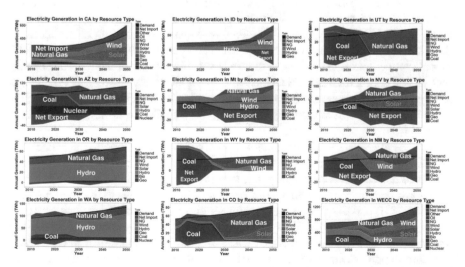

Fig. 5 Electricity generation by resource type in the Western US states in the GHG scenario. Negative import means export

(Idaho, Montana, New Mexico and Wyoming). As a result of California's carbon policy, there is not any new investment in coal plants after the existing ones are retired in all regions and also we have lots of investment in renewables. In this scenario, natural gas generates 23% of the total electricity production in the entire western US, solar plants contribute to 25% of total generation and wind plants generate 30% of electricity in this region in 2050. It is important to know that these results are obtained under the assumption that electricity can freely flow between any two points and the only constraints are physical and economic constraints. On top of significant electrification in the transportation sector, we have aggressive electrification in all others sectors including the residential and commercial sectors.

3.3   Zero Emission Scenario

In the Zero emission scenario, with a 2050 zero emission cap across all of the Western US in 2050, we only model the electricity sector for the states other than California. Therefore, the scope of our GHG emissions is California's emissions plus electricity related emissions for other states. Our results show that it is not possible to become carbon neutral in 2050 without using CCS technology; therefore, we allow CCS technology under this scenario.

Due to CCS technology, electrification is not as aggressive as in the GHG scenario and electricity demand increases by about 84% in 2050 compared with 2010 rather than 100% in GHG scenario (Fig. 6). The residential and commercial sectors in California use less electricity under this scenario since the model can offset their emissions using CCS technology. Battery electric vehicles and fuel cell vehicle adoption increases slightly compared with the GHG scenario, however, internal combustion vehicles still have more than 50% of VMT. Under this scenario, we can produce more cost-effective hydrogen using CCS technology, therefore, the model can adopt more FCVs compared with other scenarios. The model can neutralize emissions in the transportation sector as well as in all other sectors in California using CCS technology. In the electricity sector, more than 80% of in-state generation comes from solar and wind in 2050 in California and the rest of electricity comes from natural gas power plant with CCS, geothermal, hydro and imports from the other regions.

In all of the Western US states other than California, coal production phases out after 2030, natural gas as well as renewables including solar and wind pick up after 2030 and natural gas power plants start to retire after 2040 (Fig. 7). The electricity network is carbon free in 2050. California is still importing electricity in all periods between 2010 and 2050, however, imports decrease from about 30% in 2010 to 16% of total demand in 2050 in California. This is due to the cost of expanding transmission lines and the cost-effectiveness of solar and wind power in California.

Idaho imports electricity between 2010 and 2030, and it exports electricity after 2030 by investing in natural gas power plants. Montana, Nevada, New Mexico and Wyoming roughly export electricity in all periods. Colorado, Washington and

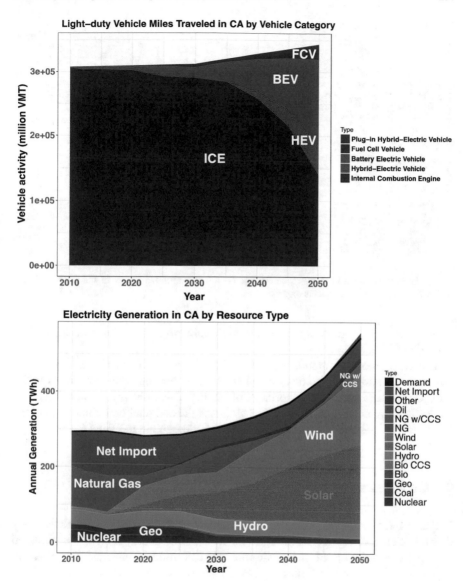

Fig. 6 Light-duty vehicle miles traveled by vehicle category (top) and electricity generation by resource type in California (bottom) in the Zero emission scenario

Oregon do not import or export and they produce electricity to fulfill their own demand. Arizona is a net exporter in the first periods (2010–2030), after retirement of coal power plants in 2030 Arizona becomes net importer of electricity.

The results in different scenarios highlight the importance of decarbonization of the electric grid along with significant adoption of ZEVs in reaching low-carbon

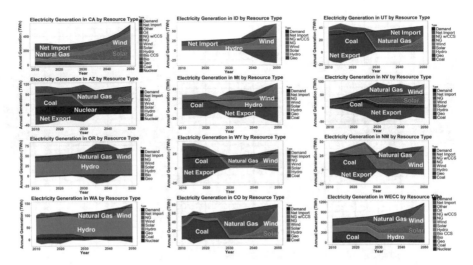

Fig. 7 Electricity generation by resource type in the Western US states in the Zero emission scenario. Negative import means export

future. Moreover, carbon capture and storage technology is vital in reaching carbon neutrality in 2050.

Currently, most of imports/exports in WECC are based on bilateral contracts; however, we assumed electricity could freely flow between any two points in the grid. Our results show the importance of having a well-integrated electricity network. Consequently, to take advantage of the spatial and temporal differences of the Western US, it is very important to completely deregulate the WECC grid.

4 Conclusion

In this chapter, we developed a modeling framework that can be used to study many different elements that are essential for the low-carbon energy future. We increased the spatial resolution of the California TIMES model, included the Western Electricity Coordinating Council (WECC) electricity network and analyzed the impact of California's climate policies on the evolution of the WECC electricity grid. It is shown that even in the absence of climate policy in the rest of the western US states California's climate policies can deeply decarbonize the WECC grid. Our results also indicate the importance of having a well-integrated electricity grid. Moreover, it is essential to adopt alternative fuel vehicles along with a carbon free electricity network to meet California's climate targets. In order to reach zero emissions in 2050, which is in line with the well below 2° scenario, the role of carbon capture and storage is vital. In addition, CCS reduces the demand for electricity since we can decarbonize the energy system without electrification.

In future endeavors, a comprehensive cost analysis is necessary to determine the cost-effectiveness of various policies. Moreover, it is essential to do some sensitivity analysis to see how changing different factors such as electricity demand in other states will impact the results. Moreover, we would like to do a sensitivity analysis on the impact of smart charging. We can use the benefits of the electric grid with high penetration of renewables to optimally charge electric vehicles. On the top of these transportation sector related results, it is also crucial to study how the implementation of smart grid and flexible demand in the building sector can benefit the grid and utility companies.

References

Axsen J, Kurani KS, McCarthy R, Yang C (2011) Plug-in hybrid vehicle GHG impacts in California: integrating consumer-informed recharge profiles with an electricity-dispatch model. Energy Policy 39:1617–1629. https://doi.org/10.1016/J.ENPOL.2010.12.038

EIA (2010) United states energy information administration. Washington, DC

FERC (2012) Annual transmission planning and evaluation report, 2012. Washington, DC, 2009. Washington, DC

Fripp M (2012) Switch: a planning tool for power systems with large shares of intermittent renewable energy. Environ Sci Technol 46:6371–6378. https://doi.org/10.1021/es204645c

Greenblatt JB (2015) Modeling California policy impacts on greenhouse gas emissions. Energy Policy 78:158–172. https://doi.org/10.1016/j.enpol.2014.12.024

Hadley SW, Tsvetkova AA (2009) Potential impacts of plug-in hybrid electric vehicles on regional power generation. Electr J 22:56–68. https://doi.org/10.1016/J.TEJ.2009.10.011

Jansen KH, Brown TM, Samuelsen GS (2010) Emissions impacts of plug-in hybrid electric vehicle deployment on the U.S. western grid. J Power Sources 195:5409–5416. https://doi.org/10.1016/j.jpowsour.2010.03.013

Lenox C, Dodder R, Gage C et al (2013) EPA US Nine-region MARKAL Database: database documentation. US Environmental

Loulou R, Remme U, Kanudia A et al (2005) Documentation for the TIMES Model Part I. Energy Technology Systems Analysis Programme

Mileva A, Nelson JH, Johnston J, Kammen DM (2013) SunShot solar power reduces costs and uncertainty in future low-carbon electricity systems. Environ Sci Technol 47:9053–9060. https://doi.org/10.1021/es401898f

Nelson J, Johnston J, Mileva A et al (2012) High-resolution modeling of the western North American power system demonstrates low-cost and low-carbon futures. Energ Policy 43:436–447. https://doi.org/10.1016/j.enpol.2012.01.031

Wei M, Nelson J (2013) Deep carbon reductions in California require electrification and integration across economic sectors. Environ Res Lett 8(1):014038

Wei M, Nelson JH, Ting M et al (2012) California's carbon challenge: scenarios for achieving 80% emissions reductions in 2050. Lawrence Berkeley National Laboratory, UC Berkeley, UC Davis, and Itron to the California Energy Commission

Williams JH, DeBenedictis A, Ghanadan R et al (2012) The technology path to deep greenhouse gas emissions cuts by 2050: the pivotal role of electricity. Sci 335:53–59. https://doi.org/10.1126/science.1208365

Yang C, Yeh S, Zakerinia S et al (2015) Achieving California's 80% greenhouse gas reduction target in 2050: Technology, policy and scenario analysis using CA-TIMES energy economic systems model. Energy Policy 77:118–130. https://doi.org/10.1016/j.enpol.2014.12.006

Towards Zero Carbon Scenarios for the Australian Economy

Luke J. Reedman, Amit Kanudia, Paul W. Graham, Jing Qiu,
Thomas S. Brinsmead, Dongxiao Wang and Jennifer A. Hayward

Key messages

- Supporting strong climate ambition of well below 2 °C and successfully managing a high share of VRE generation will be important in ensuring Australia's electricity sector achieves deep emission reductions
- The Australian implementation of TIMES is well suited to examine energy sector abatement, providing a detailed techno-economic representation of the relevant energy sub-sectors
- Further research and development is required to unlock the potential of additional sources of low emission energy such as hydrogen and solar thermal heat to ensure emissions can be completely eliminated without the need to purchase potentially higher cost emission credits from other domestic sectors or the international market.

L. J. Reedman (✉) · A. Kanudia · P. W. Graham · J. Qiu · T. S. Brinsmead
D. Wang · J. A. Hayward
CSIRO Energy, Newcastle, Australia
e-mail: luke.reedman@csiro.au

A. Kanudia
e-mail: amit@kanors.com

P. W. Graham
e-mail: paul.graham@csiro.au

J. Qiu
e-mail: qiujing0322@gmail.com

T. S. Brinsmead
e-mail: thomas.brinsmead@csiro.au

D. Wang
e-mail: dongxiao.wang@csiro.au

J. A. Hayward
e-mail: jenny.hayward@csiro.au

© Springer International Publishing AG, part of Springer Nature 2018
G. Giannakidis et al. (eds.), *Limiting Global Warming to Well Below 2 °C: Energy System Modelling and Policy Development*, Lecture Notes in Energy 64,
https://doi.org/10.1007/978-3-319-74424-7_16

1 Introduction

There is evidence that greenhouse gas (GHG) emissions from human activities have accelerated in recent years (IEA 2016) implying that significant emissions reduction may be required to limit the chances of dangerous climate change. This has significant implications for Australia given its emissions intensive economy (Denis et al. 2014). In late 2016, Australia ratified the Paris Agreement, committing to achieve a 26–28% reduction in GHG emissions below 2005 levels by 2030. The Paris Agreement also requires signatories to strengthen their abatement efforts over time with the overarching goal of limiting the increase in global average temperature to well below 2 °C above pre-industrial levels, with efforts to limit the temperature increase to 1.5 °C.

Australia's high ranking in GHG emissions per capita reflects its relatively high proportion of fossil fuels in energy consumed, high usage of relatively less efficient private transport and relatively high production of non-ferrous metals per capita. The energy sector (electricity, transport and direct combustion) is the single largest source accounting for around 71% of the total 540 megatonnes (Mt) of carbon dioxide-equivalent (CO_2-e), with electricity generation accounting for the majority at 195 Mt CO_2-e (Commonwealth of Australia 2016).

The high share of GHG emissions from the energy sector is mainly due to coal-fired electricity generation which accounted for 63% of Australia's electricity generation in 2015 (OCE 2016a). The dominance of coal in power generation masks Australia's rich diversity of renewable energy resources (i.e., wind, solar, geothermal, hydro, wave, tidal, bioenergy). Except for hydro and wind energy which currently account for most renewable generation, these resources are largely undeveloped and could contribute significantly to Australia's future energy supply (AEMO 2013; Geoscience Australia, ABARE 2010).

Previous analyses in the Australian context have found that deep decarbonisation of the electricity supply enables other sectors (e.g. buildings and transport) to decarbonise their activities through fuel switching, (Commonwealth of Australia 2008, 2011; Graham and Hatfield-Dodds 2014; Campey et al. 2017). With carbon capture and storage (CCS) technologies yet to be commercially deployed, and nuclear power generation currently prohibited by legislation, this suggests that significant deployment of renewable electricity generation will be required to achieve this goal.

The chapter assesses how the key energy sub-sectors respond to alternative global climate ambition scenarios, particularly in a well below 2 °C world. The scenarios also explore the importance of successfully managing high variable renewable (VRE) generation shares on the electricity sector and thus energy sector emission abatement as a whole of energy sector abatement strategy.

2   Methodology for Modelling Energy Systems for Deep Decarbonization Assessment

2.1   Overview of the Modelling Framework

Our modelling framework uses several linked models to derive key results (Fig. 1). In particular, energy technology capital cost projections are determined by global scale energy sector models, the Global and Local Learning Model (GALLM) given that technological learning that leads to cost reductions is a global phenomenon. National energy demand projections are determined by the Victoria University Regional Model (VURM) of the national economy. National energy technology uptake projections are determined by a national model of the energy sector, AUS-TIMES (the Australian version of The Integrated MARKAL-EFOM System).

VURM and GALLM are run first in the modelling framework sequence, independently and in parallel. AUS-TIMES is run using energy and transport demand information from VURM and energy technology costs from GALLM and uses the same global carbon and fuel prices after currency adjustments.[1]

Given the focus of this work is largely on electricity generation technological pathways rather than policy impacts or cross industry interactions, this framework was considered adequate. However, it is possible to include a broader range of model interactions and these models have been employed in more ambitious integrated assessment modelling frameworks in other research (Smith et al. 2017).

We briefly describe each model in the next subsections.

2.1.1   Global and Local Learning Model (GALLM)

The Global and Local Learning Model (GALLM) has been used to project capital cost of electricity generation technologies, which is then used as input in AUS-TIMES. GALLM is solved as a mixed integer linear program in which total costs of electricity supply are minimised while meeting electricity demand. The model features endogenous technological learning at both the global and local scale. This is through the use of experience curves which provide learning rates based on historical data for commercial technologies. For emerging technologies the learning rates have been based on those of similar technologies.

The uptake and cost of the technologies is solved simultaneously. This approach provides an objective and transparent methodology for assigning a timeline to technology cost reductions, although it is unfortunately not able to identify what improvements need to be made to bring the cost down. Cost projections can also become unrealistically low and thus a cost floor based on knowledge of the technology needs to be applied.

---

[1]Conversion from USD to AUD real 2015 dollar was assumed to be constant 1 USD = 1.331 AUD based on the average 2015 exchange rate.

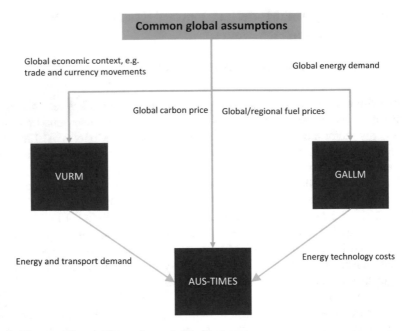

Fig. 1 Diagram of model interaction and data inputs

In the early to mid-2000s, the price of supply-constrained technologies (e.g. wind turbines and solar photovoltaics) was higher than that projected from the experience curves. In order to endogenously deal with this issue, GALLM features a "penalty constraint", which increases the price of technologies above the experience curve when new installed capacity in any year exceeds a threshold. This negative feedback counters the positive feedback of the experience curves in the model, and helps to ensure that a variety of technologies are installed.

GALLM has 13 regions based on OECD regional definitions and 27 electricity generation technologies. A full description of an earlier version GALLM can be found in Hayward and Graham (2013).

2.1.2 VURM

VURM (The Victoria University Regional Model) is a dynamic computational general equilibrium (CGE) economic model of the Australian economy. For this study, VURM represented an economic disaggregation of 72 industrial sectors, mostly corresponding to categories in the Australian and New Zealand Standard Industrial Classification, and a spatial disaggregation of Australia's six states and two territories, and was calibrated to the Australian National Accounts for the financial year 2015–16.

The key results from VURM comprise of projected demand quantities for electricity generation and transport services demand, reflecting the impact of a carbon tax in the economy, modelled as recycled through the economy via the household sector. In addition to the carbon price trajectory as a key input, other significant economic assumptions driving the results include international trade market conditions (prices for imports and exports, especially international fuel prices).

For this exercise, international market conditions were derived from an international trade model (the Global Analysis of Trade Project—GTAP) with global economic scenario conditions equivalent to Representative Concentration Pathway 2.6 (van Vuuren et al. 2011; Fricko et al. 2017). Oil prices were based on the medium scenario projections from the EIA (2016), and gas and coal price projections from IEA (2015).

2.1.3 AUS-TIMES

TIMES (The Integrated MARKAL-EFOM system) is a linear optimization energy model generator developed by ETSAP-IEA. The model satisfies energy services demand at the minimum total system cost, subject to physical, technological, and policy constraints. Accordingly, the model makes simultaneous decisions regarding technology investment, primary energy supply and energy trade. Extensive documentation of the TIMES model generator is available in Loulou et al. (2016).

The authors have created an Australian version of the TIMES model (AUS-TIMES). The model has a spatial resolution to State and Territory level (8 regions) and sub-state scale for the power generation sector (20 transmission zones). At its current stage of development, AUS-TIMES has a detailed representation of the electricity and road transport sectors, and a moderate representation of end-uses in the industrial, commercial and residential sectors. For the electricity sector, existing generation assets are vintage tracked to individual units, with plant specific data on fuel efficiency, and operating and maintenance and fuel costs. Highly temporal and spatial data sets on renewable resource availability are also incorporated into the model (AEMO 2016b).

AUS-TIMES has been calibrated to a base year of 2015 based on the latest available energy balance (OCE 2016b), national inventory of greenhouse gas emissions (DoEE 2017) and stock estimates of electricity generation plants (ACIL Allen 2014a, b; AEMO 2016a; ESAA 2016) and vehicles in the transport sector (ABS 2015). Cost and performance data on future technologies are mainly sourced from Hayward and Graham (2017) and Reedman and Graham (2016).

2.2 Scenario Definition

The scenarios are designed to explore two questions with regard to the feasibility of Australia's electricity sector achieving very low carbon emissions:

- What is the impact on Australia of the level of global ambition with respect to limiting changes in average global temperature?
- Given all low emissions generation technologies present either social, commercial or technological risks which could prevent their deployment, what is the impact of pursuing a narrower set of technological options?

The wide range and vast quantity of energy resources available in Australia is a source of uncertainty in determining the likely technology pathways for reducing GHG emissions. The exploration of technology options via scenarios builds on Campey et al. (2017), which conducted an audit of the technological pathways for reducing GHG emissions from the Australian energy sector; and by ClimateWorks Australia and ANU (2014), which developed alternative deep decarbonisation pathways for Australia, in coordination with other country studies (SDSN and IDDRI 2014).

2.2.1 Global Climate Policy Ambition

The three levels of ambition with respect to global temperature changes that we include in the scenario assumptions are well below 2, 2 and 4 °C. The alternative levels of ambition in global climate policy lead to different assumptions with respect to the global context. The 4 °C world is assumed to have moderate carbon prices, relatively higher fossil fuel prices and higher growth in the demand for Australia's fossil fuel exports, compared to a 2 °C and well below 2 °C world which have higher carbon prices and more subdued global markets for fossil fuels. We also assume that while there is already significant improvement in the costs of generation technologies under a 4 °C world, cost reductions are greater and occur earlier under a 2 °C world and greater still in the well below 2 °C world.

2.2.2 National Climate Policy

While there has been significant volatility in Australian climate policy in the past decade, current policy proposals tend towards imposing an average emission intensity constraint on the electricity sector in the period between 2020 and 2030 (Finkel et al. 2017; Energy Security Board 2017). This decade is the focus period of Australia's nationally determined commitment for a 26–28% reduction in GHG emissions on 2005 levels. We assume the design of the emission intensity scheme, which operates to 2030, allows for generators to choose not to reduce emissions but rather purchase emission credits from other sectors at the prevailing global carbon price. As such, the emission intensity target is not a hard constraint. Rather it operates as a subsidy to low emission technologies up to a given level of emission reduction and a tax on high emission technologies capped at the prevailing carbon price. In the period beyond 2030, we assume that electricity sector abatement will be driven by global and nationally linked carbon prices through emissions trading.

2.2.3 Generation Technology Risk

Australia is rich in almost all types of energy resources and as a consequence there are many different pathways by which the electricity sector could be decarbonised. Campey et al. (2017) emphasised the diversity of risks faced by competing low emissions electricity generation technologies. For example, nuclear and CCS face significant risk of public opposition (also called achieving a 'social license'). Enhanced geothermal technology, for which there are potentially vast resources in Australia (Geoscience Australia, ABARE 2010), is yet to reach technical and commercial maturity (International Geothermal Expert Group 2014). VRE technologies such as solar photovoltaics and wind are commercially mature but face coordination risks in ensuring sufficient enabling technologies such as storage and demand management as their level of penetration increases. Solar thermal with storage lacks commercial maturity. The first commercial plant in Australia, a 150 MW project, is due for completion in 2020 (Government of South Australia 2017).

We explore a technology risk scenario that limits VRE to 45% of electricity supply in Australia. This represents a world without the necessary enabling technologies, such as batteries, to make use of VRE at higher penetration rates. The limit of 45% was found by Campey et al. (2017) to be the upper limit before enabling technologies would be needed to support higher shares. Instead dispatchable low emission generation technologies such as CCS, nuclear, enhanced geothermal and solar thermal with storage are given the priority.

2.2.4 Scenario Summary

We have designed 5 scenarios based on the research questions to be explored (Table 1). All scenarios assume the continuation of current policies including the national Large-scale Renewable Energy Target (LRET), the Victorian Renewable Energy Target (VRET), and the New South Wales and Queensland biofuel mandates. The 4 °C *scenario* serves as a reference scenario (Ref); it includes moderate abatement, moderate technological change and no technological restrictions. The second scenario, 2 °C *scenario unconstrained (2DS-Unc)*, explores the impact of a 2 °C global climate policy ambition including a higher carbon price, faster technological change and no constraints on technology deployed. The *2DS-HD scenario* caps the generation share of VRE (solar photovoltaics and wind) at 45%, which may favour higher cost but more dispatchable low emission generation

Table 1 Summary of scenarios

Setting	Ref	2DS-Unc	2DS-HD	B2DS-Unc	B2DS-HD
Technology costs	Consistent with 4 °C climate ambition	Consistent with 2 °C climate ambition		Consistent with well below 2 °C climate ambition	
Emission intensity scheme	Available subsidies apply 2020 to 2030				
Carbon pricing	Commence 2030, consistent with 4 °C climate ambition	Commence 2030, consistent with 2 °C climate ambition		Commence 2030, consistent with well below 2 °C climate ambition	
Low emission generation technology constraints	Nil		The share of VRE technologies is capped at 45%	Nil	The share of VRE technologies is capped at 45%

technologies. The final two scenarios, *B2DS-Unc and B2DS-HD*, explore the impact of a well below 2 °C climate ambition with carbon prices and technological change greater than either of the other scenarios. Similar to the 2 °C world scenarios, the well below 2 °C scenarios include an unconstrained technology scenario and a scenario which imposes a limit on VRE.

Each climate scenario corresponds to a specific carbon price (Fig. 2) based on IEA (2017).

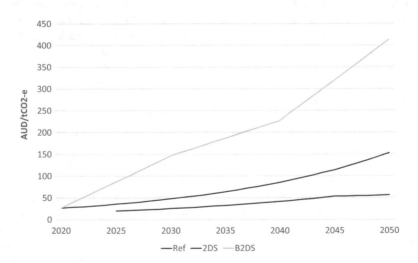

Fig. 2 Carbon price trajectories

3 Results and Discussion

3.1 Capital Cost Projections

GALLM was run under two climate scenarios: Ref and 2DS-Unc. We did not run GALLM for B2DS-Unc but rather inferred further technological improvements from differences in the global generation mix between a 2 °C and well below 2 °C world reported by IEA (2017) which indicated slightly greater deployment of renewables, gas with CCS and biomass with CCS in a well below 2 °C world.

The differences in assumptions under the two scenarios relate to the level of electricity demand, exogenous fuel prices, carbon price and specific regional climate policies. These scenario assumptions were sourced from (IEA 2016).

All costs are in AUD 2017. The costs for Gas with CCS are lower under the 2DS-Unc and B2DS-Unc scenarios (Fig. 3), as this technology is installed in greater quantities at higher carbon prices and thus the costs reduce due to learning-by-doing. The renewable technologies show a negligible difference in cost across the three scenarios, which illustrates that higher carbon prices are not a strong driver for further deployment. These technologies are competitive alternatives to conventional fossil generation technologies, even at low carbon prices.

By 2050, the cost of lithium-ion (Li-ion) batteries is projected to be $68/kWh, small-scale PV $665/kW and wind $1654/kW under the Ref and 2DS scenarios (Fig. 3).

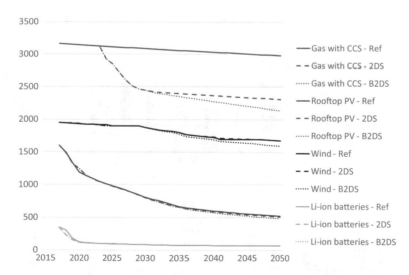

Fig. 3 Projected capital cost of selected electricity generation technologies in Australia in $/kW and batteries in $/kWh under all scenarios

3.2   Electrification in End-Use Sectors

Due to the energy efficiency gains and GHG emission reduction opportunities, particularly with near-zero emission electricity, electrification intensifies for more aggressive GHG emission reduction scenarios (Fig. 4). In B2DS-Unc (2DS-Unc), electricity consumption is 30% (5%) higher than the Ref scenario by 2050. There is increased electrification of all end-use sectors, especially the industrial subsectors of non-ferrous metals, gas mining and alumina production.

The electrification of the commercial and residential sectors is mainly in space heating and hot water provision, displacing natural gas. In transportation, there is a significant uptake of electric vehicles in the passenger and light commercial vehicle segments, and minor uptake in heavy vehicles. There is some increased electrification in rail transport (fuel switching from diesel), however no uptake in shipping or aviation.

3.3   Electricity Generation Sector

With electricity generation technology cost projections from GALLM and demand drivers from VURM, the projected electricity generation mix for Australia is expected to change significantly from a system that is currently dominated by coal-fired power generation.

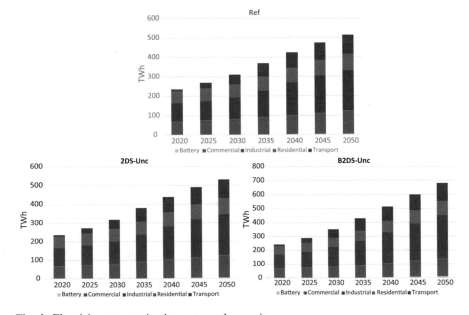

Fig. 4 Electricity consumption by sector and scenario

In the early years of the projection period of the Ref scenario, national and state renewable policies lead to significant uptake of wind generation as the lowest cost renewable technology. However, over time as the costs of large-scale solar photovoltaics decline and gas prices stabilise, the wind share stabilises at around 45% by 2035 while increased investment in solar and gas-fired generation takes place. Towards the end of the projection period in 2050, solar accounts for 25%, wind 50%, and gas 15% of electricity generation. Due to the vintage tracking of existing plant in AUS-TIMES, as coal and lignite plants near the end of their technical life, their fixed operating costs start to rise. In combination with an increasing carbon price, the share of coal and lignite declines significantly, nearing 3% by 2050.

In the 2DS-Unc scenario, coal and lignite phase out more rapidly from the electricity generation mix as more plants become uneconomic, in favour of wind and gas initially, and an increased share in wind power in the long-term due to a lower levelized cost of electricity compared to the Ref scenario (Fig. 5).

Recall that the 2DS-HD scenario caps the generation share of VRE at 45%, which favours the higher cost dispatchable low emission generation technologies. Under such a scenario, there is an increased deployment of gas-fired open cycle and combined cycle plants, as well as some deployment of coal-fired power with CCS towards the end of the projection period.

In the B2DS-Unc scenario, coal-fired generation is virtually phased-out by 2040 (<1%) mainly due to aggressive deployment of non-hydro renewable generation (wind and solar), accounting for around 75% of generation by 2050. Near-zero emission gas with CCS is also deployed. In the high dispatch scenario, VRE is constrained, resulting in much greater deployment of gas with CCS.

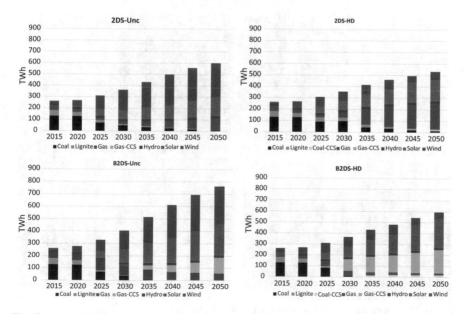

Fig. 5 Australian electricity generation mix by selected scenario

3.4   Transport

The road transport task is expected to grow by 90% between 2015 and 2050 mainly driven by population growth and economic activity. The fuel mix in Australia is currently dominated by diesel in the heavy vehicle segment and petrol in the light vehicle segment (Fig. 6), with some consumption of liquefied petroleum gas (LPG), mainly in taxis and utility vehicles, a small amount of compressed natural gas (CNG) in buses, and low blend ethanol predominantly in light vehicles (E10–10% ethanol blended with petrol).

In the Ref scenario, growth in road fuel consumption up to the early 2020s growth in kilometres travelled (average growth 2% p.a.) which more than offsets the assumed improvements in the efficiency of new internal combustion (ICE) vehicles. However, given the assumed price parity of electric vehicles (EVs) with internal combustion light vehicles in 2025 (Reedman and Graham 2016), there is significant electrification in light vehicles and moderate electrification in rigid trucks and buses and uptake of low blend biodiesel (B20–20% biodiesel blended with diesel) in articulated vehicles, leading to a decline in road transport sector fuel consumption to 2035. The growth in the passenger transport task means there is nevertheless significant consumption of petrol, particularly in larger vehicles.

The dynamics in the 2DS are very similar to the REF scenario, however the stronger carbon price signal leads to more high blend ethanol in the light vehicle segment, reducing petrol consumption. Under B2DS, these changes are accelerated.

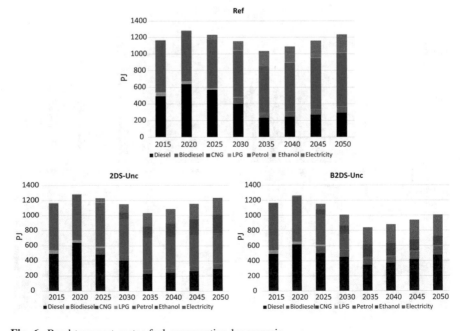

Fig. 6 Road transport sector fuel consumption by scenario

Petrol is virtually phased out by 2035, only remaining in high blend ethanol. The degree of electrification is greater, adding around 80 TWh to electricity consumption by 2050, and all diesel use is in the form of B20.

Domestic aviation has moderate uptake of bio-jet fuel in B2DS reaching one-quarter of fuel consumption, with rail having significant electrification away from diesel fuel use.

3.5 Greenhouse Gas Emissions

The projected GHG emissions indicate that failure to coordinate the management of high VRE shares and instead relying on traditional dispatchable generation technologies in the electricity sector is a risk to low GHG abatement (Fig. 7). Electricity sector GHG emissions under the Ref scenario are lower than the 2DS-HD scenario. However, if the carbon price signal is strong enough, as in B2DS-HD, this problem is overcome, with CCS being able to drive emission reduction instead of VRE.

Electricity GHG emissions do not reach zero under any of the scenarios, but are 50–82% below 2005 levels by 2050 in the 2DS-Unc and B2DS-Unc scenarios respectively. A key limiting factor is the availability of zero emission peaking plant for supporting renewables, fuelled for example by biogas or hydrogen. Biomass with CCS was included in the technology set but was not cost effective under the assumed carbon prices (although most available biomass is used elsewhere in the energy sector under the B2DS scenario).

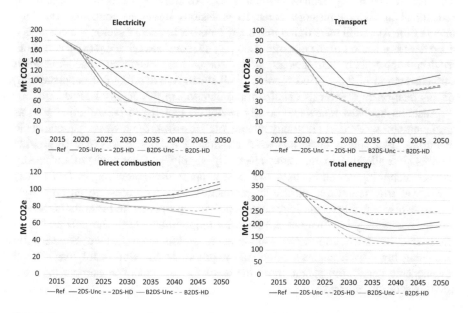

Fig. 7 Energy and energy subsector greenhouse gas emissions by scenario

In the transport sector, due to higher electricity system costs under the 2DS-HD and B2DS-HD, GHG emissions are slightly higher than in 2DS-Unc and B2DS-Unc reflecting lower incentives for electrification of transport fuels. Otherwise higher transport GHG abatement is driven by higher carbon prices. Abatement peaks around 2035 and then GHG emissions begin rising again. This reflects the fact that most available and cost effective abatement options, mainly electrification and use of biofuels, have saturated. There are insufficient biomass resources in Australia to decarbonise all liquid fuel consuming sectors, and electrification is limited to only some modes (short haul road and some rail). Improvements in hydrogen or other low emission options could remove this limit on transport abatement. Transport sector GHG abatement is between 30 and 71% below 2005 levels by 2050.

The direct combustion sector is least successful in reducing GHG emissions. As discussed, this sector takes advantage of opportunities for electrification but there remain a wide range of activities that are reliant on natural gas. Biogas is completely consumed by other sectors. Hydrogen and solar thermal heat are options to be included in future research. Across the scenarios, direct combustion emissions are 31% above to 16% below 2005 levels by 2050. Overall, the energy sector is between 29 and 64% below 2005 levels by 2050.

4 Conclusion

To meet its Paris Agreement commitments, Australia must achieve significant reductions in GHG emissions given its emissions intensive economy. Previous analysis for Australia have found that a key element in energy sector abatement is a deep decarbonisation of the electricity supply, as it enables other energy sub-sectors to decarbonise their activities as well through fuel switching. The Australian implementation of the TIMES model is well suited to extending our understanding of this relationship given its detailed representation of the energy sub-sectors.

Australia has vast renewable energy resources which can be exploited/used as zero emission sources of electricity generation. However, the lowest cost renewable resources, solar photovoltaics and wind, provide VRE. High shares of VRE provide additional challenges in matching electricity supply and demand without costly storage technologies. Ultimately, low cost support for generation in (even rare) periods of low renewable electricity production requires access to an emissions intensive dispatchable technology such as peaking gas plants. In the situation where storage technology is constrained, reliability requirements limit the potential for a high share of VRE and hence emissions reduction even further. Successfully managing high VRE shares without the requirement for significant fossil based dispatchable generation technologies increases the amount of abatement the electricity sector can achieve.

Under the scenarios examined, the modelling found that the electricity and transport sectors can achieve the greatest GHG emissions reductions of around 70–80% by 2050. The direct combustion sector has a harder abatement task owing to fewer directly substitutable low emission energy sources. GHG abatement in the electricity sector, and, through electrification, the remainder of the energy sector, was greatest when VRE could achieve high shares as they have lower emission intensities and are a lower cost solution than relying more heavily on fossil based dispatchable technologies. Under the well below 2 °C scenario, abatement across the total energy sector is 64% below 2005 levels by 2050.

It is important to note that none of the scenarios impose hard emission constraints, but rather carbon prices consistent with different global climate ambitions up to a well below 2 °C world. A hard constraint may have forced additional energy sector GHG abatement through adoption of higher cost abatement options. However, the energy sector's response to carbon pricing indicates that, given the choice, the energy sector would purchase emission credits. While least cost in this scenario, this strategy involves some risk. To offset the risk that abatement credits are at a higher cost than expected, it would be prudent to invest in further research and development into additional sources of low emission energy in those areas where abatement is higher cost. These include, for example, hydrogen and solar thermal heat to replace gas and liquid fuels in long haul transport and industrial process heat where electrification is not appropriate. Biomass is useful for these applications but there is not enough domestic biomass resources to cover all low emission energy requirements for complete energy decarbonisation.

Acknowledgements The authors acknowledge the contributions of ClimateWorks Australia in the database development of the Australian TIMES model presented in this chapter.

References

ABS (2015) Survey of motor vehicle use, Australia, 12 months ended 31 October 2014, Catalogue No. 9208.0. Australian Bureau of Statistics, Canberra

ACIL Allen (2014a) Fuel and technology cost review, Report to the Australian Energy Market Operator, June

ACIL Allen (2014b) RET Review modelling: market modelling of various RET policy options, Report to the RET Review Expert Panel, August

AEMO (2013) 100 percent renewables study—modelling outcomes. Australian Energy Market Operator, Melbourne

AEMO (2016a) National transmission network development plan. Australian Energy Market Operator, Melbourne

AEMO (2016b) 2016 NTNDP (National Transmission Network Development Plan) Database Input Data Traces. Australian Energy Market Operator, Melbourne

Campey T, Bruce S, Yankos T, Brinsmead T, Deverell J (2017) Low emissions technology roadmap. CSIRO, Canberra

Commonwealth of Australia (2008) Australia's low pollution future: the economics of climate change mitigation. Australian Government, Canberra

Commonwealth of Australia (2011) Strong growth, low pollution: modelling a carbon price. Australian Government, Canberra

Denis A, Graham P, Hatfield-Dodds S et al (2014) Introduction. In: Sue W, Ferraro S, Kautto N, Skarbek A, Thwaites J (eds) Pathways to deep decarbonisation in 2050: how Australia can prosper in a low carbon world. ClimateWorks Australia, Melbourne, pp 5–21

DoEE (2017) National inventory Report 2015, Three volumes, Commonwealth of Australia, Canberra

EIA (2016) Annual energy outlook 2016. Energy Information Administration, Washington DC

ESAA (2016) Electricity gas Australia 2016. Energy Supply Association of Australia, Melbourne

Finkel F, Moses K, Munro C, Effeney T, O'Kane M (2017) Independent review into the future security of the national electricity market: blueprint for the future. Commonwealth of Australia, Canberra

Fricko O, Havlik P, Rogelj J, Klimont Z, Gust M, Johnsona N, Kolpa P, Strubegger M, Valin H, Amanna M, Ermolieva T, Forsell N, Herrero M, Heyes C, Kindermann G, Krey V, McCollum D, Obersteine M, Pachauri S, Rao S, Schmid E, Schoepp W, Riahia K (2017) Shared socioeconomic pathway 2 (The marker quantification of the Shared Socioeconomic Pathway 2: A middle-of-the-road scenario for the 21st century. Glob Environ Change 42: 251–267

Geoscience Australia, ABARE (2010) Australian energy resource assessment, Canberra

Government of South Australia (2017) News releases—Jay weatherill: port augusta solar thermal to boost competition and create jobs. Government of South Australia, Adelaide

Graham P, Hatfield-Dodds S (2014) Transport sector. In: Sue W, Ferraro S, Kautto N, Skarbek A, Thwaites J (eds) Pathways to deep decarbonisation in 2050: how Australia can prosper in a low carbon world. ClimateWorks Australia, Melbourne, pp 23–52

Hayward JA, Graham PW (2013) A global and local endogenous experience curve model for projecting the future uptake and cost of electricity generation technologies. Energy Economics 40:537–548

Hayward JA, Graham PW (2017) Electricity generation technology cost projections 2017–2050, CSIRO Report No. EP178771

IEA (2015) World energy outlook. International Energy Agency, Paris

IEA (2016) World energy outlook. International Energy Agency, Paris

IEA (2017) Energy technology perspectives 2017. International Energy Agency, Paris

International Geothermal Expert Group (2014) Looking forward: barriers, risks and rewards of the Australian Geothermal Sector to 2020 and 2030, Australian Renewable Energy Agency (ARENA), Canberra

Loulou R, Goldstein G, Kanudia A, Lettila A, Remme U (2016) Documentation for the TIMES model (5 parts). International Energy Agency, Paris

OCE (2016a) Australian energy update. Office of the Chief Economist, Canberra

OCE (2016b) Australian energy statistics. Office of the Chief Economist, Canberra

Reedman L, Graham P (2016) Transport greenhouse gas emissions projections 2016, CSIRO Report No. EP167895, December

Smith K, Hatfield-Dodds S, Adams P, Baynes T, Brinsmead TS, Ferriera S, Harwood T, Hayward JA, Lennox J, Martinez RM, Nolan M (2017) Assessing risks and opportunities for Australia's future in a novel integrated assessment framework: the GNOME.3 suite for the Australian national outlook. Paper presented at 22nd international congress on modelling and simulation (MODSIM2017), The Hotel Grand Chancellor Hobart, Tasmania, Australia, 3–8 December 2017

Sustainable Development Solutions Network (SDSN), Institute for Sustainable Development and International Relations (IDDRI) (2014) Pathways to Deep Decarbonization. Report. SDSN, New York

van Vuuren DP, Stehfest E, den Elzen MGJ et al (2011) RCP2.6: exploring the possibility to keep global mean temperature increase below 2 °C. Clim Change 109:95

Economic Assessment of Low-Emission Development Scenarios for Ukraine

Maksym Chepeliev, Oleksandr Diachuk and Roman Podolets

Key messages

- Maintaining the current highly inefficient energy system is more expensive than a transition towards a high renewables share in the energy mix.
- Under the current energy policy set up there is an inconsistency between targets of different strategic energy documents (e.g. Energy Strategy, Low-carbon Development Strategy etc.).
- A number of additional incentives should be implemented in order to enable efficient market transformation: measures towards efficient pricing of fossil fuels, in particular price signals for industrial users, more transparent and market-oriented approach to residential consumers, elimination of cross subsidization in the electricity sector, move to competitive energy markets (in particular, fully implement the Third Energy Package), as well as proceed with further integration to the ENTSO-E.
- A successful coupling of TIMES-Ukraine and Ukrainian general equilibrium model helped to identify double dividends (economic and environmental) of the policies under consideration.

M. Chepeliev (✉)
Center for Global Trade Analysis, Purdue University, West Lafayette, USA
e-mail: mchepeli@purdue.edu

O. Diachuk · R. Podolets
Institute for Economics and Forecasting, NASU, Kiev, Ukraine
e-mail: oadyachuk@ukr.net

R. Podolets
e-mail: podolets@gmx.de

© Springer International Publishing AG, part of Springer Nature 2018
G. Giannakidis et al. (eds.), *Limiting Global Warming to Well Below 2 °C: Energy System Modelling and Policy Development*, Lecture Notes in Energy 64,
https://doi.org/10.1007/978-3-319-74424-7_17

1 Introduction

Global environmental pressure on the Earth System has overpassed safe operating boundaries on many directions (Steffen et al. 2015). One of such dimensions includes climate change, with both advanced and transition economies facing challenges with low-emission development (LED) and the fossil fuels depletion. And while transition economies often face much more severe environmental pressure than the developed countries, they have better opportunities for energy efficiency improvements and emissions reduction with potential for "double dividend" effects (Parry and Bento 2000). Ukraine is a good candidate for such a case. According to the International Energy Agency (IEA 2017), it has 6th highest GDP carbon intensity in the world and its energy intensity is 2.6 times higher than the OECD average. Significant opportunities for technological and environmental improvements are coupled with inefficient market regulatory framework and a poor investment environment.

In recent years, the Ukrainian government has implemented a number of initiatives in energy sector transformation, energy efficiency improvements and renewable energy development, which serve as a good starting point, but are not stringent enough to put the country on the LED path. In particular, the National Renewable Energy Action Plan up to 2020 (CMU 2014), sets an 11% target share for renewable energy (RE) sources in GFEC by 2020. As a comparison, the share of RE in GFEC was 3.9% in 2009 (CMU 2014). Ukraine's nationally determined contribution (NDC) states that it will not exceed 60% of 1990 GHG emissions level in 2030 (GOU 2015), which is still 40% higher than the 2012 emission level. While Ukraine's NDC indicates that this target will be revised after the restoration of country's territorial integrity and approval of post-2020 economic strategies, climate change experts consider current goal as a "critically insufficient" in terms of limiting global warming below 2 °C (CAT 2017).

In this study, we provide an assessment of LED scenarios for the Ukrainian economy. We take into account both current energy policy developments, more stringent than Ukrainian NDC contribution and consistent with the national low-carbon emission strategy initiative (USAID MERP 2017), as well as even more ambitious environmental targets. In this contribution, we mainly focus on the GHG emissions, and we consider this indicator as an integral representation of the energy consumption patterns.

2 Methodological Approach

This section describes the methodology, which we use for the assessment of Ukrainian LED scenarios. We start with an energy system TIMES-Ukraine model, which we use for the assessment of policy impacts on the energy sector and further proceed with an overview of the Ukrainian general equilibrium model (UGEM), which is used to estimate macroeconomic and sectoral effects. We also discuss the soft-linkage of TIMES-Ukraine and UGEM models, as well as some limitations of the applied methodology.

2.1 TIMES-Ukraine Model

TIMES-Ukraine is a typical linear optimization energy system model of MATKAL/ TIMES family (Loulou et al. 2004), which provides a technology-rich framework for estimating energy dynamics in the long-run (Podolets and Diachuk 2011). The Ukrainian energy system in the model is divided into seven sectors: energy supply, electricity and heat generation, industrial users, transport, agriculture, households and services (Fig. 1).

Industrial users are further disaggregated into two categories. Energy-intensive subsectors are represented by product-specific technologies. For other industrial subsectors, we use a standard representation according to the four types of general processes: electric engines, electrochemical processes, thermal processes and other processes. Energy consumption by households and commercial sector is defined by the most energy intensive categories of consumer needs (Fig. 1).

Energy system models, like TIMES-Ukraine, are usually used for long-term analysis of energy system development paths. By changing the assumptions on useful energy demands, technologies, prices or other exogenous variables baseline

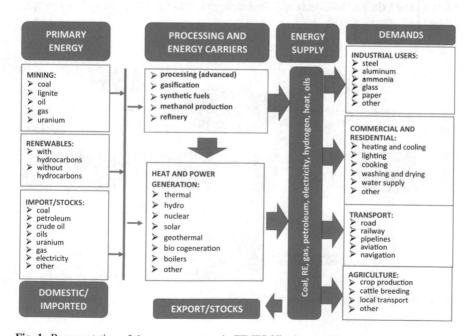

Fig. 1 Representation of the energy system in TIMES-Ukraine model

scenarios are developed. For the next step, policy scenarios are designed by imposing additional constraints or targets to the energy system. In this study, we develop one baseline scenario (BaU) and two policy scenarios. Differences between baseline and policy scenarios are further analyzed.

2.2 Dynamic UGEM Model

With energy policy impacts going beyond energy sector, we also need a modelling tool that provides a top-down view of the national economy. For this reason, in addition to the TIMES-Ukraine, we use the dynamic UGEM model. The Current version of the model is based on the static model described in Chepeliev (2014) and dynamic mechanisms introduced in TRPC (2014). It is a typical single-country recursive dynamic computable general equilibrium (CGE) model with producers divided into 40 sectors and households disaggregated into 10 groups according to their income level. Figure 2 represents key circular flows in the UGEM.

UGEM is formulated as a static model and solved sequentially over time. The Energy sector in the UGEM is represented by 7 subsectors: coal mining, extraction of the natural gas and oil, coke and oven products, petroleum products, electricity production and distribution, distribution of natural gas, heat and hot water supply. Key input data for the model is sourced from Input-Output tables, households' surveys and National accounts. It is organized in the form of Social Accounting Matrix based on the 2013 data and further updated to the 2015 using RAS method (Trinh and Phong 2013).

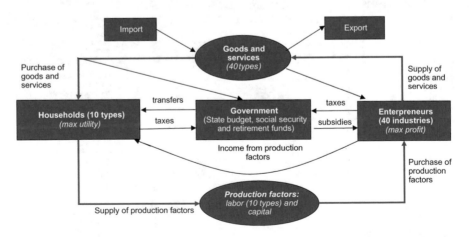

Fig. 2 Circular flows in the UGEM model

2.3 Model Linkage

To provide an assessment of LED policies in Ukraine we use a soft-linkage of TIMES-Ukraine and UGEM models (Fig. 3).

On the *first step*, we calibrate both models to match the assumptions of the BaU scenario, in particular, GDP and population projections. On the *second step*, we provide an assessment of the LED policies using TIMES-Ukraine model. As a result, we analyze policy impact on the energy sector and estimate additional investments required to reach the energy policy targets. We also estimate specific energy consumption changes relative to BaU scenario. For the *third step,* we prepare shocks for the UGEM. We map TIMES-based changes in additional investments and specific energy consumption to the UGEM classification of economic activities. *Finally*, we introduce these shocks to the UGEM and provide an assessment of economic impacts (Fig. 3).

Such an approach is not without limitations. In particular, in terms of the additional investments for the UGEM simulation, we do not assume any external sources (e.g. foreign borrowings), but use only domestically available resources. We also do not explicitly represent renewables in the UGEM and account for changes in the generation structure by altering composition of the intermediate inputs in the corresponding sectors (e.g. share of coal used for electricity generation). Finally, we make only one-way linkage from TIMES-Ukraine to UGEM, while a more consistent approach should include multiple iterations. Nevertheless, such approach can be considered more inclusive than a stand-alone application of TIMES-Ukraine or UGEM.

As many studies have shown, results of the CGE modelling can significantly depend on the values of exogenous parameters, in particular, elasticities of substitution and transformation. Sometimes, variation of these parameters can even

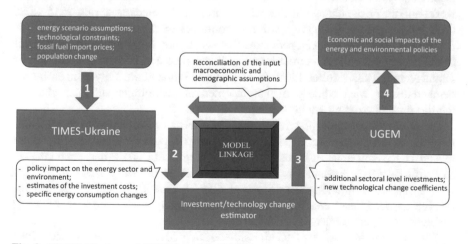

Fig. 3 TIMES-Ukraine and UGEM models linkage

change the results qualitatively, for instance, by turning net welfare gain into loss under the trade policy experiment (Taylor and von Arnim 2006). Therefore, in this study we accompany all UGEM-based estimates with error bars. We follow a Systematic Sensitivity Analysis (SSA) approach developed by Arndt and Pearson (1998) and apply a 50% variation to all 12 groups of substitution and transformation elasticities in the UGEM (Chepeliev 2014). We further derive 95% confidence intervals and indicated them using error bars. We use a triangular distribution for parameters variation and Strouds quadrature for approximation. To derive confidence intervals we assume normal distribution of the elasticities.

3 Energy Policies and Environmental Impacts

In this section, we discuss the BaU scenario and low emission development scenarios.

3.1 Business as Usual Scenario

Before moving to the discussion of long-term energy and environmental policy scenarios in Ukraine, we should consider a benchmark situation and develop a BaU path, based on the main drivers of energy demand. We assume an average 4% GDP growth rate over 2016–2050 based on the long-term forecasts of the macroeconomic model for Ukraine (Diachuk et al. 2017). For demographic forecast, we assume an −0.4% annual average population change over the same period, which corresponds to the central scenario provided by the Ukrainian Institute of Demography and Social Studies (PIDSS 2014). It is also assumed that domestic and world energy prices change in line with World Bank forecasts during 2015–2035 (WB 2017a) and 2030–2035 growth rate continues until 2050.

The BaU scenario is developed under the assumption of no fundamental changes in the energy system, i.e. current trends will continue and no new policies will be implemented. Thus resulted fuel mix and energy demand are fully defined by the demand drivers. Meanwhile gradual replacement of technologies still takes place, as the life time of existing equipment terminates.

GHG emissions under the BaU grow by almost 68% relative to 2012 levels (Fig. 4). In this study we consider only industrial processes and energy sector related GHG emissions in terms of IPCC definitions. In 2012, they amounted to 88% of total GHG emissions in Ukraine (Diachuk et al. 2017). According to the BaU scenario, Ukraine's share in global GHG emissions would almost double relative to 2015 level and reach 1.4% by 2050 under the assumption of world Reference CO_2 emissions (US EIA 2016).

Significant reductions in GHG emissions during 2012–2015 is associated with a severe economic recession and violation of Ukraine's territorial integrity. Within

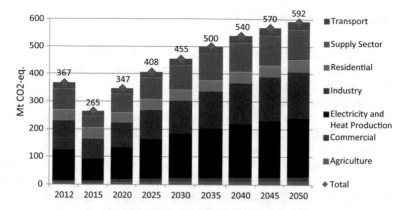

Fig. 4 Forecast of the GHG emissions in Ukraine according to the BaU scenario

the BaU scenario we assume that state sovereignty would be restored by 2020, therefore GHG emissions grow faster during this transition phase, relative to post-2020 period. Maintenance of the highly energy intensive economy like Ukraine would require significant expenses in the long-run as energy system costs may quadruple in 2050 relative to 2012 (Fig. 5). Due to the high level of depreciation, significant investments are required to replace obsolete technologies. Further depletion of fossil-fuels would even more inflate these costs and put higher pressure on the national economy.

The BaU scenario suggests that implementation of LED policies in Ukraine would primarily serve its national economic, social and environmental interests. Furthermore, any lag in moving towards this direction is only increasing cumulative energy system costs, as obsolete plants and equipment would require significant funds for their maintenance and update over time.

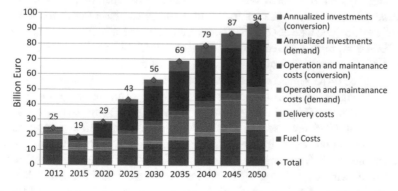

Fig. 5 Energy system costs under the BaU scenario (constant 2012 EUR)

3.2   Low-Emission Development Scenarios (LED)

LED scenarios are developed using additional assumptions imposed on the BaU path, i.e. they are based on the same macroeconomic and demographic forecasts, but differ in energy policy measures.

In accordance with Article 4, paragraph 19, of the Paris Agreement, all Parties should design long-term low GHG emission development strategies and communicate them by 2020 to the secretariat mid-century (UNFCC 2017). *ULCDS scenario* (Ukraine's low-carbon development strategy) is designed in the scope of Ukrainian low-carbon development Strategy initiative within the framework of USAID project (USAID MERP 2017). All energy policies are grouped into four streams: energy efficiency improvements, modernization and innovation, promotion of renewable energy (including biomass) and market transformations (such as national GHG emission trading scheme and improvements of emission taxation system). ULCDS scenario does not explicitly include national GHG emissions reduction target, but is based on the contributions of energy experts and their views on the economically feasible LED path for Ukraine.

The main goal of the second energy policy scenario is to analyze the benefits and challenges of moving towards high shares of renewables (*RE scenario*) more in line with the need to keep the global average temperature to well below 2 °C. This scenario has been developed in collaboration with the Heinrich Böll Foundation's Office in Ukraine (Diachuk et al. 2017), which supports the Greenpeace initiative in performing conceptual studies on energy system transformations with RE domination, the so called "Energy [R]evolution" (Greenpeace International 2016).

In contrast to ULCDS scenario, where the set of diverse targets and conditions are imposed, in RE scenario the largest possible share of renewables is key driving force for energy system transformation. RE scenario also assumes implementation of the Energy Community acquis (EC 2017). With such a stringent constraints, this scenario should be considered more as an exploratory assessment of RE potential and energy system flexibility, rather than guideline for specific policy measures.

Although most similar studies focus on a complete phase out of fossil fuels, in case of Ukraine, we take a 92% RE share as an economically feasible target. As our analysis shows, further increase of the RE share in GFEC results in the exponential growth of additional investments and total system costs.

In terms of GHG emissions, both LED scenarios are much more stringent than current Ukrainian NDC contribution (Fig. 6). To estimate the correspondence between LED scenarios and temperature paths a CI (2017) approach is applied. We take Ukrainian NDC level as a peak of GHG emissions in 2030 and apply required emission reduction rates for developing country afterwards. Under such an approach the ULCDS scenario corresponds to the 2.0 °C target, while RE scenario is consistent with 1.5 °C target (Fig. 6). Although the Ukrainian NDC level may seem high compared to the actual 2012 (pre-war) emissions, relative to the 1990 GHG emissions it is almost 40% lower. In addition, a significant emissions drop during 1990–2000 was achieved almost solely by sharp economic recession, therefore in case of rapid economic recovery without significant structural shifts NDC may serve as a reasonable constraint for possible emission peak.

3.3 Energy System Effects

Before moving to the comparison of ULCDS and RE scenarios, one particular point should be discussed in terms of results interpretation. While by 2050, RE scenario has much higher share of renewables than ULCDS, main decoupling between these two paths takes place only after 2035–2040. Therefore, in terms of cumulative changes over the whole 2012–2050 time horizon, differences between these two scenarios may not seem so large. This point can be best illustrated using GHG emissions reduction (Table 1). While in the case of the RE scenario cumulative emissions reduction is only 9% higher than in ULCDS, in the former case 2050 GHG emissions are 70% lower (85 Mt CO_2-eq. in RE vs. 285 Mt CO_2-eq. in ULCDS).

In general, both scenarios have negative total system costs, which means that additional capital expenditures are offset by associated energy efficiency improvements and reduction in fuel consumption. While the RE scenario has slightly higher total system costs than ULCDS, it is still more attractive from the economic point of view than BaU path. In other words, further maintenance of the

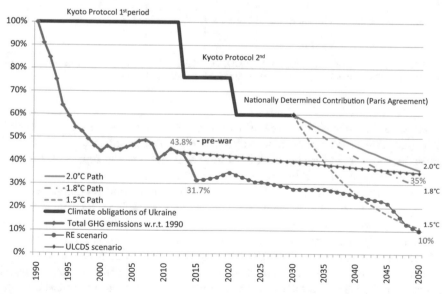

Fig. 6 LED scenarios emission and climate obligations of Ukraine. *Note* Estimates of the temperature paths for Ukraine are based on the CI (2017) methodology. Ukrainian NDC level is taken as a peak emission in 2030, afterwards annual reduction rates of 3.5% (in case of 1.8 °C path) and 8% (for 1.5 °C path) are applied to derive corresponding emission levels by 2050

Table 1 Cumulative 2012–2050 LED policy results

Indicators	Units	BaU	Change w.r.t. BaU			
			ULCDS		RE	
			Absolute	%	Absolute	%
GHG emissions	Mt CO$_2$-eq.	17,502	−7169	−41.0	−8718	−49.8
Total primary energy supply	Mtoe	5354.2	−1348.3	−25.2	−1820.5	−34.0
Import	Mtoe	1923.4	−815.8	−42.4	−972.0	−50.5
Final energy consumption	Mtoe	2684.7	−524.2	−19.5	−634.0	−23.6
Electricity generation	TWh	9901	−1675	−16.9	−501	−5.1
New power plant capacities	GW	101	14	13.5	133	132.2
New storage capacities	GW	0	59	–	139	–
Fuel expenditures	billion EUR (*constant 2012*)	629.4	−262.0	−41.6	−295.5	−46.9
Power plant investments	billion EUR (*constant 2012*)	120.3	−13.1	−10.9	82.9	68.9
Final demand technologies investments	billion EUR (*constant 2012*)	911.0	63.5	7.0	110.5	12.1
Total system costs	billion EUR (*constant 2012*)	734.8	−73.0	−9.9	−46.7	−6.4

Note "Mt CO$_2$-eq" stands for million tons of CO$_2$ equivalent; "Mtoe" stands for million tons of oil equivalent; "TWh" stands for Terawatt-hours; "GW" stands for Gigawatts

existing, highly inefficient energy system in the long-run, is even more expensive than transition towards 92% renewables share. As in case of BaU scenario, fuel expenditures account for almost 86% of total system costs and represent the most attractive "low hanging fruits" in terms of costs reduction.

Both scenarios positively contribute to national energy and economic security, by significantly reducing expenditures on imports, as well as final energy consumption (Table 1). While demand for electricity decreases, there is a need for new generation capacity, which is a key driver behind large additional investments in RE scenario. At the same time, ULCDS requires even lower power plant investments than BaU. However, both scenarios need additional investments to boost changes in final demand technologies, including expenditures on domestic appliances, vehicles, industrial machines, as well as insulation. In both cases, this category accounts for the largest share of additional investments.

Both LED scenarios assume significant change in generation mix (Fig. 7). Even under ULCDS path, nuclear energy would lose its dominating share after 2035. Existing nuclear plants would be gradually decommissioned, while new nuclear units (under current forecasts) would not be able to compete with other generation technologies. As a result, new coal power plants become a basis of the load curve, while 2/3 of the demand is covered by renewables. As RE scenario shows (Fig. 7), apart from hydro energy, other renewables (solar, wind, biomass) have even higher technically feasible potential. Implementation of their potential would require additional measures and technical solutions. In particular, they should include development of the grid and long-term electricity storage technologies, implementation of the demand control system for the load curve smoothing, further development of the transmission capacities and elimination of operational constraints that renewable generation faces for the integration to the national energy system. Additional measures should include introduction of tariff incentives for the renewable energy co-generation and development of the attractive funding opportunities. The latter one is especially important for Ukraine, as under the current conditions, high level of the green tariffs is almost fully offset by unaffordable funding options.

Changes in total final consumption include a decrease in energy use due to the implementation of energy efficiency (EE) measures, penetration of renewables on the energy market and significant increases in electricity consumption (Fig. 8). Due to the transition to market prices for natural gas in Ukraine, which started in 2009 for industrial users and in 2015–2016 for households, natural gas consumption gradually reduces and is substituted by other sources within ULCDS scenario. One of the methodological reasons behind this is that LED scenarios do not include

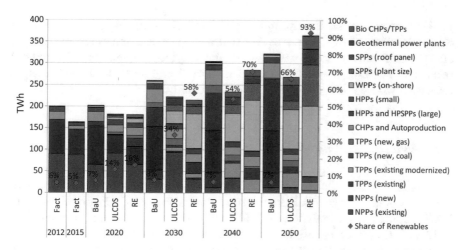

Fig. 7 Electricity generation in LED scenarios for Ukraine. *Note* CHPs—combined heat and power plants; TPPs—thermal power plants; SPPs—solar power plants; WPPs—wind power plants; HPPs—hydro power plants; HPSPPs—hydro pumped storage power plants; NPPs—nuclear power plants

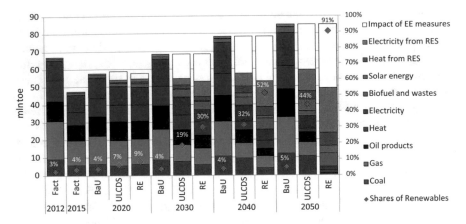

Fig. 8 Total final consumption in LED scenarios in Ukraine

stringent ecological constraints (in case of RE scenario share of renewables is the only target), as a result there is no impact on the price of carbon-intensive fuels and technologies. At the same time, the renewables share doubles by 2050 and the corresponding emissions reduction es not put significant financial pressure on the final consumers. In particular, each additional percentage point of the renewable energy share requires 1 bn Euro investments on aggregate over 2015–2050.

4   Economic Assessment of LED Scenarios

In this section, we use the dynamic UGEM model to provide an assessment of economic effects of LED scenarios. We start with the discussion of macroeconomic effects and further proceed with the sectoral impacts. All results are estimated relative to the BaU scenario discussed in Sect. 3 and show additional changes associated with the implementation of LED scenarios.

4.1   Macroeconomic Effects

According to our results, both LED scenarios are associated with positive macroeconomic effects. While in the short run, additional GDP growth (w.r.t to BAU scenario) is relatively low (1–2%), in the mid- and long-term it can reach up to 12–16% (Fig. 9). Key insights from this result is that over time energy efficiency improvements overstate associated investment costs. This is one of the

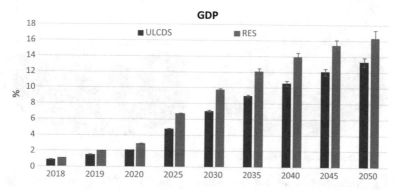

Fig. 9 Macroeconomic effects of LED policies implementation (w.r.t. BaU, %). *Note* Error bars indicate 95% confidence intervals for substitution and transformation elasticity changes under the SSA approach. See Sect. 2.3 for more details

"low hanging fruits" of the Ukrainian energy sector, as an introduction of new technologies can substantially reduce both energy and carbon intensity, as well as benefit economic development.

Overall, the RE scenario results in a higher GDP and output growth rates than ULCDS path, mainly due to the much higher additional investments and corresponding energy efficiency improvements. Although ULCDS scenario exploits most of the "low hanging fruits", there is still some space for environmentally friendly economic development. At the same time, a continuously growing necessity to accumulate additional investments can be seen as a main challenge towards implementation of the RE scenario. In this context, a sectoral contribution of the additional investments differs significantly by scenarios. In case of ULCDS policies, key contributions are provided by households, which account for over 60% of additional investments. At the same time, under the RE scenario, residential users' contribution does not change in value terms, but much higher investments are required from the industrial users, in particular, electricity producers (Fig. 10).

According to the National accounts (SSSU 2016), Ukrainian households' expenditure on capital goods are around 11% of the total national investments (1,3 bn EUR in 2015), while only implementation of the ULCDS scenario additionally requires 31 bn EUR over the 2017–2050 period. This may seem to require a substantial change in the households' consumption patterns and a significant increase of the marginal propensity to accumulate. At the same time, estimates of the additional residential investments, provided in this chapter, include not only capital goods, but also some final consumption expenditures (electrical appliances, motor vehicles etc.). In this context, additional residential investments (in case of both LED scenarios) represent a relatively small share of the total final consumption expenditures—less than 0.5% over the 2017–2050 period.

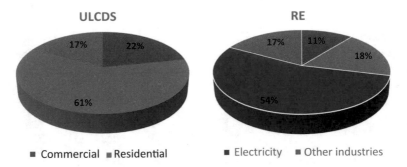

Fig. 10 Distribution of the additional investments by sources in LED scenarios, 2017–2050 (w.r. t. BaU, %)

With growing GDP, households also experience positive impacts of LED policies. In the short run, they are relatively insignificant, and even slightly negative for the RE scenario, but in the mid and long term, additional income growth can reach up to 12–14% (Fig. 11). During initial years, RE scenario has lower real income growth rates as much higher additional investments are needed for its implementation (compared to ULCDS), while efficiency improvements are distributed over time and have larger impact in the mid-term (by 2020–2025).

Households of the lower income deciles experience higher income growth rates than the richer households. Higher decile households have bigger share of services in their final consumption, and as domestic prices for services grow quicker than aggregate consumer price index, lower income deciles benefit more.

Both GDP and households' income estimates seem to be robust under elasticity changes, as deviations within 95% confidence intervals in most cases do not exceed 5% of the estimated values (Figs. 10 and 11).

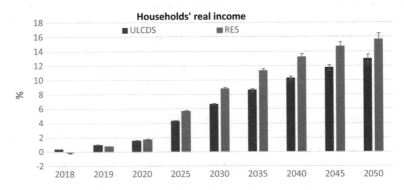

Fig. 11 Impacts of LED policies implementation on households real income (% w.r.t. BaU). *Note* Error bars indicate 95% confidence intervals for substitution and transformation elasticity changes under the SSA approach. See Sect. 2.3 for more details

4.2 Sectoral Effects

Implementation of LED scenarios is accompanied with significant structural changes, driven by the impacts of energy efficiency and fossil fuel substitution on productivity growth and price effects.

In general, both LED scenarios lead to reductions in fossil fuels mining and processing industries (Fig. 12). This is especially representative in the case of RE scenario, as coal mining, gas distribution and coke production decline by over 40% relative to BaU in 2050. Energy efficiency improvements and fossil-fuel prices reduction contribute to the output growth in some energy intensive industries, including basic metals, electricity and heating. As the share of biomass increases, agriculture and wood processing sectors raise their output to meet the growing demand. Some resources from mining and fossil fuel production shift towards services, as a result their output share slightly increases.

On average, estimates of the output changes show a higher degree of uncertainty compared to aggregate macroeconomic effects. For some sectors, they even indicate the possibility of sign change, like in case of electricity industry under ULCDS scenario (Fig. 12). Nevertheless, in most cases results do not change qualitatively, while high uncertainty around quantitative estimates is justifiable considering the role of substitution elasticities.

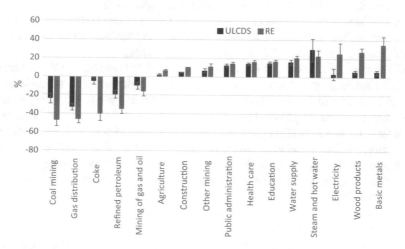

Fig. 12 Changes in the sectoral output under LED scenarios in 2050 (w.r.t. BaU, %). *Note* Error bars indicate 95% confidence intervals for substitution and transformation elasticity changes under the SSA approach. See Sect. 2.3 for more details

5 Policy Recommendations

Implementation of both LED scenarios is associated with significant challenges for Ukraine, especially in terms of current energy market structure, institutional capacity, social, economic and political conditions. In this context, policies and measures required for the implementation of ULCEDS scenario (consistent with 2 °C target) are essential since they serve as a basis for further movement towards RE paths (consistent with 1.5 °C target). Indeed, the ULCEDS scenario itself would require the whole set up of new policy mechanisms, significant market and institutional transformations etc., and the transition from the ULCEDS to the RE path would include scaling the intensity of policy measures (e.g. emission taxation level). In addition, most RE path-oriented measures are implemented after 2035–2040, which enables the possibility of a smooth transition from ULCEDS to RE during this period.

While implementation of both scenarios provides double dividends and is socially acceptable in the long run, existing institutional environment and inefficient market framework can pose significant challenges. In particular, this includes a necessary transition to the new institutional model of the Ukrainian energy market. The basics of such market set up (innovative development, networks' integration, consumers' empowering etc.) are already identified in the Ukrainian Energy Strategy (CMU 2017). At the same time, target indicators of the Energy Strategy are inconsistent with goals of some other strategic energy documents, such as Ukrainian low-carbon development strategy initiative. This identifies one of the key challenges of the institutional transformation in the Ukraine, which includes consistency and coherence of the energy strategic planning.

Elimination of the market disparities should readily contribute to the solution of another challenge, in particular, energy efficiency improvements, which may significantly reduce fuel costs. While all final consumption sectors have high energy efficiency potential, key contribution is associated with industry and households. Therefore, main policy efforts and government support should be focused on these two sectors. Transformation of the energy system under RE scenario would require significant additional investments—over two times higher than under ULCEDS. Even further reduction of fuel costs would not compensate increasing total system costs within RE scenario.

In terms of more specific challenges and policy measures, there are several high-priority steps that should be implemented. While Ukraine has successfully implemented a tariff reform in the energy sector (OECD 2016) and moved from uniform to targeted subsidies, over half of the households still do not have incentives and resources for energy efficiency improvements, since they benefit from the subsidized price rather than receiving any direct monetary transfer which could stimulate energy savings. Even with a more efficient subsidization system, due to the low level of households income, it would take a long time to substantially reduce volumes of subsidization. Over that period, National budget may experience significant pressure, as in 2016 subsidy-related transfers reached over 1.6 billion EUR or 7% of the National budget expenditures (STSU 2017).

Another potential challenge includes incentives for industrial users. As of the end of 2017, CO_2 emitters are facing an emission tax of 0.01 EUR per ton of CO_2, which is too small for achieving environmental targets.

In a broader context, national energy markets transformation towards transparency and competitiveness becomes a necessary condition of LED policies success. Remaining highly ineffective, Ukrainian energy markets are not able to sufficiently perform their key functions and fully integrate into the global environment. As a result, excessive government interventions continuously take place, which leads to the price distortions, as well as creates high risks for potential investors. Ukraine's access to the European legal framework through joining the Energy Community has defined a more specific direction and set up a way forward for efficient sectoral reforms. This form of integration should be further reasonably exploit to establish effective energy market design and technological renovation of the Ukrainian energy system.

6 Conclusion

With one of the highest levels of energy and emission intensities in the world, Ukraine has a high potential to exploit the "low hanging fruits" of the energy sector transformation by implementing LED policies, which can benefit both economy and environment.

According to our results, further maintenance of the existing highly inefficient energy system in the long-run is even more expensive than transition towards 92% renewables share. Key differences between ULCEDS and RE scenarios, both in terms of policy measures and results, arise after 2035–2040, which enables the possibility of smooth transition from ULCEDS to RE during this period. Only the RE scenario provides sufficient national contribution in terms of limiting global warming well below 2 °C.

With initially low level of energy efficiency in Ukraine, both LED policies result in positive macroeconomic and sectoral effects, both in terms of GDP and households real income growth, with better perspectives in case of the RE scenario, which at the same time requires 3 times higher investments. In this context, Ukraine benefits from double dividends under both policy options, while RE scenario also provides an economically acceptable way of going from relative to absolute decoupling.

Finally, adjustments in the institutional environment and market framework are identified. First, the energy strategic planning set up must be improved to avoid inconsistencies between strategic energy documents (e.g. Energy Strategy, Low-carbon Development Strategy etc.), including a social and political consensus around key strategic targets. Second, an efficient pricing of fossil fuels is required, in particular price signals for industrial users, as well as more transparent and market-oriented approach to residential consumers, with more targeted subsidies and elimination of cross subsidization in the electricity sector. Finally, Ukraine should move to competitive energy markets (in particular, fully implement Third Energy Package), as well as proceed with further integration to the ENTSO-E.

Acknowledgements The authors acknowledge the financial support from the Heinrich Böll Foundation in Ukraine and the USAID "Municipal energy reform in Ukraine" Project. We thank for assistance with TIMES-Ukraine modelling to DWG experts Gary Goldstein and Pat DeLaquil.

References

Arndt C, Pearson KR (1998) How to carry out systematic sensitivity analysis via Gaussian Quadrature and GEMAPCK. GTAP technical paper no. 3. https://www.copsmodels.com/ftp/gpssatp3/tp3.pdf

Cabinet of Ministers of Ukraine (CMU) (2014) National renewable energy action plan. Approved by the Cabinet of Ministers of Ukraine Executive order no. 902-p of 1 Oct 2014. https://www.iea.org/policiesandmeasures/pams/ukraine/name-131666-en.php

Cabinet of Ministers of Ukraine (CMU) (2017) Energy strategy of Ukraine until 2035. http://mpe.kmu.gov.ua/minugol/doccatalog/document?id=245239554

Chepeliev M (2014) Simulation and economic impact evaluation of Ukrainian electricity market tariff policy shift. Econ Forecast 1(1):1–24. https://papers.ssrn.com/sol3/papers.cfm?abstract_id=2608980

Climate Action Tracker (CAT) (2017) Ukraine. http://climateactiontracker.org/countries/ukraine

Climate Interactive (CI) (2017) Scoreboard science and data. https://www.climateinteractive.org/programs/scoreboard/scoreboard-science-and-data/

Diachuk O et al. (2017). Transition of Ukraine to the renewable energy by 2050. Heinrich Boll Foundation in Ukraine, Kyiv, Ukraine. https://ua.boell.org/sites/default/files/transition_of_ukraine_to_the_renewable_energy_by_2050_1.pdf

Energy Community (EC) (2017) Energy Community acquis. https://www.energy-community.org/legal/acquis.html

Government of Ukraine (GOU) (2015) Intended Nationally-Determined Contribution (INDC) of Ukraine to a New Global Climate Agreement. http://www4.unfccc.int/ndcregistry/PublishedDocuments/Ukraine%20First/Ukraine%20First%20NDC.pdf

Greenpeace International (2016) The Energy [R]evolution 2015. http://www.greenpeace.org/international/en/campaigns/climate-change/energyrevolution/

International Energy Agency (IEA) (2017) Key world energy statistics. http://www.iea.org/publications/freepublications/publication/KeyWorld2017.pdf

Loulou R et al. (2004) Documentation for the MARKAL family of models. ETSAP. https://iea-etsap.org/MrklDoc-I_StdMARKAL.pdf

Organisation for Economic Co-operation and Development (OECD) (2016) Inventory of energy subsidies in the EU's Eastern Partnership countries: Ukraine. https://www.iisd.org/gsi/sites/default/files/ffs_ukraine_draftinventory_en.pdf

Parry IWH, Bento AM (2000) Tax reductions, environmental policy, and the "double dividend" hypothesis. J Environ Econ Manage 39(1):67–96

Podolets RZ, Diachuk OA (2011) Strategic planning in fuel and energy complex based on TIMES-Ukraine model. Scientific report. Kyiv, Ukraine. http://ief.org.ua/docs/sr/NaukDop(PodoletsDiachuk)2011.pdf

Ptoukha Institute for Demography and Social Studies (PIDSS) (2014) Population projection for Ukraine. 2014 revision (2014 to 2060). Released Aug 2014. http://idss.org.ua/forecasts/nation_pop_proj_en.html

State Statistics Service of Ukraine (SSSU) (2016) Capital investment in Ukraine, 2010–2015. http://www.ukrstat.gov.ua/

State Treasury Service of Ukraine (STSU) (2017) Annual report on the state budget of Ukraine for 2016. http://www.treasury.gov.ua/main/uk/doccatalog/list?currDir=359194

Steffen W et al (2015) Planetary boundaries: guiding human development on a changing planet. Science 347:736–746

Taylor L, von Arnim R (2006) Modelling the impact of trade liberalisation: a critique of computable general equilibrium models. Oxfam research report, July. Oxfam International, London

The World Bank (WB) (2017a) Commodity markets outlook. http://pubdocs.worldbank.org/en/820161485188875433/CMO-January-2017-Full-Report.pdf

The World Bank (WB) (2017b) CO_2 emission (kg per PPP $ of GDP). https://data.worldbank.org/indicator/EN.ATM.CO2E.PP.GD?view=map&year_high_desc=true

Thompson Reuters Point Carbon (TRPC) (2014) Improving the existing carbon charge in Ukraine as an interim policy towards emissions trading—detailed report. http://www.ebrdpeter.info/uploads/media/report/0001/01/9705b0af32bc096636a81554c185d9181b49d916.pdf

Trinh B, Phong NV (2013) A short note on RAS method. Adv Manage Appl Econ 3(4):133–137. http://www.scienpress.com/Upload/AMAE/Vol%203_4_12.pdf

United Nations Framework Convention on Climate Change (UNFCC) (2017) Communication of long-term strategies. http://unfccc.int/focus/long-term_strategies/items/9971.php

United States Energy Information Administration (EIA) (2016) International energy outlook 2016. https://www.eia.gov/outlooks/ieo/ieo_tables.php

USAID Municipal Energy Reform Project (USAID MERP) (2017) Low emission development strategy of Ukraine. http://www.merp.org.ua/index.php?option=com_content&view=article&id=388&Itemid=1052&lang=us

Long-Term Climate Change Mitigation in Kazakhstan in a Post Paris Agreement Context

Aiymgul Kerimray, Bakytzhan Suleimenov, Rocco De Miglio, Luis Rojas-Solórzano and Brian Ó Gallachóir

Key messages

- The energy targets of the Strategy 2050 and the Green Economy Concept of Kazakhstan are compatible with the least-cost 25% emissions reduction pathway. In other words, a 25% emissions reduction target, rather than a 15% reduction target as currently proposed, is feasible for Kazakhstan.
- Renewable energy reaches 50% of the electricity generation mix, the rest is attributed to gas-fired power plants.
- A coal ban is not sufficient; emission reduction strategies must be also supported by carbon pricing and market mechanisms to promote zero emission sources.
- A TIMES model disaggregating the energy system of Kazakhstan in 16 sub-national regions is used for this study.

A. Kerimray (✉)
School of Engineering, National Laboratory Astana, Nazarbayev University,
Astana, Republic of Kazakhstan
e-mail: aiymgul.kerimray@nu.edu.kz

B. Suleimenov
National Laboratory Astana, Astana, Republic of Kazakhstan
e-mail: bakytzhan.suleimenov@gmail.com

R. De Miglio
E4SMA Srl, Turin, Italy
e-mail: rocco.demiglio@gmail.com

L. Rojas-Solórzano
School of Engineering, Nazarbayev University, Astana, Republic of Kazakhstan
e-mail: luis.rojas@nu.edu.kz

B. Ó Gallachóir
MaREI Centre, Environmental Research Institute, University College Cork,
Cork, Ireland
e-mail: b.ogallachoir@ucc.ie

© Springer International Publishing AG, part of Springer Nature 2018   297
G. Giannakidis et al. (eds.), *Limiting Global Warming to Well Below 2 °C: Energy
System Modelling and Policy Development*, Lecture Notes in Energy 64,
https://doi.org/10.1007/978-3-319-74424-7_18

1 Introduction

Kazakhstan ratified the Paris Agreement and its nationally determined contribution (NDC) is a 15% reduction in greenhouse gas emissions (GHG) as an unconditional target and 25% reduction as conditional target by 2030 compared to the level of 1990 (UNFCCC 2016a). The 25% conditional target is subject to additional international investments, access to the low carbon technologies transfer mechanism, green climate funds and flexible mechanism for economy in transition countries. Historical trends show a steadily increasing level of emissions over the last decade, with an average annual growth rate of 3.7%, already exceeding in 2014 the net GHG emissions of the unconditional NDC 15 target by 7% (Fig. 1). From the recent GHG emissions trends it can be concluded that progress towards achieving even the unconditional NDC 15 target is not sufficient and the mitigation actions are inadequate.

The Climate Action Tracker (2017) indicates that Kazakhstan's unconditional NDC commitment (−15%) in 2017 is not consistent with holding the increase in average global temperature to below 2 °C and is instead consistent with warming between 2 and 3 °C. According to the World Energy Outlook 2016, Eastern Europe/Eurasia region (where Kazakhstan belongs to) will be required to reduce its CO_2 emissions by 50% by 2030 and 58% by 2040 compared to 1990 level in the 450 Scenario (IEA 2017). This points out that Kazakhstan may be required to take a more ambitious emissions reduction target, depending on the burden sharing method. The implications of the conditional NDC (−25%) target have not previously been studied for Kazakhstan. This chapter addresses this knowledge gap. More ambitious targets (−50, −100%) have not been discussed previously in the country.

Kazakhstan has experienced a rapid economic development since the year 2000, and in 2016 the country was rated as an upper-middle-income country; its GDP per capita reached 7700 USD (World Bank 2017). Due to poor building insulation, low coverage by energy infrastructure (gas and district heating) in some of its regions

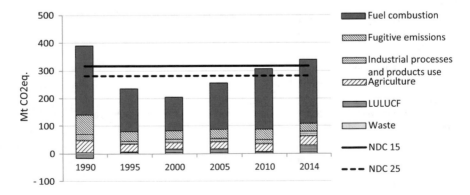

Fig. 1 Historical GHG emissions trend in Kazakhstan and the NDC cap (UNFCCC 2016b)

and large income inequalities, households are affected by energy poverty, more particularly energy affordability and lack of access to clean fuels. 28% of surveyed households spent more than 10% of their income on energy in 2013 (Kerimray et al. 2017a). There is high reliance on unsustainable fuels: 40% of households used coal, mainly for heating purposes in low-efficiency stoves (Kerimray et al. 2017a). There could be large incidence of insufficient thermal comfort values, but there are no data to quantify these values. Economic development (reduction of income inequality) and satisfaction of the demand for energy services have been higher priority in the country, compared to the climate change mitigation (Government of the Republic of Kazakhstan 2014; Tengrinews 2017). In this regard, taking into account current policies and long-term strategic documents, as well as the need to increase con-tribution to climate change mitigation, we considered the NDC 25 target as a possible main target for Kazakhstan rather than the NDC 15 target.

Previous energy system modelling studies for Kazakhstan focused on the unconditional target (NDC 15) and focused mostly on the mid-term analysis (2030) (Kerimray et al. 2016; Sarbassov et al. 2013; Kerimray et al. 2015; Suleimenov et al. 2016; PMR 2016). This study explores for the first time, the more ambitious the NDC 25 reduction target, under a longer time horizon (2050). As there is no official national long term GHG emissions reduction pledge by 2050, we assume in this chapter that it extends beyond 2030–2050. These scenarios serve as a bench-mark for how the NDC targets can be achieved at least cost.

Coal is currently widely used in power plants (74% of electricity generated with coal) and in the domestic heating appliances (40% of households use coal) in Kazakhstan. Therefore, emissions reductions are not possible without urgent actions on substantially reducing coal use across all sectors of the economy. This chapter goes further than previous work by analysing the implications of phasing-out coal (with a coal-ban scenario) in Kazakhstan by 2050 as an additional contribution from Kazakhstan towards achieving "well below 2 degrees world".

The TIMES-based sub-nationally disaggregated 16-region energy system model for Kazakhstan was employed in this study. The need for regionally disaggregated analysis for Kazakhstan is mainly driven by spatially heterogeneous conditions of the national energy system and different dynamic factors of its regions. Due to the absence of data on thermal comfort and unmet demand values, energy poverty was not explicitly modelled. The results of the study can be used for development of a low-carbon development program for Kazakhstan and to inform actions to be undertaken for fulfillment of the NDC targets.

2 Policies for Energy Transition and GHG Emissions Reduction in Kazakhstan

Kazakhstan has introduced many policies and measures domestically over the last 5–7 years to promote penetration of renewable energies and to improve energy efficiencies (Kerimray et al. 2017b, c). The law 'On Energy Saving and Improving Energy Efficiency' was adopted in 2012. Since its enactment, many industrial and buildings energy audits have been conducted and a regulatory framework for energy efficiency has been introduced. A law promoting the use of renewable energy resources was introduced in 2009, with fixed feed-in tariffs adopted initially, and renewable energy auctions later in 2017.

Kazakhstan is the first country in Central Asia that has launched an emissions trading scheme (ETS) in 2013. It involves 140 big companies (including oil and gas), mining and chemical industry, covering around 50% of country's CO_2 emissions. However, industry involvement and trading activity has been very weak, which could be partly due to the economic recession. Thus, in the first year of ETS (2014), just 32 transactions were completed with an average price of 1.67 USD/tonne and in its second year, 40 transactions took place for a total of 1.98 Mt (<1% of all tonnes capped under the system) with the average price of USD2.06 (IETA 2016). In 2016, trading and penalties for non-compliance were suspended until January 2018, to give time to make amendments to the ETS, although annual reporting and verification requirements remain in place. The ETS suspension was due to high resistance of industries and concern of the negative impact of the ETS on the economic growth associated with lack of flexibility of "historical" allocation methods to production levels. It is planned to restart ETS with new allocation methods (benchmarking) and trading procedures.

Despite existence of policies and measures, changes in the energy mix are slow: renewable energy penetration is still low (less than 1% without accounting for biomass) and there were no significant energy intensity reductions for any sectors between 2010 and 2014 (except for oil and gas), resulting in the positive trend in growth of GHG emissions (Kerimray et al. 2017c).

Strategic documents on energy development of Kazakhstan suffer from the lack of integrated approach and consistency. According to the "Strategy-Kazakhstan-2050": the new political course of the state" by 2050, 50% of energy consumed in Kazakhstan should be supplied through renewable and alternative energy sources. While the strategic document "Fuel-energy complex development concept till 2030", adopted in 2014, implies that coal will remain to be the main fuel for power generation in Kazakhstan (Government of the Republic of Kazakhstan 2014). Reduction of coal share in the fuel mix, actions towards achieving ambitious 2050 goals, as well as emissions reduction measures, are not indicated in the "Fuel-energy complex development concept till 2030".

Energy system models provide a comprehensive description of possible scenarios for development of the energy system by considering intertemporal, interregional and intersectoral relations and thus, may assist decision makers to take systemic interdependencies of the energy sector into consideration.

3 TIMES-Kazakhstan Multi-regional Model

The TIMES-Kazakhstan multi-regional model represents all steps of an energy chain region by region: from an extraction of primary resources to their supply to primary energy markets, from transformation of primary energy carriers to their transmission and distribution to the final energy-use sectors, from use of final energy commodities to satisfaction of end-users demand for energy services (Suleimenov et al. 2016).

The optimisation paradigm used here is energy system cost minimisation with perfect foresight. The modelling time horizon is from base year (2011) to (2050). The model for Kazakhstan is calibrated for the year 2011 with the data provided by the regional Energy Balances (Kazmaganbetova et al. 2016; Kerimray et al. 2017c). Regional representation corresponds to administrative division of the country: 14 regions and 2 cities: Astana (capital) and Almaty (ex-capital, financial centre).

The model is one multi-regional model with 16 regions which are allowed to trade energy forms through the existing and new infrastructures (pipelines for crude oil and natural gas), through electrical grids and via land transport (oil products and coal) depending on regional demand for energy. Capacities of the existing infrastructures are used to describe the maximum level of "tradable" energy between pairs of regions. New capacities of energy infrastructure between regions is endogenous to the model, investment costs and possible routes were described. The capital cost for new gas pipeline infrastructure (described in the model as one of the technology investment options) was obtained from the recently constructed gas pipeline Beineu-Bozoi-Shymkent at 7 mln USD/(TJ*km).

CO_2 emissions from combustion are tracked using the emission coefficients per "fuel", according to the IPCC guidelines (based on the carbon content of each fuel) and the national inventory of emissions.

The technology database was inherited from national (single region) TIMES-Kazakhstan model, using the latest updated version by Nazarbayev University Research and Innovation System (Kazakhstan) under the Project funded by Partnership for Market Readiness (2015–2016).

3.1 Electricity and Heat Generation Sector

The electricity and heat generation sector in the base year (2011) represents all power and combined heat and power plants, region by region, by input fuel type and calibrated according to data from KEGOC (Kazakhstan Electricity Grid Operating Company) for 2011. The existing stock is dominated by coal-fired plants, low efficiencies, and low shares of renewables (hydro). For the future years, the retirement of existing stock and new capacities is fully endogenous to the model. Electric power transmission and distribution losses of Kazakhstan's grid was around 7%.

In Kazakhstan the renewable energy potential is high and it exceeds the projected energy demand (Karatayev et al. 2016). Hydro, solar and wind technologies were assumed to have three levels of costs, depending on the region. The regions with the highest wind speed and high solar insolation were assumed to have the lower end of technology cost. The regions with lower renewable potential were assumed to have high medium and high technology cost. As there were no studies on renewable energy technology costs for Kazakhstan, it was assumed as described in the Table 1.

3.2 Demand Projections

The model includes various demands for energy services categorised by sector (for example, industries, types of transport, household and commercial processes: washing, drying, cooking, heating, hot water supply, lighting, etc.). Each energy service demand has its own macro-economic or physical output driver. Correlation factors for associating energy service demand to their drivers were inherited from the national model (PMR 2016) and assumed to be the same across the regions. Energy service demands are assumed here to be "inelastic" to prices for energy, due to the absence of data on price elasticities. Projections of drivers of energy demand is described by Suleimenov et al. (2016).

3.3 Export and Import Assumptions

Kazakhstan has significant fossil fuel resources and is a net exporter of energy and energy products. One of the pillars of the national strategy is to minimize energy imports. According to the energy balance of Kazakhstan, most of the energy commodities consumed are supplied from domestic production, with the exception of oil products (due to insufficient capacity of domestic oil refineries).

Existing import/export from/to abroad are also taken into consideration in the model and projected over the time horizon based on the following assumptions:

Table 1 Investment cost of solar, wind and hydro, USD$_{2013}$/W

	Low cost	Medium cost	High cost
Hydro	3	6.6	9.6
Solar	1.57	1.62	2.22
Wind	1.44	1.68	2.76

(a) Minimum level of crude oil export is equal to the base year net export (2000 PJ) till 2030; (b) Natural gas export level decreases twice in 2030 from the level of the base year, allowing the system to supply gas to domestic users which is consistent with the adopted Law "On gas and gas supply" (2012), setting priority of gas supply to domestic users; and (c) Export level for coal decreases by 25% from the base year (540 PJ). Trade of oil products, electricity and biomass is endogenous to the model, with exogenously defined import and export prices.

4 Scenarios

TIMES 16-region model employed in this study cover only fuel combustion related emissions. Thus, projections for sectors not related to fuel combustion (e.g. waste; industrial processes and products use; land use, land-use change and forestry, fugitive emissions from fuels) were obtained from the previous study funded by Partnership for Market Readiness (PMR 2016). The results of "with measures" scenario was taken. Upper limits on fuel combustion emissions take into account these other emissions.

The model was run for four scenarios: Business as usual (BaU), unconditional and conditional NDC targets, and coal ban (Table 2). The NDC scenarios inherit all the key characteristics of a reference case (key assumptions about the gas production profile, energy self-sufficiency plan, gas network development, etc.), and add an emission reduction target to the decision problem. The system can respond to the constraint through investing in higher efficiency technologies, and/or different fuels, energy infrastructure (gas pipeline, electricity network, district heating system) in some or in all regions. The underlying assumption is that there is an (endogenous) allocation of an emission reduction among regions and sectors based

Table 2 Scenario runs

Scenario	Description
BaU	The BaU (business as usual) scenario is the least cost solution of the energy system without any specific environmental target
NDC 15	Least cost solution of the energy system with the constraint on total GHG emissions from fuel combustion to the amount of 199 Mt CO_2 equivalent for the entire time horizon (until 2050), which is equivalent to 80% of emissions in the year 1990. With GHG emissions for sectors not related to fuel combustion accounted, total GHG emissions correspond to 85% relative to 1990 levels
NDC 25	Least cost solution of the energy system with the constraint on total GHG emissions from fuel combustion in the amount of 174 Mt CO_2 equivalent for the entire time horizon (until 2050), which is equivalent to 70% of 1990 levels. With GHG emissions for sectors not related to fuel combustion accounted, total GHG emissions correspond to 75% relative to 1990 levels
Coal-ban	Total coal consumption is reduced by 20% in 2020, by 60% in 2030 and by 100% in 2050 compared to the base year

on a cost-effectiveness approach to equalize the marginal cost of CO_2 eq. emissions across the regions.

5 Results

5.1 High Abatement Potentials, Coal Ban Alone Is Insufficient

Without any targets (the BaU Scenario), the GHG emissions from fuel combustion increases by 47% by 2050 compared the base year level due to rising demand for energy services and low technology/fuel improvements (Fig. 2). In 2050, GHG emissions in BaU scenario exceed the NDC 25 by 184 $MtCO_2$ eq. Cap on GHG emissions has resulted in the fuel switch and improving efficiencies in technologies and processes. Coal ban has resulted in GHG emissions reductions by 153 $MtCO_2$ eq. in 2050 (compared to BaU), however, it exceeds the NDC 25 since coal is substituted by fossil fuels rather than renewable, pointing out the need for additional actions for achieving GHG emissions reductions. Most of the GHG emissions reductions are realized through abatement of CO_2, in particular in the power sector (Fig. 2).

Due to necessity for investments in the energy system, CO_2 price reaches 36–59 USD per tonne in 2030 and 209–281 USD per tonne in 2050 (Table 3). These prices apply to all sectors. ETS sectors (upstream, industry and power generation sectors) are collectively responsible for 75% of total GHG emissions reductions in 2050 in the NDC 25 compared to BaU. This demonstrates that sectors covered by ETS have the highest abatement potential at the lowest cost.

Coal continues to dominate in the BaU scenario (from the current share in the TPES of 55–63% in 2050). In the NDC 25 scenario, share of coal in the TPES reduces to 12% and share of gas increases from the current share of 20–52% in 2050 (Fig. 3). There is an investment to the gas pipeline construction to the regions, which currently do not have access to network gas in alternative scenarios. Total final consumption reduces by 27% in the NDC 25 in 2050 compared to the BaU case.

As a result of the replacement of existing inefficient coal-fired power plants with new gas and renewable based generation, the efficiency of energy transformation processes (measured as TFC/TPES) increases and reaches 80% in 2050 in the NDC 25 scenario (compared to the existing 54%).

5.2 Focus on Power Generation: Less Coal, Less CHP

Without any climate policy actions (BaU), coal continues to dominate the fuel mix for electricity and heat generation (70% in 2050), while in the NDC 25 scenario it is almost fully phased out (0.3%). Share of gas in electricity generation reaches 50%,

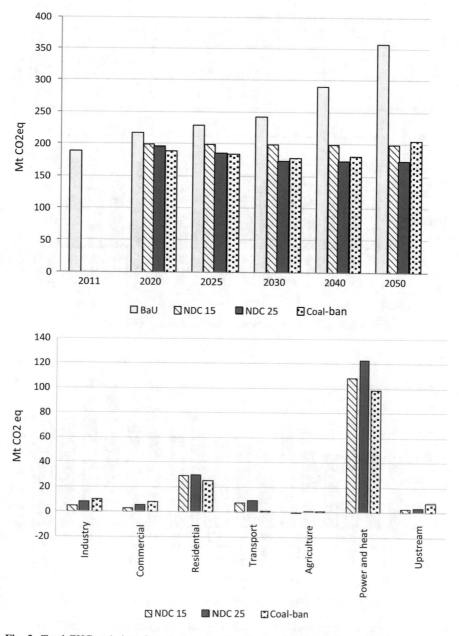

Fig. 2 Total GHG emissions from fuel combustion (top) and CO_2 emissions reduction by sectors compared to the BaU scenario in 2050 (bottom)

Table 3 Marginal CO$_2$ eq. price, USD/tonne

Scenario	2020	2025	2030	2035	2040	2045	2050
NDC 15	11.5	25.1	36.0	40.6	62.7	137.3	208.5
NDC 25	12.0	46.2	58.6	89.5	171.7	181.3	280.5

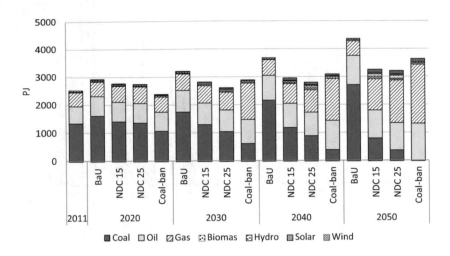

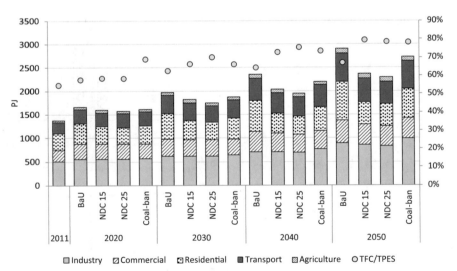

Fig. 3 Total primary energy supply (TPES), PJ (top) and total final consumption (TFC), TJ by sectors and TFC/TPES (bottom)

with the remaining 49% provided by the renewable energy sources in 2050 in the NDC 25.

The total installed capacity of coal fired power plants is 13 GW in 2050 in the NDC 25, while 5.9 GW only of which is utilized (new coal-fired CHP plants generate heat and electricity). By 2050 there will be a substantial installation of renewable energy sources reaching to 14 GW of wind, 10 GW of hydro and 9 GW of solar in the NDC cases. Renewable energy potential at the lowest costs is fully utilized in the NDC scenarios. Additional emissions reduction needed for the NDC 25 (compared to the NDC 15) is achieved by replacement of remaining coal-based capacity with gas. The cost for building gas infrastructure depends on the region (distance). Thus, the regions located at the longest distance from gas production regions are provided with gas in the NDC 25 and the coal ban scenario. In the coal ban scenario most of the electricity is generated with gas (73%), indicating that gas is the most economically viable substitution for coal when the emissions constraint is not imposed (Fig. 4).

Current production of electricity in Kazakhstan is largely dependent on coal-fired power plants with limited possibility of quick start-up and hot standby. In all alternative scenarios, the gas network is constructed in the northern and central regions, with installed capacity of gas power plants reaching 16 GW in northern and central Kazakhstan in 2050, thus providing necessary back up for variable renewable energy sources (wind, solar).

Due to high demand for heating (cold climate conditions), currently, CHP generation capacity makes up a large share of total installed capacity, providing 42% of the total electricity generation and 55% of the total district heating generation. In the BaU scenario, CHP (mainly coal based) provides up to 72% of electricity generation in 2050 and satisfies most of the heat demand (65%). While in the NDC cases CHP plants (mainly gas based) generate up to 47% of the electricity (due to utilisation of renewable energy sources). Production of district heat by CHP plants increases in all scenarios, by 79–87% compared to base year level.

In the NDC scenarios, heat only plants reduce district heat generation substantially (by 92%) compared to the base year level. This occurs as a result of significant heat savings in the residential sector (up to 77% in the NDC 25) and the switch to individual heating systems (natural gas, electricity) in the residential and commercial sectors.

Biomass is not deployed in any scenarios, because of limited biomass potential in Kazakhstan (low level of large agriculture industries, limited stock of forests available only in certain regions).

5.3 Focus on Final Consumption Sectors: Large Changes in the Residential Sector

The residential sector is affected the most among end-use sectors by the emissions constraint, with total final consumption of this sector reducing by up to 43% (compared to the BaU scenario). There is a complete phase-out of coal in the residential sector in the NDC cases (from the base year level 104 to 0 PJ in 2050). In the NDC scenarios, the least cost energy system incorporates significant building

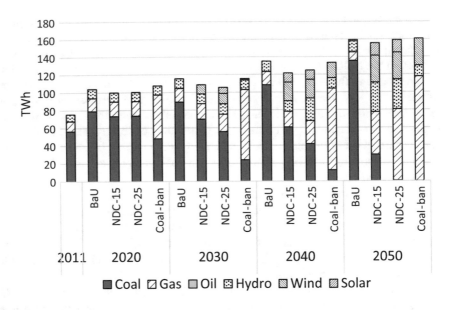

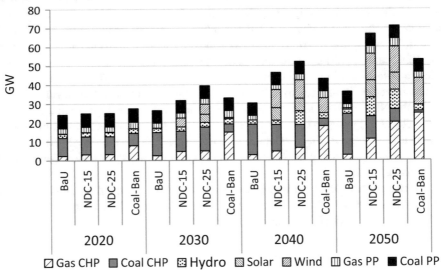

Fig. 4 Electricity generation by fuel type (top) and total installed capacity of power and CHP plants (bottom)

energy retrofit measures in the residential sector (e.g. insulation of walls, replacement of windows), which result in the reduced demand for heating by 32–46% in 2030 and by up to 77% in 2050. Along with building energy retrofit measures, coal is replaced by natural gas, electricity and LPG in the NDC scenarios. Additional emissions reductions in the NDC 25 (compared to the NDC 15) are achieved by additional heat savings, which are higher by 43% and 24% in 2030 and 2040, respectively.

In the transport sector, total consumption reduces by 22% by 2050 in the NDC 25 compared to the BaU scenario, as a result of switch to more efficient heavy trucks, light trucks and light duty vehicles. Thus, diesel oil consumption reduces 29% in the NDC 25 compared to the BaU scenario in 2050. Gasoline consumption reduces by 39% in the NDC 25 compared to BaU scenario in 2050.

In the industry sector, total consumption reduces in the NDC 25 by 6% compared to the BaU scenario in 2050. There is a fuel switch from coal to natural gas, electricity and district heating.

In the commercial sector there is up to 15% reduction in the total final consumption. The share of coal in the commercial energy consumption (which is mainly used for heating) in 2050 decreases from the 25% in the BaU to 9% in the NDC 25. Coal is mainly replaced by natural gas and electricity. The share of gas increases from 7% in the BaU to 19% in the NDC 25 in 2050. The share of electricity rises from 27% in the BaU to 42% in the NDC 25.

5.4 Implications for Energy Poverty

Energy poverty (or unmet demand) indicator is not explicitly tested in the model due to lack of data. However, marginal prices for heating were analyzed to account for the affordability dimension of energy poverty, as heating is the highest end-use service in the residential sector in Kazakhstan. The NDC 25 scenario has resulted in an increase in the marginal price of useful energy for heating by 125% and 288% by 2030 and 2050 respectively compared to the BaU (Fig. 5). This clearly indicates that additional investment in technology and more expensive fuel results in increased costs for heating, which can have negative impact for population, particularly on low income and vulnerable households.

Despite the negative impact on energy affordability (which can be mitigated by appropriate policies), there is an improved access to energy infrastructure in the NDC 25. Gas pipelines are constructed to all regions without access to gas in the NDC 25. Coal is fully phased out in the residential sector.

6 Recommendations for Policy Makers

The overall target as set by Kazakhstan's "Strategy 2050" and the Green Economy Concept to reach 50% of renewable and alternative energy sources by 2050 is very close to the least-cost emissions reduction pathway (NDC 25) as demonstrated by modeling results here (49% of renewable energy sources by 2050). Comparing values for installed capacity by energy source in the Green Economy Concept with this study demonstrate similar installed capacities for gas (23–24 GW) and wind

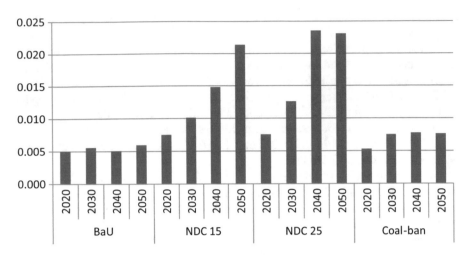

Fig. 5 Marginal price of useful energy for heating, USD/MJ

(14–15 GW). Capacity of coal power plants is also similar (5–6 GW) with the Green Economy Concept, as only 6 GW (out of 13 GW installed) is utilized in the NDC 25 scenario in 2050. The values for solar, hydro and nuclear are different (Table 4). According to the modeling results, nuclear power generation was not selected, while the renewable energy potential available at the lowest cost is fully utilized. Methodology, input data and assumptions used for the preparation of the Green Economy Concept and the Fuel-energy complex development concept till 2030 are not available for the public, thus making it challenging to compare assumptions and input data.

Compared to the results of the WEO 2017 for Eastern Europe/Eurasia region in the 450 scenario, Kazakhstan may need to reduce its fossil fuel power generation substantially more compared to the current study. Eastern Europe/Eurasia region have the following fuel mix for electricity generation in the 450 scenario: 30% nuclear, 21% gas, 27% hydro, 3% coal, with the remaining (46%) provided by renewable energy sources in 2040.

Table 4 Total installed capacity in 2050, GW

	The green economy concept	This study (NDC 15)	This study (NDC 25)
Gas	23	15	24
Wind	15	14	14
Solar	15	9	9
Coal	5	18	13 (5 GW utilized)
Nuclear	2	0	0
Hydro	4	11	11
Total	63	68	72

Coal has been used as an inexpensive and abundant resource for energy generation in the country. Due to the reduction of demand for coal globally and low export levels, the coal industry of Kazakhstan is mainly oriented for domestic use. The results presented here depict that the plan to increase coal consumption as set by the "Fuel-energy complex development concept till 2030" is not consistent with Kazakhstan's climate pledge. The modeling results demonstrated that in 2030 there is a 25% reduction in the use of coal in the NDC 25 compared to the base year level. Fulfilling the NDC target would mean inevitable reduction/elimination of the consumption of coal. Carbon pricing and market mechanisms (such as ETS) can serve as an effective tool for achieving emissions reduction targets. Modeling results suggest that the ETS sectors contribute to the highest emissions reductions at the lowest cost, indicating that it can be an effective tool if it is appropriately designed and operated.

Kazakhstan has considerable gas supply potential, as it has its own gas reserves. Several possible routes for providing a gas to capital Astana city (with further extension to northern and eastern locations) have been discussed by the Government of Kazakhstan in the past. However, to date (2017), the investment decision for construction of a gas pipeline to northern and central regions has not been made yet. The results of this study suggest that the construction of gas pipelines to the northern, central and eastern regions of Kazakhstan is a necessary action to achieve emissions reduction targets as natural gas is the most economically feasible alternative to coal. Gas fired power plants can serve as a back-up capacity for balancing the system with high shares of renewable energy sources. "Coal to gas" strategy alone is not sufficient, as additional mitigation actions such as deployment of large shares of renewable energy sources (by up to 50%), energy efficiency improvements (heat savings in buildings, efficiency of transport technologies and early retirement of inefficient power plants) are necessary.

Despite the official announcements of Kazakhstan on its contribution to climate change mitigation (Ministry of Energy of the Republic of Kazakhstan 2017), high level officials stated that Kazakhstan will continue to rely on coal (as the least expensive fuel) and will not deploy alternative energy sources (due to its high cost) (Tengrinews 2017). Thus, compared to the current energy policies and mitigation actions which appeared to be inefficient, not fully implemented and sometimes contradictory, the actions proposed by these modeling results (NDC 25) are quite ambitious.

Negative implications of climate mitigation actions on energy affordability (due to higher prices) can be mitigated by providing subsidies on building energy retrofits, on clean technologies, with targeted support for low income and vulnerable population.

Analyses based on integrated energy and emissions modeling should be promoted and deployed in the country, to provide different views and pathways on the energy system development and to provide verification and comparison of the results.

7 Conclusion

Fulfilling the 25% emissions reduction target requires not only a coal ban but also the promotion of zero emission sources, which can be achieved by extending the gas network to the non-gasified regions, accelerated retirement of existing old and inefficient power plants and deployment of renewable energy sources. The share of renewable energy (including hydro) could represent half of the electricity generation mix. The remaining share should be attributed to gas-fired power plants. In other words, the overall target as set by Kazakhstan's "Strategy 2050" and "Green Economy Concept" to reach 50% of renewable and alternative energy sources by 2050 is very close to the least-cost 25% emissions reduction pathway.

Carbon pricing and market mechanisms (such as ETS) can be effective tools for climate change mitigation. Due to regulated (relatively low) energy prices, the future of the construction of gas pipelines largely relies on a strong political will to implement pricing reforms and/or allocation of funding from the Government. Additionally, a gradual coal ban across all sectors of the economy is a fundamental step towards achieving emissions reductions.

Coal is also fully phased-out in the residential sector in both NDC scenarios, in favor of natural gas, electricity and LPG, with substantial building energy retrofit measures. Mitigation actions in the transport sector include utilization of more efficient heavy trucks, light trucks and light duty vehicles, while in the commercial and public sector there is a substantial reduction of coal use in favor of gas and electricity.

Meeting the NDC target is technically possible, however, the corresponding abatement costs appear to be rather high at around 36–59 USD per tonne of CO_2 eq. by 2030 and 209–281 USD per tonne of CO_2 eq. by 2050 (if the NDC target is applied for 2050). Marginal price for heating increases substantially in the NDC 25, which can worsen energy affordability of households (if no actions supporting energy poor are taken).

The results of this work can be used by policy makers in formulating and justifying a climate mitigation roadmap.

Future studies need to be conducted to explore 40%, 50% or even higher emissions reductions compared to 1990 level by 2050. To provide more arguments necessary for energy transition, future studies are needed to quantify all external damage and costs of the existing energy system, consequences of climate change to the national economy and the health of population. Future studies should also be conducted to better quantify "unmet demand" based on Households Survey data of thermal comfort, indoor air temperature and behavioral issues. Finally, a finer representation of operating parameters of the power system (e.g. ramping rate, minimum up and down times) would be necessary to explore the full implications of the integration of renewable energy sources.

References

Government of the Republic of Kazakhstan (2013) Concept for transition of the Republic of Kazakhstan to green economy. http://gbpp.org/wp-content/uploads/2014/04/Green_Concept_En.pdf. Accessed 10 Oct 2017

Government of the Republic of Kazakhstan (2014) Fuel-energy complex development concept till 2030. https://tengrinews.kz/zakon/pravitelstvo_respubliki_kazahstan_premer_ministr_rk/promyishlennost/id-P1400000724/. Accessed 11 Oct 2017

IEA (2017) World energy outlook 2016

IETA (2016) Kazakhstan: an emissions trading case study. http://www.ieta.org/resources/2016%20Case%20Studies/Kazakhstan_Case_Study_2016.pdf. Accessed 7 Oct 2017

Karatayev M, Hall S, Kalyuzhnova Y, Clarke M (2016) Renewable energy technology uptake in Kazakhstan: policy drivers and barriers in a transitional economy. Renew Sustain Energy Rev 66:120–136

Kazmaganbetova M, Suleimenov B, Ayashev K, Kerimray A (2016) Sectoral structure and energy use in Kazakhstan's regions. In: 4th IET clean energy and technology conference (CEAT 2016), Kuala Lumpur, p 9(7). https://doi.org/10.1049/cp.2016.1266

Kerimray A, Baigarin K, Bakdolotov A, De Miglio R, Tosato GC (2015) Improving efficiency in Kazakhstan's energy system. In: Lecture notes in energy, vol 30, pp 141–150. https://doi.org/10.1007/978-3-319-16540-0_8

Kerimray A, Baigarin K, De Miglio R, Tosato GC (2016) Climate change mitigation scenarios and policies and measures: the case of Kazakhstan. Clim Pol 16:1–21

Kerimray A, De Miglio R, Rojas-Solórzano L, Ó Gallachóir BP (2017a) Causes of energy poverty in a cold and resource rich country. Evidence from Kazakhstan. Local Environ. https://doi.org/10.1080/13549839.2017.1397613

Kerimray A, Rojas-Solórzano L, Amouei Torkmahalleh M, Hopke PH, Ó Gallachóir BP (2017b) Coal use for residential heating: patterns, health implications and lessons learned. Energy Sustain Dev 40C:19–30

Kerimray A, Kolyagin I, Suleimenov B (2017c) Analysis of the energy intensity of Kazakhstan: from data compilation to decomposition analysis. Energ Effi. https://doi.org/10.1007/s12053-017-9565-9

Law "On gas and gas supply" (2012). https://online.zakon.kz/Document/?doc_id=31107618. Accessed 28 Nov 2017

Ministry of Energy of the Republic of Kazakhstan (2017) Kazakhstan was represented by the Minister of Energy at the climate summit in Paris. http://energo.gov.kz/index.php?id=16460

PMR (2016) Assessment of economic, social and environmental effects of different mitigation policies using combined top-down (CGE-KZ) and bottom-up (TIMES-KZ) models. World Bank project: development of policy options for mid- and long-term emissions pathways and role of carbon pricing. Final report

Sarbassov Y, Kerimray A, Tokmurzin D, Tosato G, De Miglio R (2013) Electricity and heating system in Kazakhstan: exploring energy efficiency improvement paths. Energy Policy 60:431–444

Suleimenov B, De Miglio R, Kerimray A (2016) Emissions reduction potential assessment in regions of Kazakhstan using TIMES-16RKZ model. IEA-ETSAP Workshop, Madrid. https://www.slideshare.net/IEA-ETSAP/emissions-reduction-potential-in-regions-of-kazakhstan-using-times16rkz-model. Accessed 20 Sept 2017

Tengrinews (2017) Cheap coal is available for 300 years, said Deputy. https://tengrinews.kz/kazakhstan_news/deshevogo-uglya-nam-hvatit-na-300-let-deputat-331922/

The Climate Action Tracker (2017) Kazakhstan. http://climateactiontracker.org/countries/kazakhstan.html. Accessed 5 Oct 2017

UNFCCC (2016a) Intended Nationally Determined Contribution—Submission of the Republic of Kazakhstan. http://www4.unfccc.int/submissions/INDC/PublishedDocuments/Kazakhstan/1/INDCKz_eng.pdf. Accessed 25 Jun 2017

UNFCCC (2016b) National Inventory Submissions 2016. http://unfccc.int/national_reports/
   annex_i_ghg_inventories/national_inventories_submissions/items/9492.php.   Accessed   25
   Aug 2017
World Bank (2017) Kazakhstan. Available from http://www.worldbank.org/en/country/kazakhstan.
   Accessed 17 Aug 2017

Mexico's Transition to a Net-Zero Emissions Energy System: Near Term Implications of Long Term Stringent Climate Targets

Baltazar Solano-Rodríguez, Amalia Pizarro-Alonso, Kathleen Vaillancourt and Cecilia Martin-del-Campo

Key messages

- Our modelling suggests that deep decarbonisation of Mexico's power system to 2050 is techno-economically feasible and cost-optimal through renewables
- An over-investment in gas infrastructure in the next 15 years may delay the power sector's transition to lower carbon sources and put at risk either meeting carbon targets cost-effectively or leaving some gas assets stranded
- A novel TIMES energy systems model for Mexico has been used to explore the implications of a whole energy system decarbonisation on the power sector, considering energy efficient technologies in end-use sectors; these results have been used as input to the Balmorel-Mexico model to simulate our scenarios.

B. Solano-Rodríguez (✉)
UCL Energy Institute, London, UK
e-mail: b.solano@ucl.ac.uk

A. Pizarro-Alonso
Energy Systems Analysis Group, Systems Analysis Division, DTU Management Engineering, Technical University of Denmark, Lyngby, Denmark
e-mail: aroal@dtu.dk

K. Vaillancourt
ESMIA Consultants, Blainville, QC, Canada
e-mail: kathleen@esmia.ca

C. Martin-del-Campo
Facultad de Ingeniería, UNAM, Mexico City, Mexico
e-mail: cecilia.martin.del.campo@gmail.com

© Springer International Publishing AG, part of Springer Nature 2018
G. Giannakidis et al. (eds.), *Limiting Global Warming to Well Below 2 °C: Energy System Modelling and Policy Development*, Lecture Notes in Energy 64, https://doi.org/10.1007/978-3-319-74424-7_19

1 Introduction

The United Nations Framework Convention on Climate Change (UNFCCC) 21st Conference of the Parties (COP), celebrated in Paris in 2015, resulted in a climate change agreement to achieve global emission reductions in order to keep climate change well below 2 °C above pre-industrial levels, with the aim of limiting it to as close to 1.5 °C as possible (UNFCCC 2015). The agreement also pledges a 'net-zero' emissions energy system by 2100. While the long-term aim is highly ambitious, the current short-term pledges are less demanding up to 2030. There has been a small number of analyses on the Intended Nationally Determined Contributions (INDCs), mostly conducted in the months prior to the Paris agreement, using the commitments of the major emitters and those who had already submitted their INDCs (IPCC 2014; Fawcett et al. 2015; Boyd et al. 2015; Admiraal et al. 2015; Ekholm and Lindroos 2015). These analyses suggest the INDCs effort puts the average global temperature increase on a course to somewhere between 2.7 and 3.7 °C by the end of the century. While the INDCs are a useful interim step towards decarbonisation, the international community recognises the need to increase these ambitions in order to achieve Paris long-term goals. Mexico's contribution to global GHG emissions is below 2% (Federal Government of Mexico 2015); however, Mexico is considered to be highly vulnerable to the negative impacts of climate change (Federal Government of Mexico 2013a, b). It is therefore in Mexico's interest to show leadership and commitment towards climate change mitigation targets. This is a significant challenge, since Mexico is an emerging country and reducing emissions without jeopardizing socio-economic growth and competitiveness will require international collaboration, political commitment, a broad consensus across sectors and significant capital investments in order to transform energy production and consumption; including increased electrification in end-use sectors, and for this electricity to be supplied by clean sources. All these goals also have implications on the way electricity is transmitted. According to the National Electric System Development Programme (PRODESEN) (Secretaría de Energía 2017) developed by Mexico's Ministry of Energy (SENER), in Mexico 79.7% of the electricity is generated by fossil fuel power plants, and about 54.2% of the total electricity in 2016 is based on natural gas. The projections in PRODESEN (2017) suggest that the investment required in the power sector for generation, transmission and distribution projects over the next 15 years is USD 105 billion. Most of this investment (81%) will be allocated to power generation projects, of which 37.8% will come from conventional technologies (21.6 GW) and 62.2% from clean technologies (35.5 GW). There is a global debate around the role of natural gas as a fuel bridge to a cleaner energy system. Natural gas may replace more polluting fossil fuels-based generation and provide support to an increasing share of variable renewables; however, the emission factor of any gas technology is higher than the average carbon intensity of the grid needed, for a well below 2 °C global target to be reached, unless carbon capture and storage becomes commercially available. The risk of stranded natural gas assets may increase further under deep decarbonisation policies without the commercial availability of CCS, or if energy system end-use sectors, fall short in their contribution to decarbonisation. Therefore, public decision makers and private sector investors would do well to consider the consequences of ambitious long-term decarbonisation pathways to increase the carbon risk resilience of their natural gas-based assets. Making investment decisions based on moderate INDC ambitions to 2030, that reflect an inadequate long-term ambition, could lead to poor investment choices in energy infrastructure (Pye et al. 2017). In this chapter, we explore the near-term implications of increasing climate policy ambitions in 2050 to better understand the extent to which power sector investments may lead to technological lock-in, or remain consistent with long-term stringent climate targets. This study will consider the whole energy system but focus on the electricity sector implications since, after transport, this is the largest contributor to emissions in Mexico; and given its key role as enabler for the decarbonisation of end-use sectors, which are considered to be more difficult to decarbonise than the power sector.

2 Mexico's Climate Policies and Energy Reform

Mexico was the first developing country to publish a Climate Change Law with stringent mitigation targets to 2050. Mexico's GHG emissions targets are stated in the General Law on Climate Change (DOF 2012), in the National Strategy on Climate Change (Federal Government of Mexico 2013a, b) and in the Intended Nationally Determined Contributions (INDC) (Federal Government of Mexico 2015) presented to the United Nations Framework Convention on Climate Change (UNFCCC) as part of the preparations for the Paris Agreement (UNFCCC 2015). To achieve these ambitious targets, the transformation of the electricity sector and an increase in renewables in the system are crucial. The renewable energy market in Mexico is shaped by the General Climate Change Law, which published Mexico's intent to increase electricity generated from clean energy sources, including efficient co-generation and nuclear energy, to 35% by 2024 and to 50% by 2050. The country approved in 2014 an Energy Reform bill whose main drivers are to boost domestic oil and gas production and to decrease electricity production costs (Federal Government of Mexico 2013a, b). However, the energy reform was not only designed for the oil and gas sector; it was also created to liberalise the electricity generation market and open future development to private firms, thereby creating competition among energy producers. The reform package created an independent grid operator (CENACE), which controls a new, wholesale market and enables customers to purchase power directly from generators. The creation of CENACE has established an independent power producer (IPP) market for the first time in Mexico. In order to comply with national goals of sustainable development and emissions reduction outlined in Mexico's General Climate Change Law, the Mexican Government created Clean Energy Certificates (CELs). Mexico's Energy Regulatory Commission (CRE) administers the system known as DECLARACEL, which grants a CEL per each MWh of electricity produced by a generator using clean energy technologies. Large consumers of electricity, mainly industrial and commercial, also known as Qualified Consumers, are required to consume electricity generated from clean energy sources. The requirements will begin in 2018 with a five percent share gradually increasing over the next few years to 30% by 2021 and 35% by 2024. Large consumers of electricity will obtain the CELs they need to comply with this requirement from Qualified Service Suppliers and they will apply the DECLARACEL system to submit these certificates and thereby avoid sanctions. The Mexican Government has also introduced long-term auctions to provide new and existing clean electricity generation projects with a certain return on investment for 15–20 years. The main objectives of the long-term auctions are to attract investment, promote competition amongst all technologies, and ensure efficiency for the buyer. This new regulatory environment has resulted in increased investment during the second power auction held in Mexico in 2016, contracting 8.9 TWh of power, mostly wind and solar (CENACE 2016) at an average tender price of $33.5/MWh, a wind price of $32/MWh and a solar price of $27/MWh. These prices compare favourably with results from auctions which took place in other countries over the past few years, in particular on solar prices. A decline can also be observed in onshore wind auction prices, although not as steeply as for solar PV. According to the IEA (2017), over the period 2017–22 global average generation costs are expected to further decline for solar PV by 25 and 15% for onshore wind.

3 Decarbonisation Pathways to 2050: What Do Other Studies Say?

In recent years, there have been a number of studies that explored different decarbonisation pathways for Mexico using a wide range of models. Studies point to an important mitigation potential in the country, with a wide variety of pathways available for Mexico's energy system transition. Tovilla and Buira (2015) explored deep decarbonisation pathways towards 2050 based on the 2050 Energy and Emissions Calculator model for Mexico. This study explored three scenarios: central, no CCS and limited CCS, concluding that deep decarbonisation pathways are feasible with increased energy efficiency in all sectors, CCS, zero emission vehicles, energy storage technologies and smart grids. Elizondo et al. (2017) modelled a number of low-carbon scenarios using the same tool to assess current energy policy strategies. The assessment found that industrial efficiency, cities and transport have the largest GHG emissions mitigation impact. In Mexico's Climate Change Mid-Century Strategy (2016) the Government carried out a modelling exercise using the EPPA Model and Balmorel to evaluate two scenarios; one considering unconditioned NDC emissions reduction goal of 22% reductions from baseline by 2030, with the goal of 50% reduction by 2050. The second scenario explores a more ambitious policy of 36% reduction by 2030 with additional policies agreed at the regional level with USA and Canada. In both models renewable energy plays an important role as well as cogeneration and natural gas technologies. The CLIMACAP-LAMP project (Veysey et al. 2016) undertook a cross-modelling exercise which included a wide-range of modelling techniques such as general equilibrium models, energy systems models and market equilibrium models. This exercise involved three scenarios: current energy and climate policies, 50% GHG emissions abatement by 2050 and 50% abatement of fossil fuels and industry. Results from the six models involved, showed different decarbonisation pathways, but shared common decarbonised electricity supply and mitigation actions in the transport sector. Using IEA's World Energy model, Mexico's Energy Outlook (IEA 2016) also shows a strong reduction in the GHG emissions intensity of the power sector in their new policies scenario, with solar PV and wind accounting for half of capacity additions over the period to 2040. As far as we are aware, the only study that has explored increased climate ambition scenarios to 2050 beyond targets existing in current national policy using an optimisation model is the Balmorel model-based analysis carried out by Togeby and Dupont (2016). Three high CO_2 price scenarios where modelled in myopic mode, resulting in decreased emissions (20–50%) from the electricity sector at an additional cost of up to 11% in the most ambitious scenario. It is worth noting that some of the tools used in these studies may not have a detailed enough representation of Mexico's energy system due to their global or top-down nature, and some of the pathways described in the studies may not necessarily be cost-optimal, given the accounting nature of some of the tools used. However, these exercises are indicative of the size of the challenge and contribute to our understanding of the trade-offs and energy-economy-environment dynamics of different pathways. These studies are also in general agreement with the proposed short-term steps to limit warming to 1.5 °C (Climate Action Tracker 2016). In terms of the electricity sector, this study highlights the need to develop a power system consisting largely of renewables and other zero and low carbon sources, with rapid and sustained growth in the coming decade. It also suggests that no new coal-fired power plant can be built and that fossil fuels should incur externalities and these ought to be included in the cost/price of electricity. Additionally, it recommends higher electrification in the transport, industrial and residential sectors to reduce GHG emissions.

4   Methodology and Scenarios

We have used two soft-linked models in this chapter: an energy systems optimisation model (TIMES MX-Regional) and a power systems model (Balmorel-Mexico), both commissioned by SENER; the former was developed in collaboration with University College London (UCL) and the latter with the Technical University of Denmark (DTU). While the energy systems optimisation model includes a representation of the power sector, it does not have the spatial and temporal disaggregation of a power systems model which is better placed to find the optimal dispatch of variable renewable sources such as wind and solar. On the other hand, the power systems model does not have the bottom-up, technology-rich sectorial detail of an energy systems model to better understand how different decarbonisation pathways and their related uptake of low-carbon technologies may impact electricity generation. TIMES (an acronym for The Integrated MARKAL-EFOM System) is a bottom-up, least cost optimisation, techno-economic, partial equilibrium model generator for local, national or multi-regional energy systems (Loulou et al. 2005). The TIMES MX-Regional model (TMXR) (Solano-Rodriguez 2017) represents Mexico's energy system tracking energy flows, greenhouse gas emissions (carbon dioxide, methane and nitrous oxide) and related costs from resource supply through conversion and distribution to end-use demands. As a partial equilibrium energy system and technologically detailed model, it is well suited to investigate the techno-economic, trade-offs between long-term divergent decarbonisation scenarios. The model represents the energy systems of 5 Mexican regions; each region is described and modelled to include its supply sector (fuel mining, primary and secondary production, exogenous import and export), its power generation sector and its demand sectors (residential, commercial, industrial, agriculture and transport). Balmorel-Mexico is a partial equilibrium model, with perfectly competitive markets, that allows the simultaneous optimisation of investments and power dispatch, including transmission capacity, with a high detail of spatial and temporal resolution (more information on the Balmorel platform at www.balmorel.com). Technologies are modelled as unique plants, without any aggregation. The model is deterministic and assumes full foresight, and refurbishments or shutdown of existing capacity can be part of the investment options. All information is available for the optimisation in Balmorel, e.g. hourly demand, hydro inflow, wind and solar profiles. Also, the availability of power plants is known. The Balmorel-Mexico model includes 53 regions, which are interconnected through power transmission lines (Fig. 1).

The soft-link approach is summarised as follows:

(1) The total electricity demand, as well as consumption of fossil fuels and biomass for power generation are endogenously calculated by the TMXR model under different decarbonisation scenarios.

(2) The consumption of fossil fuels and biomass calculated from TMXR is set as an upper boundary limit in the Balmorel model; this consumption is consistent with the contribution of the power sector to achieve the decarbonisation target imposed on the whole energy system.

(3) The electricity demand in Balmorel is exogenously defined as the electricity demand calculated by TMXR but disaggregated geographically to the 53 regions of Balmorel, according to the allocation factors defined in PRODESEN (2017), which evolve between years.

(4) Resulting from this approach, Balmorel shows the optimal investment made in the power system portfolio subject to the same emission constraints as TMXR.

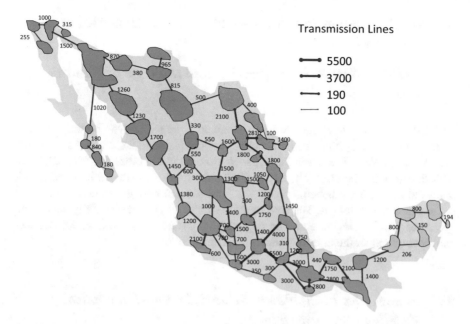

Fig. 1 Power transmission lines in Balmorel-Mexico (the colours represent the 5 regions in TMXR)

We have applied this soft-link approach to three reduction scenarios using an assessment horizon of 2014–2050. The trajectory of the key drivers of GDP, population growth, and number of households, which influence energy service demand in the economy, is the same in each scenario.

The scenarios modelled include:

- **Current Policy (CP)**: The clean energy goals as formulated in the Mexican energy and climate laws are fulfilled, including a 50% reduction in GHG emissions relative to the year 2000 by 2050. New coal-based capacity cannot be added, as this is not in line with current policy.
- **Deep Decarbonisation (DD)**: This is a hypothetical scenario with a 75% reduction in GHG emissions relative to the year 2000, used to find the implications of a 2050 emissions target consistent with a well below 2 °C scenario.
- **Net-zero (NZ)**: This scenario has the same GHG targets as the Current Policy scenario up to 2030, but increases its decarbonisation ambition towards a net-zero GHG emission energy system by 2050.

5 Decarbonisation Pathways to 2050: What Does Our Study Say?

5.1 End-Use Sectors Electrification and the Role of Natural Gas

Our TMXR results show that electricity demand is largely driven by the industrial sector towards 2050, followed by the transport and residential sectors. As expected, there is larger electrification in the NZ scenario than in the other two (Fig. 2). Even with this growth, the electricity sector GHG emissions are reduced in all scenarios thanks to a significant penetration of wind and solar in new generation capacity, at the expense of natural gas. The consumption of natural gas for electricity generation decreases under stringent emission targets (Fig. 3); over 40% less in the NZ scenario compared to the current policy scenario by 2050.

5.2 Renewables in the Power Sector: Techno-Economically Feasible and Cost-Optimal

Results from the CP scenario in Balmorel, with the restrictions given by TMXR, concerning the availability of resources, including the natural gas consumption, and the limits associated with the contribution of the electricity sector to the Mexican decarbonisation goals (Figs. 4 and 5). Investments in renewable energy such as hydropower, geothermal, wind and solar in areas with high capacity factors are

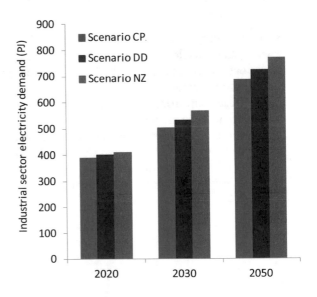

Fig. 2 Industrial sector electricity demand under scenarios CP, DD and NZ

Fig. 3 Natural gas consumption for electricity generation

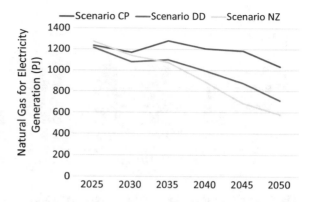

economically preferable to the use of existing gas plants. However, once the integration of renewables is high, a larger share of them in the energy matrix would require significant expansions of the infrastructure for power transmission and/or of storage technologies, such as pumped-hydro or batteries, which are more costly than the use of natural gas; therefore, the use of natural gas for electricity generation is favoured from an economic perspective, unless there are decarbonisation targets that limit its consumption.

5.3 Continued Investment in Gas Risks Stranded Assets

Stranding assets might not be a feasible or cost-effective solution for an optimal power sector transition. At present, a more secure framework for energy investments might be a preferred option, even if long-term decarbonisation targets should be considered when planning the future energy system. Therefore, in order to ensure that the existing capacity is not stranded, a constraint associated with a

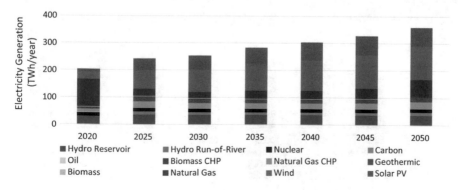

Fig. 4 Electricity generation in Mexico–CP scenario

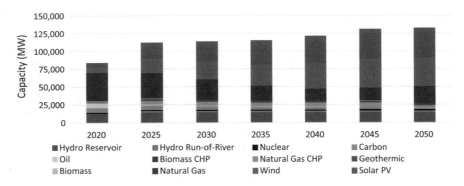

Fig. 5 Installed capacity for electricity generation in Mexico—CP scenario

minimum use of 40% of each natural gas combined cycle plant is defined (Figs. 6 and 7). Investments in renewable energy technology are postponed, compared to the CP scenario (Figs. 4 and 5), but from 2035 onward, when a large capacity of the existing combined cycle (CC) plants achieves the end of their lifetime, investments in renewable energy are similar to the ones without any limitation about the use of existing CC plants. By 2050, the integration of renewable energy and the level of GHG emissions are almost equal in both approaches; however, an early retirement of some CC plants could allow for a faster decrease in GHG emissions in a cost-efficient way, although if they remain in the system they would not hinder the accomplishment of the decarbonisation targets.

The NZ scenario, which has the most ambitious decarbonisation targets with the constraints that avoids having stranded assets for existing CC natural gas plants (minimum use of 40% of each natural gas CC plant) has a higher electricity demand, due to a larger electrification of the transport and industrial sectors, and a lower natural gas consumption allocated to the power sector (Fig. 8). In NZ scenarios the power sector is not fully decarbonised by 2050 (over 80% of electricity generation has zero emissions, with most of the remaining generation being

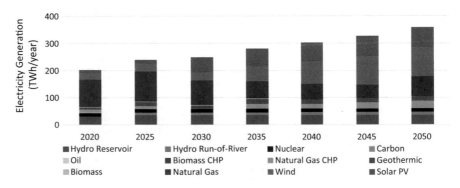

Fig. 6 Electricity generation in Mexico—CP Scenario, no stranded assets

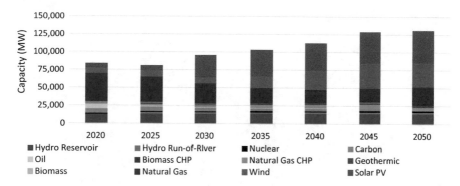

Fig. 7 Installed capacity for electricity generation in Mexico—CP scenario, no stranded assets

gas-based), since we are allowing some energy system emissions to be sequestered through removal options such as afforestation.

5.4 The Electricity System Can Operate with High Levels of Variable Renewable Energy

A large integration of renewable technologies is observed: the feasible potential defined in the scenarios for hydropower and geothermal plants is reached, and there are investments in wind and solar technologies in areas with high potential (AZEL 2017) (Figs. 9 and 10).

In the NZ scenario for 2050, there are also some investments in natural gas plants with CCS, because they play a very important role in providing flexibility and ensuring the stability of the system (Fig. 11). Unless the generation from hydropower from reservoirs or from biomass increases, storage is integrated in the

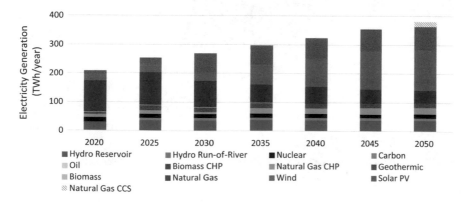

Fig. 8 Electricity generation in Mexico—NZ scenario, no stranded assets

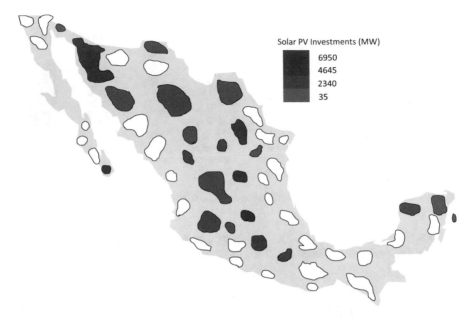

Fig. 9 Cumulative investments in solar PV plants from 2020 to 2050—NZ scenario

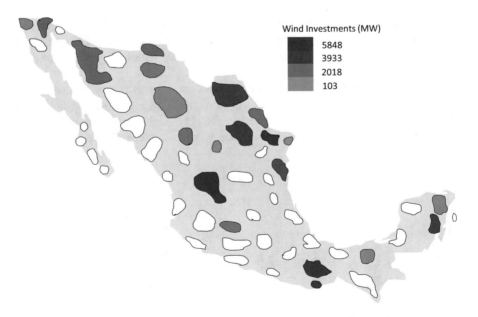

Fig. 10 Cumulative investments in wind plants from 2020 to 2050—NZ scenario

energy system, or there are some measurements that promote demand side management, it is not possible to satisfy the total energy demand without the use of natural gas in periods where the generation from variable energy sources is low and the electricity demand is high.

Mexico's power system is highly flexible due to geographical (spatial) differences, wind and solar patterns, the existence of hydropower and a strong transmission grid. Nonetheless, due to the limitations provided by the GHG emissions target, carbon capture and storage technologies are required for achieving the decarbonisation targets of the NZ scenario. Therefore, the integration of shares higher to 86% of clean energy in the system (clean energy as defined by the Mexican government in the Energy Transition Law), and of 61% of variable renewable energy for power generation (wind, solar and hydropower run-of-river) would require larger investments in storage (e.g. pumped-hydro or batteries) and a higher degree of flexible electricity demand if CCS technologies are not to be implemented in the electricity sector.

5.5 Investment Versus Fuel Costs

As would be expected, the total power generation system costs for generating electricity will increase when the GHG emissions are reduced; Balmorel-Mexico shows that both the DD and NZ scenarios are more costly than the CP scenario in terms of total costs. Compared to the CP scenario, the DD scenario is between 125–312 Million USD/year more expensive over the 2020–2050 time period

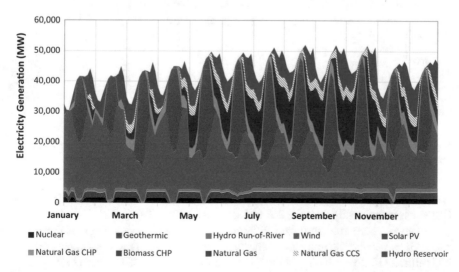

Fig. 11 Electricity generation balance—scenario NZ, 2050 (Each month is depicted selecting a representative day for graphical purposes)

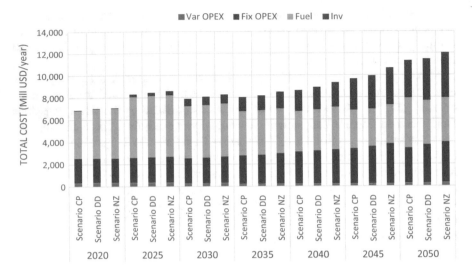

Fig. 12 Electricity sector costs by component—CP, DD and NZ "no stranded assets" scenarios

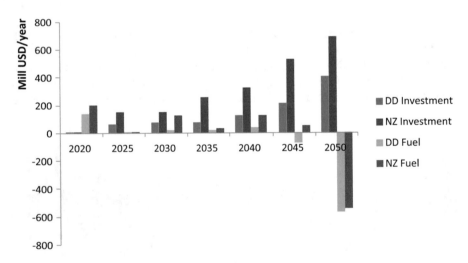

Fig. 13 Relative difference in investment and fuel costs of DD and NZ scenarios versus the CP scenario

(1.5–3.2% of total costs), while the NZ scenario is between 216 and 980 Million USD/year more costly (3.2–10.2% of total costs) (Fig. 12).

However, there are significant differences in the cost components of these scenarios, in particular the investment and fuel-related costs. Investment costs are up to 405 million USD/year higher in the DD scenario and up to 687 million USD/year in the NZ scenario, both compared to the CP scenario (12–20% higher respectively) (Fig. 13). In contrast, both CP and DD scenarios are up to half a billion USD/year more economical than the NZ scenario towards 2050, as more natural gas capacity is replaced by renewables; hence having higher fuel costs but lower investment costs.

6 Conclusion

This chapter focuses on the near-term implications of a deep decarbonisation of the energy system in the electricity sector, exploring the potential future role of natural gas in Mexico's power sector. Our modelled scenarios suggest that a higher ambition in the decarbonisation of the electricity system is possible through a significant growth post-2030 in renewable electricity generation as well as investment in hydro reservoirs, storage (e.g. pumped-hydro or batteries) or flexible electricity demand if CCS technologies are not available or adopted. The modelled scenarios considered only clean energy and GHG emission targets; further reduction of emissions could be achieved by combining these mechanisms with CO_2 prices. While the three decarbonisation scenarios modelled are feasible, there are caveats that should be taken into consideration. Firstly, the electricity demand levels used in this study are lower than those projected in PRODESEN largely due to the optimal use of energy efficient technologies in end-use sectors, not accounted for in these Government projections; higher electricity growth would likely be more costly, as it may require additional investment in CCS or for some renewable capacity to be installed in areas of lower potential. This is also true in the NZ scenario if no removal options were available to decarbonise the energy system. Secondly, if other sectors fail to decarbonise timely, the carbon budget for the electricity sector would need to be reduced accordingly. A historical analysis of Mexico's electricity system makes it clear that coal-to-gas and fuel oil-to-gas substitution has already played a major role in reducing the sector's emissions. Therefore, the potential of natural gas as a fuel bridge to further decarbonise the electricity system is limited. Investors in new gas-based power plants will expect to maximise the use of their assets throughout their lifetime. Hence, it could be argued that the "no stranded assets" scenarios represent a more realistic variant of the modelled decarbonisation targets; otherwise special payments would need to be made by regulators in order for gas-based capacity to be used solely as back up to renewables, or to run at low load factors post-2030 unless they are retrofitted with CCS. This would likely discourage private investment in new gas-based power plants. Our "no stranded assets" scenarios show a delay in significant renewables growth until post-2030, while our "stranded assets" scenarios consider it more cost-effective to invest in renewables 5–10 years earlier. Our study suggests that short-term gains through low-priced natural gas generation may delay renewables penetration into the system and result in a more costly electricity system, as global renewables growth trends are expected to continue decreasing their capital cost. Our results show that investments in renewable electricity generation in areas with high capacity factors are economically preferable than the use of existing gas plants.

They also suggest that the decarbonisation path of the electricity system post-2030 is largely dependent on the investment decisions made in the 2020s; it is therefore essential that Mexico's energy planning decision-makers avoid a natural gas "lock-in" that would either cause carbon targets to be missed or risk leaving some natural gas infrastructure stranded.

References

Admiraal A, den Elzen M, Forsell N, Turkovska O, Roelfsema M, van Soest H (2015) Assessing intended nationally determined contributions to the Paris climate agreement—what are the projected global and national emission levels for 2025–2030? PBL Netherlands Environmental Assessment Agency, The Hague. PBL publication number: 1879

Atlas de Zonas con Energias Limpias (AZEL). https://dgel.energia.gob.mx/AZEL/. Accessed 15 Sept 2017

Boyd R, Turner J, Ward B (2015) Intended nationally determined contributions: what are the implications for greenhouse gas emissions in 2030? Grantham Research Institute on Climate Change and the Environment, policy paper, Oct 2015

CENACE (Centro Nacional de Control de Energía) (2016) Resultados de la Segunda Subasta de Largo Plazo. https://www.gob.mx/cenace/prensa/con-precios-altamente-competitivos-se-anuncian-los-resultados-preliminares-de-subasta-de-largo-plazo-2016. Accessed 18 Sept 2017

Climate Action Tracker (2016) The ten most important short-term steps to limit warming to 1.5 ° C. NewClimate Institute, Ecofys and Climate Analytics

Diario Federal de la Federación (DOF) (2012) Ley General de Cambio Climático. http://www.diputados.gob.mx/LeyesBiblio/pdf/LGCC_010616.pdf. Accessed 2 Oct 2017

Ekholm T, Lindroos TJ (2015) An Analysis of Countries' Climate Change Mitigation Contributions Towards the Paris Agreement, VTT Technical Research Centre of Finland

Elizondo A, Pérez-Cirera V, Strapasson A, Fernández JC, Cruz-Cano D (2017) Mexico's low carbon futures: an integrated assessment for energy planning and climate change mitigation by 2050. Futures 93:14–26

Fawcett AA, Iyer G, Clarke L, Edmonds J, Hultman N, McJeon H, Rogelj J, Schuler R, Alsalam J, Asrar G, Creason J, Jeong M, McFarland J, Mundra A, Shi W (2015) Can Paris pledges avert severe climate change? Science 350(6265):1168–1169

Federal Government of Mexico (2013a) Reforma Energética. http://cdn.reformaenergetica.gob.mx/explicacion.pdf. Accessed 4 Oct 2017

Federal Government of Mexico (2013b) National Climate Change Strategy (NCCS). 10-20-40 Vision Mexico: Federal Government of Mexico

Federal Government of Mexico (2015) Intended nationally determined contributions. Available at http://www4.unfccc.int/Submissions/INDC. Accessed 4 Oct 2017

International Energy Agency (2016) Mexico's energy outlook. OECD/IEA, Paris, France

International Energy Agency (2017) Renewables 2017. OECD/IEA, Paris, France

IPCC (2014) Climate Change 2014: Synthesis report. Contribution of working groups I, II and III to the fifth assessment report of the Intergovernmental Panel on Climate Change. In: Core Writing Team, Pachauri RK, Meyer LA (eds). IPCC, Geneva, Switzerland

Loulou R, Remne U, Kanudia A, Lehtila A, Goldstein G (2005) Documentation for the TIMES model—part I. Energy Technology Systems Analysis Programme

Pye S, Li FGN, Price J, Fais B (2017) Achieving net-zero emissions through the reframing of UK national targets in the post-Paris agreement era. Nat Energy 2:17024

Secretaría de Energía (2016) Sistema de Información Energética (SIE). http://sie.energia.gob.mx/bdiController.do?action=temas. Accessed 3 Oct 2017

Secretaría de Energía (2017) Programa de Desarrollo del Sistema Eléctrico Nacional (PRODESEN)

SEMARNAT-INECC (2016) Mexico's climate change mid-century strategy. Mexico City, Mexico: Ministry of Environment and Natural Resources (SEMARNAT) and National Institute of Ecology and Climate Change (INECC)

Solano-Rodriguez B (2017) Documentación del modelo TIMES MX-Regional. University College London, London, UK

Togeby M, Dupont N (2016) Renewable energy scenarios for Mexico. Ea Energy Analyses

Tovilla J, Buira D (2015) Pathways to deep decarbonization in Mexico, SDSN—IDDRI (Sustainable Development Solutions Network and Institute for Sustainable Development and International Relations)

United Nations Framework Convention on Climate Change (UNFCCC) (2015) Paris Agreement

Veysey J, Octaviano C, Calvin K, Martinez SH, Kitous A, McFarland J, van der Zwaan B (2016) Pathways to Mexico's climate change mitigation targets: a multi-model analysis. Energy Econ 56:587–599

Mitigation Challenges for China's End-Use Sectors Under a Global Below Two Degree Target

Wenying Chen, Huan Wang and Jingcheng Shi

Key messages

- Emission reduction beyond the Nationally Determined Contribution of China is feasible but challenging, especially in the end-use sectors.
- Electrification drives the CO_2 mitigation approaches for end-use sectors, resulting in deep changes in power source structure and supply capacity.
- Energy efficiency standards are essential to promote the transition of end-use sectors to low-carbon sectors. They need to be complemented by market tools such as carbon trading to motivate more initiatives in mitigation actions and innovations.
- The global model, GTIMES, is used to generate cost-optimal CO_2 mitigation pathways at global level and describe energy transition details for China.

1 Introduction

To meet the carbon mitigation requirements brought by the 2° target, every single country on this planet will have to participate actively in controlling their national emissions. It is of great importance to explore each regions' mitigation task under the global common mitigation goal as this work could help regions to propose energy transition roadmaps based on their own unique challenges.

W. Chen (✉) · H. Wang · J. Shi
Institute of Energy, Environment and Economy, Tsinghua University, Beijing, China
e-mail: chenwy@mail.tsinghua.edu.cn

H. Wang
e-mail: huan-wan15@mails.tsinghua.edu.cn

J. Shi
e-mail: sjc11@mails.tsinghua.edu.cn

© Springer International Publishing AG, part of Springer Nature 2018
G. Giannakidis et al. (eds.), *Limiting Global Warming to Well Below 2 °C: Energy System Modelling and Policy Development*, Lecture Notes in Energy 64, https://doi.org/10.1007/978-3-319-74424-7_20

As the biggest developing country, and also the biggest CO_2 emitter, China will face a severe challenge to balance economic progress and controlling CO_2 emissions. On one hand, China has just reached a GDP per capita of 8000 USD in 2015 (World Bank Group 2017), which is still far below the average level of developed countries, and a large population are still living a low standard quality of life, especially in rural areas. Both economic progress and people's life improvement may bring growing energy demand. On the other hand, China has a coal-dominated energy structure which was determined by its rich coal, deficient oil and lean gas resource endowment. The transition to a low carbon energy system is not easy— large-scale investments into low-carbon technologies would be needed, and scientific and rational overall planning involving both technology and finance should be conducted as soon as possible.

In its Nationally Determined Contribution (NDC) to the Paris Agreement on Climate Change, China has committed to peaking CO_2 emissions by 2030 at the latest, reducing the carbon intensity of GDP by 60–65% below 2005 levels by 2030, and increasing the share of non-fossil energy carriers of the total primary energy supply to around 20% by the same year, alongside other measures. Reflecting on these commitments, the 13th Five Year Plan of China integrates several policies covering energy conservation regulations and the planning on renewable energy development, several instruments including mandatory standards and targeted subsidies. However, Chinese NDC commitments, as in the case of many countries, will need to be further updated in order to be more consistent with a well below 2° limit of the global temperature increase.

Many studies have been conducted to analyze China's CO_2 mitigation, most of them only involving China's own independent emission reduction target. Liu (2015) proposed four steps to promote emissions reduction and stated that clear regional targets and mature carbon market will help significantly in peaking CO_2 emissions by 2030. Li and Wei (2015) conducted decomposition analysis on China's CO_2 emissions and found that CO_2 intensity may play an important role in CO_2 emissions reduction, and absolute reduction is hard to achieve in short term. Chen et al. (2016) compared the China TIMES model results with global models and pointed out substantial efforts in economic development mode transformation, advanced the promotion of technologies, and international cooperation reinforcement are all essential for low-carbon transition.

Socio-economic development can create significant differences in future energy systems. This research took the Shared Socio-economic Pathways (SSPs) structure as a reference and conducted the analysis under the SSP2 trajectory. SSPs were proposed by the Intergovernmental Panel on Climate Change (IPCC) in 2010 (Kriegler et al. 2012; van Vuuren et al. 2014), and several studies have been conducted to get a universal qualitative description of main structure and basic pathways (Bauer et al. 2017; Fricko et al. 2017; Riahi et al. 2017). The SSP approach proposes five basic pathways, each standing for a unique development pattern: SSP1 stands for sustainability where protection of environment and resources would be further emphasized, SSP2 stands for middle of the road which would basically follow the current trend, SSP3 stands for regional rivalry where potential regional conflicts may force the government to concern more about energy and food security, SSP4 stands for inequity where gaps between developed and developing countries would further widen, SSP5 stands for fossil-fueled development which would maintain high dependency on fossil fuels. This chapter will focus on middle of the road SSP2, in which the current trend will be roughly continued, with a slow transition to a sustainable development pattern.

This chapter conducted energy system optimization with a 14-region global TIMES model with the assumption that future socio-economic development would follow the middle of the road trajectory SSP2. China's unique mitigation challenges under global climate targets were analyzed under two emission pathways corresponding to 2.3 and 1.6 °C respectively. Although supply sectors would also make great contributions for China's mitigation, this chapter focuses on the energy transition roadmaps for end-use sectors, thus the detailed analysis will focus on the building, transportation and industry sectors. This chapter is organized as follows: a brief introduction on the methodology and key assumptions is explained in Sect. 2, the main model results and analysis are presented in Sect. 3, and the conclusion and discussions are proposed in Sect. 4.

2 Methodology

2.1 GTIMES Model

The GTIMES model was constructed by Tsinghua University based on the development of China MARKAL and China TIMES model (Chen 2005; Chen et al. 2007, 2010, 2014, 2016; Yin and Chen 2013; Shi et al. 2016; Zhang et al. 2016). It is a bottom-up model with both supply sectors and demand sectors considered: upstream, power and heat, agriculture, building, industry and transportation. Each sector contains detailed energy technologies, making it possible to describe the whole process of energy system including extraction, conversion, transmission and distribution and final use. The current version of the model only considers the CO_2 emissions emitted by fossil fuel combustion, while non-CO_2 gases and industrial process CO_2 has not been covered yet.

This model considers the time period 2010–2050 with a 5-year time step. It divides the world into 14 regions covering most developed countries, major middle-income regions and also developing countries, based on the investigation of each regions' socio-economic situation and energy supply-demand features. Each region has a detailed description on current and future energy system, and specific and reasonable adjustment may be applied according to its unique regional characteristics.

2.2 Projection of Energy Service Demand: Following a "Middle-of-the-Road" Trend

TIMES is a demand-driven modeling framework, to optimize the future energy system needs under the projection of future energy service demands (ESD). We conducted each demand sector's ESD projection separately on the basis of SSP2 trajectory modelling. The basic method (Eq. 1) is the elastic demand forecasting method usually applied in TIMES models (Loulou and Labriet 2008; Anandarajah et al. 2011).

$$Demand_t = Demand_{t-1} \times k \times driver^{elasticity} \tag{1}$$

In this equation, the driver and elasticity are two key parameters which need to be determined by the modelers. $Demand_t$ and $Demand_{t-1}$ represents the energy service demand of period t and period t − 1 respectively. k is modifying factor which usually equals to 1. The drivers are usually macro-index reflecting the main trend of societal development, which will directly trigger the changes in ESD. Typical drivers such as GDP, GDP per capita, industrial added value are used. The elasticity reflects the impact on energy service demand when driver has a unit of changes.

Each end-use sector has its unique means of meeting energy service demand and may be influenced by several different factors. For the building sector, we considered heating, cooling, lighting, cook and other appliances as the main service and chose drivers such as GDP per capita, households, population for residential building and added value of tertiary industry for commercial building. For the transportation sector, we considered several traffic modes, such as airline, shipping, road and railway and chose GDP per capita and population for passenger transportation and GDP for freight transportation. For industry, we conducted the projection considering fuel demand during production and also the feedstock demand for major industries, includes chemistry, iron, paper, etc. The drivers of industrial ESD includes industrial value-added and GDP per capita, etc.

We took the IIASA-SSP database as reference and quantified the future trends of core socio-economic indicators including population, GDP and urbanization in the GTIMES model. Several existing studies and databases provided valuable data and methodology for this work (KC and Lutz 2014; Jiang and O Neill 2015; Leimbach et al. 2015). Both global population and GDP of SSP2 will basically continue current trend in the next decades. For China, the population will roughly be kept constant and reaches 1.45 and 1.42 billion in 2030 and 2050 respectively; GDP will continue its rapid growth, and reaches 34,000 and 52,000 billion dollars in 2030 and 2050 respectively.

2.3 Scenarios and Key Assumptions

This research has designed 3 scenarios, one reference scenario SSP2 and two mitigation scenarios based on representative concentration pathways, SSP2RCP4.5 and SSP2RCP2.6 scenarios (Table 1). Although society is expected to have a slow transition toward sustainable development in SSP2 trajectory, these changes are not efficient enough to realize a 2° target. In the reference scenario SSP2, CO_2 emissions of both the world and China would continue to increase in coming decades while the emission gaps between SSP2 and mitigation scenarios are found to be large, and would get larger as time goes by, thus specific and powerful mitigation measures need to be implemented as soon as possible. The average temperature rise of RCP4.5 pathway and RCP2.6 pathway is 2.3 and 1.6 °C respectively compared with pre-industrial levels, and the likelihood to realize a 2° target of these two pathways is less than 50% and more than 66%, respectively (IPCC 2015). Global carbon budgets over 2010–2050 provided by the IPCC (2015) are used as a cumulative upper limit in the model (Table 1).

The scenario runs were conducted under the assumption of a full cooperation between all regions in order to explore the cost-optimal mitigation pathways. This is equivalent to a uniform carbon price is introduced for all regions. This approach ignores the carbon allocation principles such as equity and ability and only focuses on the efficiency dimension to study each region's mitigation potential. Thus, it is worth noting that the developing countries' mitigation burden generated by the model may differ from the expectations of other allocation plans considering equity principles.

3 Results

3.1 Overall Emission Reductions

To meet the CO_2 mitigation requirements, China would need a deep energy transition which should cover both supply-side and demand-side. Currently, a few mitigation measures have been conducted in China's power sector such as the

Table 1 Scenario descriptions

Name	Description
SSP2	Reference scenario without carbon constraint
SSP2RCP4.5	Reference scenario + cumulative carbon constraint of RCP4.5 pathway (1472 billion tons CO_2 over 2010–2050)
SSP2RCP2.6	Reference scenario + cumulative carbon constraint of RCP2.6 pathway (1095 billion tons CO_2 over 2010–2050)

development plan for renewable power and the nationwide carbon market for power industry. Furthermore the establishment of low-carbon power system has already begun. While the decarbonization of end-use sectors are also essential to achieve stringent climate goals, since the targeted strategy and planning for these sectors have not been published yet, this chapter conducts detailed analysis on the building, transportation and industry sectors to analyse their challenges in achieving mitigation goals.

3.2 Building Sector

Energy service demand (ESD) in the building sector will continue to rapidly increase in the medium-and long-term, even with the stringent CO_2 emission constraints. China is a developing country with a huge population, the urbanization process has not been completed for now, and large rural population are still living a relatively low standard of life. Besides, tertiary industry has just entered a high-speed development period—more commercial buildings will be constructed in the next decade and it is predictable that the various forms of energy service demand in both residential building and commercial building will keep increasing rapidly. Furthermore, the total ESD of China's building sector may be over 50,000 PJ by 2050, over 4 times the 2010 level. Under the carbon constraint of RCP4.5 and RCP2.6 pathways, the growth in building sector's energy service demand will slow down due to the increased price brought by additional mitigation cost. The total ESD of the building sector will reach 48,000 PJ and 37,000 PJ by 2050 in the RCP4.5 and RCP2.6 pathways respectively.

The building sector is not the biggest contributor in national emission reduction, mainly because emissions from the building sector only accounts for a small share in China's total emissions. The building sector will contribute to around 7–8% of the total cumulative emission reduction in SSP2RCP4.5 and SSP2RCP2.6 scenario compared to the reference scenario between 2010 and 2050. Interestingly, the relative contribution is the same, although the absolute reductions are of course far bigger in SSP2RCP2.6.

For all three scenarios, energy structure will be greatly changed in the building sector compared to 2010. Up until 2010, China still had a traditional biomass dominated energy structure for the rural area, with coal acting as the second biggest energy source. The combustion of both biomass and coal will generate air pollutants and bring harm to human health and also bring energy waste for their low efficiency. A more efficient, clean and diverse structure will be established gradually in the three scenarios, including the Reference case: a higher level of electrification will be achieved, more clean energy will be consumed, and traditional biomass will be phased out step by step, but to a bigger magnitude under the severe climate target. For example, in 2050, the electricity demand in the building sector will reach 14,000–17,000PJ in SSP2, SSP2RCP4.5 and SSP2RCP2.6 scenarios, which grows by over 6 times relative to the 2010 level, electricity may provide from 40 to 63% of the total energy demand in China's building sector (excluding self-production), depending on the scenario.

Renewable energy needs to be promoted at a large scale to meet the requirements of RCP2.6 pathway. In SSP2RCP2.6 scenario, coal and oil will be eliminated from China's building sector after 2040, much more solar and geothermal will be used to satisfy the growing energy service demand (Fig. 1). The promotion of roof photovoltaics has been included in the 13th Five Year Plan of solar energy development, while obstacles such as large initial investment and difficulties in grid connection impede residents' participation. Such an electricity market, where households would participate in the electricity market and make economic profits through electricity trading with the electric price differences between peak and off-peak hours utilized alongside an advanced energy management system would provide a solution. Selling extra electricity to other consumers could then become possible and provide a greater incentive for roof photovoltaics.

Another required transition in building sector is the improvement of energy efficiency. Model result shows that 24 and 40% of final energy reduction could be realized in SSP2RCP4.5 and SSP2RCP2.6 scenarios by 2050, compared with SSP2 scenario. Energy efficiency standards apply to buildings envelop and appliances in China, but they have not been sufficient. Uniform certifications for low carbon building and related financial support such as tax reduction would prove to be more efficient.

3.3 Transportation Sector

The growth of China's transportation sector is expected to be extremely high, driven by the continuous and sustainable economic development. Passenger travel demand will expand as per capita GDP grows, the requirement of comfort level may be higher and the demand of advanced means such as airlines would increase rapidly. Larger traffic volume would also occur in the freight sector for the growing

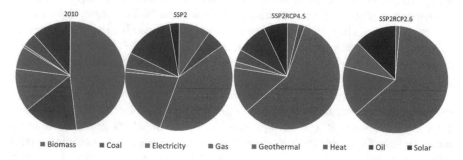

Fig. 1 Final energy structure in the building sector (shares in 2010 and 2050)

and high active economy. However, the growth trend may change in 2030 if a stringent carbon constraint is introduced. Indeed, energy service demand growth would slow down in mitigation scenarios, and even cease to grow from 2030 in SSP2RCP2.6 scenario, under the effect of the mitigation costs.

The contribution of the transportation sector to mitigation slightly increases when the carbon budget reduces. In SSP2RCP4.5 scenario, the transportation sector contributes nearly 7% of the total cumulative CO_2 emission reduction, and this share can be raised to around 10% if the RCP2.6 pathway is introduced (note that total absolute emission reductions required in RCP2.6 are also much higher). However, in both the SSP2 and SSP2RCP4.5 scenarios, the CO_2 emissions of the transportation sector will continue its increasing trend and will not reach its peak before 2050. The differences between these two scenarios is relatively small, which indicates the mitigation burden of this sector is relatively small in RCP4.5 pathway. Under the RCP2.6 scenario, a greater level of effort is realized. A further cumulative emission reduction of 9 Gt CO_2 may be required in 2010–2050, compared with SSP2RCP4.5 scenario, and the emission increase after 2030 would be rather limited.

Final energy demand in the transportation sector will increase rapidly in the reference scenario SSP2, and reaches 4.2 times the 2010 level in 2050, which is the direct result of the big increase in traffic turnover (Fig. 2).

Oil will still be the most important fuel in China's transportation sector until mid-century in all scenarios, and so to guarantee its long-term supply would be an important task. In the SSP2 scenario, oil consumption would increase rapidly. Even with powerful climate and energy policies, this indicator would still be around 500 and 400 Mt in mitigation scenarios, while China's annual oil production is expected to keep constant at 200 Mt which means transportation alone would bring at least 200 Mt of oil import demand. Based on this demand projection, a long-term strategic plan on international oil trade would be essential. Current adequate supply capacity in major oil producing countries may be a valuable opportunity for China.

Electricity and biofuels could act as substitutes for oil and gas when CO_2 emission space is limited. Targeted incentive measures and R&D support may be needed in an initial period. When oil and gas consumption decrease in mitigation scenarios, especially in the stringent scenario SSP2RCP2.6, the demand of electricity and biofuels would increase to provide a low-carbon energy source for transportation (Fig. 2). However, electric vehicles and biofuels have their unique shortages such as short running range and high cost. Incentive measures, such as subsidies, are very helpful to raise their market shares, and R&D support would be of great importance in improving battery capacity and biofuel production processes.

In other words, the stringent 2° target may be very challenging for China in the transportation sector: compared with SSP2RCP4.5 scenario, the final energy reduction of 25% may be required in SSP2RCP2.6 scenario by 2050. Improvement on energy efficiency is not enough, specific measures which can reduce transportation sector's energy service demand are essential—more transportation modes with lower energy intensity might be helpful to meet the huge travel demand, sustained efforts should be made to promote public transport and light vehicles.

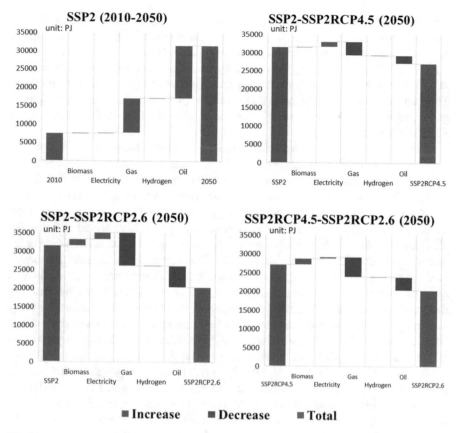

Fig. 2 Final energy changes in transportation sector

Shared bicycles are becoming increasingly popular in China's major cities, the widely promotion of new transportation mode such as subway supplemented by shared bicycles may play a critical role in balancing the growing travel demand and stringent carbon constraint.

3.4 Industry Sector

The energy demand service of China's industry may reach a saturation in near future, and mitigation goals may even bring 10–20% of demand decline in 2050. Indeed, China has experienced an explosion growth period of industry in the last 30 years, and it is expected that energy-intensive industries may reach saturation in the coming years with newly developed industries which have lower energy-intensity gradually becoming the major driving force of the industrial value-added growth. This trend is reflected in the growth rate of industrial ESD, which slows down significantly in SSP2 scenario after 2020, and may keep constant and even begin to decrease in SSP2RCP4.5 and SSP2RCP2.6 scenarios, with the reactions to high mitigation cost such as changes in behavior (Fig. 3).

As a result, the increasing trend in industrial CO_2 emissions slows down from 2015 and maintains a weak growth after that, even without the mitigation pressure. The industry sector may be one of the most important contributors in CO_2 mitigation, and accounts for 17 to 24% of the total mitigation in SSP2RCP4.5 and SSP2RCP2.6 scenarios.

The final energy demand in SSP2RCP4.5 scenario shows a small decrease compared with SSP2 scenario (around 10%), while the energy structure will not have significant change mainly caused by the reduction of coal consumption. Fossil energy will account for 65% of the total final energy supply in industry.

While to further pursue the mitigation goal to the RCP2.6 pathway, significant changes both in gross energy consumption and energy structure would be necessary alongside a rapid reduction in fossil fuel. Differing from the slight decrease in energy demand in SSP2RCP4.5 scenarios, SSP2RCP2.6 scenario would reach the energy demand peak in 2020 and then decrease rapidly, which may need nationwide total quantity control targets and regional specific policies in the near future. Besides this, the fossil fuel dominated structure is expected to change with non-fossil energy providing over 60% of the final energy by mid-century. The coal consumption will begin to decrease from now on and realize 67% of reduction by 2050, compared with 2010s coal consumption. Oil consumption would also be reduced from 2020 and achieve over 50% of reduction in 2050. Meanwhile, electricity demand may double from 2010 to 2050, to provide stable and inexpensive electricity for industrial enterprises. Service contract for direct purchase would be an efficient solution (long-term contract between electricity generation enterprises and large electricity consumers). Distributed photovoltaics could also be applied to provide certain amount of the electricity demand (not yet considered in our model).

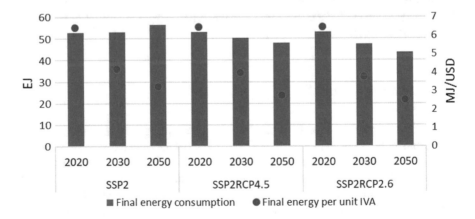

Fig. 3 Final energy and energy intensity in the industry sector

Overall energy efficiency can be improved greatly in industry, mainly because of the large-scale application of high-efficient equipment. Since the SSP2 trajectory will slowly change to a sustainable development, significant energy intensity decrease will occur from now on and the energy consumption per unit of industrial added value will reduce by 60 and 69% in 2030 and 2050 compared with 2010 respectively. With a mitigation goal, another 5 and 8 points of reduction in the energy intensity of the industry sector will be realized in 2050, driven by the penetration of more new and advanced technologies.

For energy-intensive industries such as steel and cement, the multi-stage utilization of energy needs to be realized through integrated design and energy management (e.g., in steel mills waste boilers could be used to generate electricity, and the waste heat from this boiler may still be useful to provide heating for residents). The reduction target for energy intensity of steel and cement have been included in 13th Five Year Plan; given the large-amount of heat demand in these industries, one efficient option is to make full use of the waste heat, which is capable to generate electricity using heat recovery boiler. Currently, large numbers of factories have already accepted the heat recovery idea and specific design for stage utilization should be conducted to make the most of this waste heat.

The improvement of product quality and the recycling of these products could be efficient in reducing industrial energy demand, and this method could also help to decrease the CO_2 emitted from the industrial production process. Take plastic for instance, its one-time use is common in China. If efficient recovery could be achieved, the total demand would be greatly reduced which could make contributions to CO_2 mitigation and environment protection at the same time.

4 Conclusion

The global climate governance follows the principle of common but differentiated responsibilities, which allows certain increases in emission space for developing countries. In the analysis presented in this chapter, we temporarily ignored the allocation principles involving equity and capability, and only focused on the global optimal pathway. Thus, the model results provide insights on potential challenges in mitigation strategies and do not consider the technical and financial ability of countries to implement the strategies.

The improvement on overall energy efficiency could play a critical role in meeting end-use sectors' mitigation challenges in China, and significant energy conservation is observed in the three end-use sectors in the mitigation scenarios. Barriers must be overcome to achieve this goal. Limited by the current technical level, the improvement of energy efficiency would be expensive for energy consumers and may get great dependency on policy instruments such as subsidies, which can bring large financial pressure for the government. The future domestic development progress for some key technologies may be not rapid enough to meet the requirements of stringent carbon constraint. To promote carbon mitigation in

China's end-use sectors, not only targeted policies are required, but international cooperation on technology may also be essential.

Even with rapid improvements in efficiency, deep structural transition in final energy will still be essential to meet the mitigation goal. Currently, energy sources in all three end-use sectors are relatively homogenous and mainly rely on traditional fuel as including coal and oil, and a more diversified and low-carbon final energy structure will be formed gradually. Higher shares of electricity and clean energy including natural gas will need to be achieved in next decades for all sectors, and new and advanced technology like solar photovoltaics and geothermal will also be applied in some final sectors, and this process may be accelerated under stringent carbon constraint.

To motivate mitigation actions in end-use sectors, market tools should also be introduced in addition to other policy instruments. Currently, the financial value of CO_2 emission reductions has not been fully reflected in China and most mitigation actions are conducted to meet the policy regulations or to save fuel cost, while the attentions on CO_2 mitigation are relatively lacking. Under the requirements of below 2° target, mandatory policy instruments such as efficiency standards may not be powerful enough to meet the tight carbon constraint. More incentives on mitigation measures should be provided through the internalization of environmental externality, energy consumers would pursue more efforts on carbon mitigation to reduce their cost or make financial profits and related innovations on both technology and production modes may also be motivated.

With a higher electrification level in end-use sectors, China's power sector may play an even more critical role in climate mitigation. Under the global below 2° target, electricity would gradually become a major energy source in all end-use sectors, and to meet the growing electricity demand could be challenging for the power sector. Meanwhile, the improvement on the structure of power sources would be increasingly important as more renewables are needed to be introduced into the power sector to change the coal-dominated structure. To guarantee a stable supply with increasing electricity demand and high proportion of renewables can bring new challenges for China, distributed power system may be widely used in future.

Considering the uncertainty in global warming, this chapter has assessed the consequences of two global climate mitigation pathways on the end-use sectors of China. The modelling results show that to hold the likelihood of realizing 2° target to go beyond 66% can be very challenging for China. Under the RCP4.5 pathway, the fuel substitution and efficiency improvements are effective measures, and the reduction of fossil fuels can be partly supplemented by electricity. While under the stringent carbon constraint in RCP2.6 pathway, these measures may not be effective enough to meet the mitigation requirements, and the reduction in energy service demand would become necessary, which may lead to significant welfare loss in China's end-use sectors. These reductions in energy service demand, their impacts on society change and potential solutions to ease these influences would deserve further and detailed exploration.

Acknowledgements This research is supported by the National Natural Science Foundation of China (NSFC 71690243) and Ministry of Science and Technology (2012BAC20B01).

References

Bauer N, Calvin K et al (2017) Shared socio-economic pathways of the energy sector—quantifying the narratives. Glob Environ Change 42:316–330

Chen W (2005) The costs of mitigating carbon emissions in China: findings from China MARKAL-MACRO modeling. Energy Policy 33(7):885–896

Chen W, Li H, Wu Z (2010) Western China energy development and west to east energy transfer: Application of the Western China sustainable energy development model. Energy Policy 38 (11):7106–7120

Chen W, Yin X, Ma D (2014) A bottom-up analysis of China's iron and steel industrial energy consumption and CO2 emissions. Appl Energy 136:1174–1183

Chen W, Wu Z, He J, Gao P, Xu S (2007) Carbon emission control strategies for China: a comparative study with partial and general equilibrium versions of the China MARKAL model. Energy 32(1):59–72

Chen W, Yin X, Zhang H (2016) Towards low carbon development in China: a comparison of national and global models. Clim Change 136(1):95–108

Fricko O, Havlik P et al (2017) The marker quantification of the Shared socioeconomic pathway 2: a middle-of-the-road scenario for the 21st century. Glob Environ Change 42:251–267

Anandarajah G, Pye S, Usher W, Kesicki F, Mcglade C (2011) TIAM-UCL global model documentation. University College, London

IPCC (2015) Climate Change 2014 mitigation of climate change

Jiang L, Neill BCO (2015) Global urbanization projections for the shared socioeconomic pathways. Glob Environ Change 42:193–199

Kc S, Lutz W (2014) The human core of the shared socioeconomic pathways: Population scenarios by age, sex and level of education for all countries to 2100. Glob Environ Change 42:181–192

Kriegler E, Neill BCO et al (2012) The need for and use of socio-economic scenarios for climate change analysis: a new approach based on shared socio-economic pathways. Glob Environ Change 22(4):807–822

Leimbach M, Kriegler E, Roming N, Schwanitz J (2015) Future growth patterns of world regions —a GDP scenario approach. Glob Environ Change 42:215–225

Li H, Wei Y (2015) Is it possible for China to reduce its total CO_2 emissions? Energy 83:438–446

Liu Z (2015) Steps to China's carbon peak. Nature 522(7556):279

Loulou R, Labriet M (2008) ETSAP-TIAM: the TIMES integrated assessment model Part I: model structure. CMS 5(1–2):7–40

Riahi K, van Vuuren DP et al (2017) The shared socioeconomic pathways and their energy, land use, and greenhouse gas emissions implications: an overview. Glob Environ Change 42:153–168

Shi J, Chen W, Yin X (2016) Modelling building's decarbonization with application of China TIMES model. Appl Energy 162:1303–1312

van Vuuren DP, Kriegler E et al (2014) A new scenario framework for Climate Change Research: scenario matrix architecture. Clim Change 122(3):373–386

World Bank Group (2017) The World bank data

Yin X, Chen W (2013) Trends and development of steel demand in China: a bottom–up analysis. Resources Policy 38(4):407–415

Zhang H, Chen W, Huang W (2016) TIMES modelling of transport sector in China and USA: comparisons from a decarbonization perspective. Appl Energy 162:1505–1514

The Importance of the Water-Energy Nexus for Emerging Countries When Moving Towards Below 2 °C

Gary Goldstein, Pascal Delaquil, Fadiel Ahjum, Bruno Merven,
Adrian Stone, James Cullis, Wenying Chen, Nan Li, Yongnan Zhu,
Yizi Shang, Diego Rodriguez, Morgan Bazilian, Anna Delgado-Martin
and Fernando Miralles-Wilhelm

Key messages

- Representing water in energy system planning models is essential to break thru the institutional planning silos;
- Aligning water basins and energy supply regions and representing water infrastructure supply investments rather than supply cost curves adds significant value to energy planning;

G. Goldstein (✉)
DecisionWare Group LLC, Sag Harbor, NY, USA
e-mail: DecisionWare.NY@Gmail.com

P. Delaquil
DecisionWare Group LLC, Gresham, OR, USA
e-mail: PDeLaquil@DecisionWareGroup.com

F. Ahjum · B. Merven
University of Cape Town, Rondebosch, South Africa
e-mail: mf.ahjum@uct.ac.za

B. Merven
e-mail: bruno.merven@gmail.com

A. Stone
Sustainable Energy Africa, Cape Town, South Africa
e-mail: adrian@sustainable.org.za

J. Cullis
Aurecon, Cape Town, South Africa
e-mail: James.Cullis@aurecongroup.com

W. Chen · N. Li
Institute of Energy, Environment and Economy, Tsinghua University,
Beijing, China
e-mail: chenwy@mail.tsinghua.edu.cn

© Springer International Publishing AG, part of Springer Nature 2018
G. Giannakidis et al. (eds.), *Limiting Global Warming to Well Below 2 °C: Energy
System Modelling and Policy Development*, Lecture Notes in Energy 64,
https://doi.org/10.1007/978-3-319-74424-7_21

- Examining the impact of water availability uncertainty under Climate Change needs to be part of determining viable NDC pathways and moving towards below 2 °C emission reductions.
- TIMES offers a relevant robust modeling framework for analysis of the water-energy nexus.

1 Why Examine the Water-Energy Nexus?

Water and energy are often entwined in the sense that the use of one depends on the availability of the other. Decision making in one sector significantly affects the other, but those effects often are not taken into account in traditional sector-based planning processes. The sustainable supply of services from these two interdependent resources constitutes a set of integrated challenges commonly referred to as the water-energy nexus. Improving our understanding of these complex interdependencies and developing tools to assist decision makers with future infrastructure planning are essential for sustainable development in the face of the uncertainties posed by climate change.

The World Bank has embarked on a global initiative called Thirsty Energy to help countries tackle the challenges of managing the water-energy nexus in an integrated manner. A primary aim is to demonstrate the importance of combined energy and water modeling, planning, management, and decision making and to develop practical methodologies that can be widely applied. This chapter presents the results from the development of new "water-smart" energy models in South

N. Li
e-mail: linanpro@gmail.com

Y. Zhu · Y. Shang
State Key Laboratory of Simulation and Regulation of Water Cycle in River Basin,
Institute of Water Resources and Hydropower Research, Beijing, China
e-mail: zhyn@iwhr.com

Y. Shang
e-mail: shangyz@iwhr.com

D. Rodriguez · M. Bazilian · A. Delgado-Martin
World Bank, Washington, DC, USA
e-mail: drodriguez1@worldbank.org

M. Bazilian
e-mail: mbazilian@worldbank.org

A. Delgado-Martin
e-mail: adelgado@worldbank.org

F. Miralles-Wilhelm
University of Maryland, Maryland, USA
e-mail: fwilhelm@umd.edu

Africa and China, offering proof of concept and highlighting the importance of planning for energy and water in a more integrated manner.

This chapter provides an overview of the two "water-smart" energy models realized as part of the Thirst Energy initiative, with an emphasis on the conceptual aspects that need to be considered (most notably aligning the water and energy models, developing the water supply cost curves and incorporating them into the energy model), and what the resulting framework can show that might otherwise be overlooked in traditional "silo" planning approaches. The individual case studies provide a more in-depth description of the water-energy situation in each country, the assumptions underlying the models, the process of incorporating water in each model, and the results of the analysis performed by the country teams (and are the only references provided here as each has an extensive set of source references).

1.1 South Africa

South Africa is a water-stressed country that is coming out of an electricity supply crisis that was severely impaired by financial circumstances. The sustainability of water and energy supplies is uncertain, as is the impact of shortages on social well-being, the national economy, and the environment, particularly in the context of climate change. Fully understanding the contours of the water-energy nexus is therefore particularly relevant for South Africa.

South Africa has long had processes for long-term planning related to the supply of energy and water. Planning for one has historically taken into account the cost and scarcity of the other, though to varying degrees. This has resulted in the development of an integrated system of large dams and inter-basin transfers to ensure a reliable water supply to all sectors. Therefore, South Africa provided a unique opportunity for the Thirsty Energy Initiative to develop and demonstrate methodologies for integrated water-energy planning.

Furthermore, there is a strong commitment to reduce carbon emissions in the country, as reflected by the two mitigation scenarios examined, one aligned with the country NDC aspirations (14 Gt CO_2 eq cumulative emissions) and the other more stringent one aimed at what would be the country's target as part of achieving under 2 °C (10 Gt CO_2 eq cumulative emissions), as is discussed in Sect. 3.

1.2 China

China is increasingly aware of the complex interdependencies between water and energy. China's rapid economic development has been accompanied by a similar rapid increase in energy needs, which is dominated by coal, resulting in significant air pollution and CO_2 emissions. In addition, the coal energy supply chain is very water intensive, from mining and washing to cooling power plants.

The water-energy nexus challenge is further complicated by the fact that the majority of the planned new energy projects are to be located in the water stressed Energy Bases of northern China, where significant energy resources (coal, oil and natural gas reserves) are found.

Core policies embodied in China's 13th Five Year Plan (FYP) have mandated changes in coal-fired power plants requiring that new plants built in the northern Energy Bases employ dry cooling. The 13th FYP also has ambitious targets for the deployment of renewable energy sources, as well as calling for major reductions in greenhouse gas (GHG) emissions by 2030. Therefore, the case study examined how these developments will impact the need for water, including progress in the energy sector towards meeting the "3 red lines" water policy aimed at reducing water usage, improving water efficiency and curtailing wastewater discharges. The core energy policies combined with the 3-red lines water policies are moving China towards emission reductions in line with what will be required in terms of a contribution to limiting global temperature to below 2 °C.

2 Realizing "Water-Smart" Energy System Models

The steps undertaken to realize "water-smart" energy planning models involve:

1. Identifying an appropriate energy systems planning model that can be modified to incorporate water, where both countries had advanced TIMES models (Chen et al. 2016; Huang et al. 2017; ERC 2017);
2. Use basin level hydrological models to determine likely water supplies and infrastructure needs;
3. Align the regions in the energy and water modeling platforms;
4. Incorporate this water supply (and infrastructure) information into the energy systems models, and
5. Analyze a variety of scenarios examining future water-energy trade-offs.

Aligning the regions in South Africa was done according to the water management areas (WMA, Fig. 1), while the Energy Bases drove the depiction of the regions for China (Fig. 2).

2.1 Including Water in Energy Models

In order to represent the water needs for energy production, two modifications were necessary within the energy model: representing the supply of water to each region, and determining the water withdrawal, consumption and discharge level for each energy process using water.

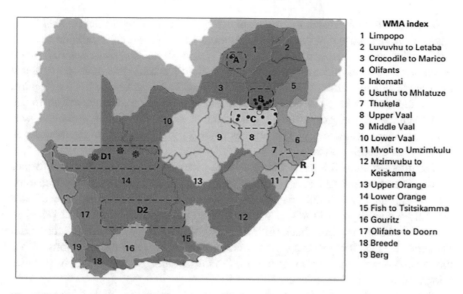

WMA index
1 Limpopo
2 Luvuvhu to Letaba
3 Crocodile to Marico
4 Olifants
5 Inkomati
6 Usuthu to Mhlatuze
7 Thukela
8 Upper Vaal
9 Middle Vaal
10 Lower Vaal
11 Mvoti to Umzimkulu
12 Mzimvubu to
   Keiskamma
13 Upper Orange
14 Lower Orange
15 Fish to Tsitsikamma
16 Gouritz
17 Olifants to Doorn
18 Breede
19 Berg

Fig. 1 Water management areas and energy producing regions in South Africa

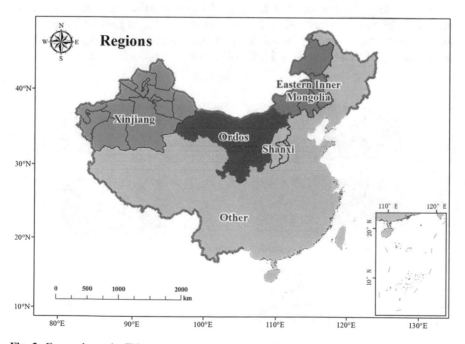

Fig. 2 Energy bases in China

With respect to representing the supply of water, two different approaches were taken. In South Africa, along with the amount of water supply available in each WMA, the associated water infrastructure build decisions (investments) were incorporated within the energy model. For South Africa this was particularly important as major water transfer schemes between basins are often the only option for getting water where it is needed for energy purposes. In China, owing to the unavailability of the project level cost data, a more traditional approach was employed using water supply cost curves (WSCC) indicating how much water is available at a certain price.

The fundamental difference between these two approaches is that the former endogenizes the cost and timing of the water infrastructure investment decisions, recognizing the need to fully pay for a project over its lifetime, whereas the latter has only a cost-supply curve that could demand more water (at a higher price) in some period then dropping back (in later periods) essentially "stranding" the initial investment. Thus, incorporating both water and energy investment decisions within the "water-smart" energy model has important advantages that result in a better integrated planning framework.

The approach taken in South Africa is portrayed in Fig. 3, where the "water-smart" South Africa model is named SATIM-W.

In order for a "water-smart" TIMES model to determine the amount of water needed for energy, the water requirements (and ideally discharges) are incorporated into the model for the energy production processes, as reflected in Fig. 4 for China, where the China model is called TIMES-ChinaW.

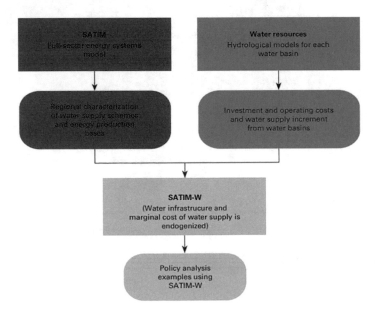

Fig. 3 "Water-Smart" energy modeling framework (South Africa)

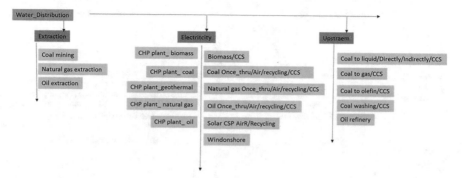

Fig. 4 Water representation in TIMES-ChinaW

2.2 Representing Water Supply Costs

The process of including water in an energy model starts by developing Unit Water Cost (UWC) curves, which show the cost and amount that can be obtained from each water delivery option. These UWC curves also show the sequencing of the projects, as some more expensive ones may need to be undertaken before less costly ones and be pursued.

In SATIM-W, the depiction of the water delivery system is done using distinct processes that include the investment, operating and other costs for each water supply scheme (Figs. 5 and 6). In addition, any costs associated with water quality either before use or afterwards for treatment is also explicitly represented as discrete processes in the water subsystem.

However, in TIMES-ChinaW, the investment decisions are built into the WSCC, providing a single cost for each increment of supply (Fig. 7). With this type of supply-cost curve, the amount of water drawn in any particular period is free to float, and thus less revenue may be available to cover the cost to build any new infrastructure.

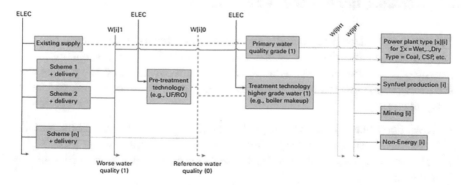

Fig. 5 Generic water supply depiction in SATIM-W

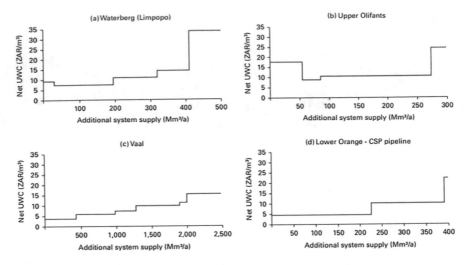

Fig. 6 Regional UWC curves for South Africa

The South Africa approach makes it much less likely that the model will choose to make an investment in a scheme (or treatment process) that won't be used later since the model optimizes using the life-cycle cost of the investment. However, with the China approach, if for example, a coal-fired plant is stranded due to pressure to reduce GHG emissions or concerns about local air quality, the model will simply lower the amount of water delivered without consideration for the need to pay off the original investment.

Thus a key recommendation arising from this research is that when possible (i.e., the necessary data is available) water infrastructure investment options should be embodied in the "water-smart" energy model, just like all the other energy sector investment options.

3   Assessing the Water-Energy Nexus

The scenarios selected for analysis using the new "water-smart" energy planning model reflect main drivers of investment uncertainty in water and energy supply that are of key importance to South Africa and China. Specifically, the models were used to examine the following questions facing each of the countries, as well as many others.

- How does accounting for regional variability in water availability and the associated cost of water supply infrastructure affect future energy planning?
- Is the current policy of dry cooling for new coal-fired power plants economically justified?
- How do stricter environmental controls affect coal investments?

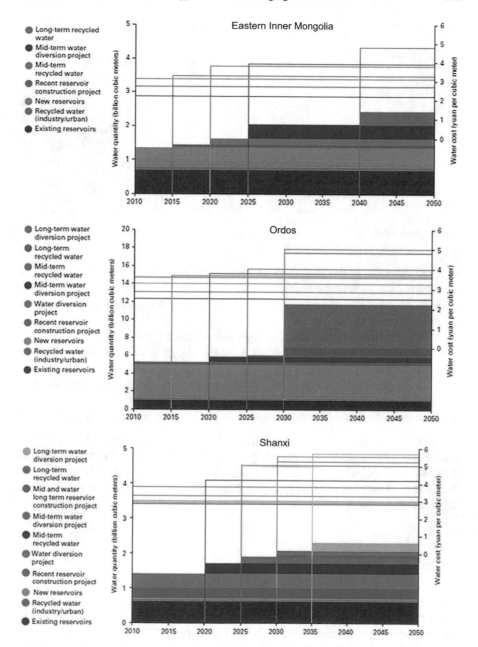

Fig. 7 Regional WSCC curves for the energy sector in China

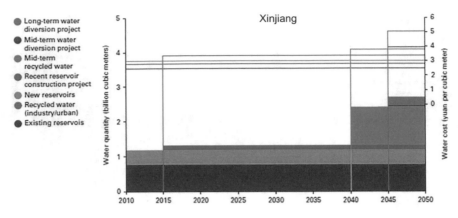

Fig. 7 (continued)

- How does the potential impact of climate change uncertainty (in particular a drier climate) affect the energy sector?
- How does the cost of water affect alternative production processes (such as shale gas, synfuels, etc.)?
- What is the impact of planned energy polices on future water needs?
- In a carbon-constrained world, what is the likelihood of stranded assets and what might that cost?
- Looking at a water limited and carbon-constrained future, what are the key technologies comprising of a robust sustainable roadmap?
- How do energy sector changes contribute to meeting water policy goals?

In the following sections, examples of scenarios examined in each country are presented. The findings are primarily illustrative and serve to showcase how "water-smart" planning models can be used to better inform decision-making and policy formulation for the energy sector. More details are available in the full Case Studies (World Bank 2017a, b).

3.1 Scenarios Examined

For each country analysis, an energy system Base case was developed reflecting current government policies, with and without considering the cost of water. The Base case with water costs, along with any other government policies in place, was established as the Reference scenario and used most of the time for comparison against the policy runs.

Table 1 shows the main scenarios analyzed using SATIM-W, which examine key areas of water and energy supply uncertainty in South Africa.

For China, the analysis focused on the various policies embodied in the 13th FYP, as summarized in Table 2. Note that a fundamental requirement of the 13th FYP is

Table 1 South Africa study scenarios

Scenario	Description
Base (no water cost)	Status quo planning continues without consideration of water
Reference (water cost)	Status quo planning continues with proper consideration of water
Dry climate	Regional water supplies and non-energy water demands are adjusted to reflect a drier climate (increasing water demand and decreasing water supply), affecting the unit water supply cost of regional schemes
Environmental compliance	• Retrofitting existing coal power plants with wet flue-gas desulfurization (FGD); • Fitting existing and new coal-to-liquids refineries with FGD technology; • Operating all combined-cycle gas turbines with wet control of nitrogen oxides; • Coal mines fully treating water discharged into the environment, and • Inclusion of Water Quality requirements
Dry climate + environmental compliance	Water demands and costs rise across sectors as Dry Climate and Environmental Compliance are combined
CO_2 cumulative cap 14GT	Cumulative national GHG emissions limited to 14 Gt by 2050 (in line with the country's NDC)
CO_2 cumulative cap 10GT	Stricter carbon budget limiting cumulative national GHG emissions to 10 Gt by 2050 (in line with what would be required for below 2 °C)

Table 2 China study scenarios

Policy issue	Description
Base	With and without cost for water, no CO_2 limit but with other 13th FYP core policies (use dry cooling, close small less efficient coal-fired power plants, cut back on the synfuels program)
Reference [NDC]	13th FYP core policies (and CO_2 limit of 9.5 billion tons in 2020 and 11.0 billion tons in 2030 (in line with China's NDC), with and without the cost of water
Non-Fossil energy	Implement government Plans for nuclear and renewables by 2020, extended until 2030
	Implement government Policy to raise the proportion of non-fossil energy to 20% by 2030
Coal Peak	Limit coal production to 2.85 Btce in 2020 and 1.5 Btce in 2050 (with the CO_2 cap)
Coal chemicals	Coal-to-liquid (CTL) capacity increases to 2400 kt/a by 2020, and the coal-to-gas capacity grows to 3.1 bcm/yr by 2030
Combined scenario	Combining Reference, Non-Fossil Plan, Coal (Peak), and other policies under consideration (aimed at moving towards what would be required as a below 2 °C contribution)

China's strong commitment to limiting CO_2 emissions in line with the country's NDC goals, which is included as part of the Reference and each of the policy scenarios examined. In addition, three climate change sensitively runs were done to look at the impact on water availability for the Representative Concentration Pathways RCP2.6, RCP4.5 and RCP8.5 proposed by the Intergovernmental Panel on Climate Change (Van Vuuren et al. 2011). However, the water modeling showed only minor variation between the Reference and RCPs levels so those scenarios are not presented here, but can be examined in the full case study (World Bank 2017b).

3.2 Key Findings

The focus is to provide a sense of why including water in an energy model is essential and how it affects the results. Some key metrics serve to highlight the key impacts (e.g., energy system cost, power plant builds, water use, CO_2 emissions) of selected scenarios (Table 3, Fig. 8).

Table 3 Key indicators for mitigation scenarios, difference from reference—South Africa

Scenario results	Units	Reference (water cost)	CO_2 cum cap 14Gt	Percent change	CO_2 cum cap 10Gt	Percent change
System cost	2010 MZAR (× 1000)	7646	7686	0.51	7865	2.86
Expenditure —supply	2010 MZAR (× 1000)	11,650	11,765	0.98	10,941	−6.90
Primary energy	PJ	333,500	284,385	−15.24	266,639	−20.52
Final energy	PJ	157,083	156,008	−0.68	154,452	−1.67
Power sector CO_2 emissions	Mt	13,756	9330	−32.18	6120	−55.51
Power plant builds	GW	134	170	26.49	189	40.88
Power plant investment	2010 MZAR (× 1000)	2670	3318	24.28	4872	82.49
Water to power plants	Mm^3	12,074	14,592	20.85	15,073	24.84
Total water for energy	Mm^3	16,265	16,941	4.16	16,753	3.00

MZAR—million of South Africa rands (1 EUR = 14.80 ZAR); PJ—petajoule (10^{15} joules); Mt—millions of tons; GW—gigawatts; Mm^3—millions of cubic meters

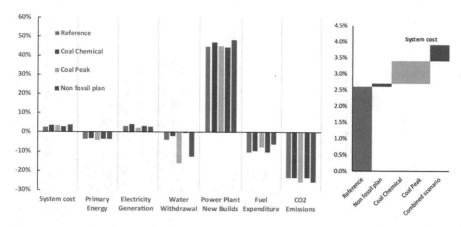

Fig. 8 Key indicators for core scenarios, difference from base—China

(a) **Striving to achieve mitigation goals now will ensure there is less infrastructure stranded**

For South Africa, not surprisingly, coal-fired and CTL generation would be directly affected by a cap on CO_2 emissions. A 14 Gt cumulative limit (in line with the country's NDC target) allows the existing CTL facility to operate at full capacity until 2025, with an increase in imports of petroleum products owing to a lack of investment in new CTL capacity. The existing and planned coal power plants are less at risk under the 14 Gt limit, as these assets remain operational for their production life, although no new coal power plants are commissioned. In contrast, a 10 Gt limit (what would be required for the country to achieve 2 °C) would reduce CTL output dramatically, resulting in an increased reliance on imported petroleum products, which replace 80% of existing CTL production by 2025. At the same time, the operating life of the committed coal-fired power plants is shortened by 15 years, with their capacity replaced by new nuclear plants. In addition, electricity production shifts from the Waterberg to the Orange River region, where concentrated solar power (CSP) is used. Illustrating the pressure on the existing generation facilities, the remaining existing power plant capacity and the associated utilization factor (CF) drop precipitously in 2035, though the one remaining coal-fired plant in the Central Basin does manage to run at 80% in the last period of the 14Gt scenario due to the "headroom" created by the substantial buildout of CSP in the country (Fig. 9).

Both CO_2 Cap scenarios also affect investment in water supply infrastructure in the Waterberg region, since the region's coal-based energy sector provides much of the funding for the water infrastructure. If the water demands for energy are greatly reduced, the associated water infrastructure may not be built. As water demand from non-energy sectors grows in this region, however, the missing investment from the power sector pushes up the costs of water supply by 2035. Meanwhile, other users have access to additional water, since the existing supply capacity is underutilized. The timing of regional water infrastructure investments under several scenarios also

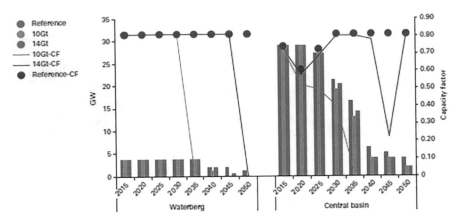

Fig. 9 Existing coal capacity and production factors (South Africa)

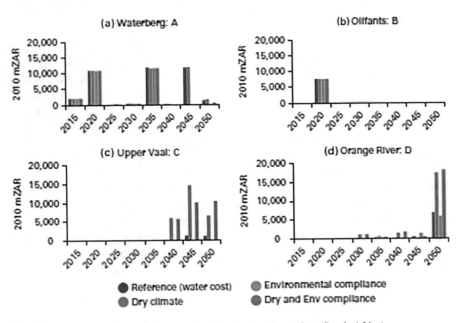

Fig. 10 Investments in new water supply infrastructure by region (South Africa)

deserves interest (Fig. 10). Olifants does not show much variation, but Upper Vaal and Orange River will require significant new investments, particularly under the Dry Climate scenario. Only an integrated approach to water-energy planning can help to ensure timely investment, proper sizing and delivery of water for the energy sector, while also reducing the likelihood of stranding major energy or water assets.

For China, the CO_2 limit policy is embedded in the Reference scenario assumptions, as well as the mandate in the 13th FYP to use dry cooling, which leads to closing small less efficient coal-fired power plants and cut backs on the Coal Chemicals (synfuels) program. There are some stranded (existing) coal assets lost (by design) and less new coal plants built in favor of more non-fossil generation. This results in a dramatic shift in the 2050 generation mix, with coal-fired power going from 60% to less than 10%, solar increasing to nearly 30% from less than 2%, wind moving from 16 to 26%, and nuclear climbing from less than 1% to over 14% (Fig. 16). At the same time, substantive progress is made on meeting the 3 red lines water goals for the energy sector with water withdrawals dropping 42% in 2050. When the Coal Peak and other policies discussed are included, CO_2 emissions are pushed 3% lower as renewable and nuclear sources account for 87.4% of the generation mix, substantially cutting water needs for the energy sector, and positioning China for what would be needed to achieve a 2 °C future, where the cost of this additional reduction is only 1.6% or $446 Billion over the 40-year planning horizon.

(b) Climate change affects regional water availability, more studies are needed

Projected changes in South Africa's regional climate (reduced water availability and increased ambient temperature) are expected to shape investment in energy and water supply, compounding the effects of a policy limiting carbon emissions. Under the Dry Climate scenario, increases in water demand traceable to a warmer and drier climate will trigger faster and larger investments in water infrastructure, which will raise the average cost of water supply. However, the ability to meet demand will not be strongly affected, owing to South Africa's integrated water supply network, which enables the transfer from areas of high rainfall (such as Lesotho) or substantial urban return flows (such as Johannesburg) to water-scarce regions such as the Waterberg.

Climate change brings with it concerns about the uncertainty of water availability in the future. However, based upon the modeling work done for China, the Energy Bases are expected to benefit from more, not less, rainfall although with greater variability, and thereby could make the energy system less water constrained. The Other (wet) region will likely experience less rainfall—though not enough to dramatically impact the energy sector. This conclusion is robust over the three climate scenarios examined, that is, ranging from modest to severe climate change there was less than a 0.01% of an impact on the overall cost of the energy system, and only modest shifts in the generation mix from the northern Energy Bases to the Other region.

However, for both countries, further analysis of the effects of climate change on the water-energy nexus is required. In particular, the use of only a single climate change model was a limitation and more analysis is warranted, in particular a water stress scenario for China.

(c) Climate change mitigation policies can help achieve water policy goals too

Some energy policies have a direct impact on water, such as the mandate to use dry cooling in both countries. Recognizing the increasing conflict between the various needs for water (agriculture, energy, industry, urban), China has taken an aggressive approach to setting strict "3 red lines" water policies. What we see in the analysis shows that properly pricing water along with current policies in the 13th FYP (including the China's NDC goal) all combine to directly help the Energy Bases comply with important aspects of the water policies including:

- Lowering water withdrawal 30% by 2030 and dropping to 42% by 2050, contributing to meeting the target to reduce total withdrawal;
- Decreasing waste water releases 48% by 2030 and 57% by 2050, helping to address water quality;
- Increasing water consumption 18% by 2030 due to the substitution of old once-through coal power plants by new recirculating plants, but decreasing by 5% in 2050 as more non-fossil generation enters, and
- Decreasing the intensity of water withdrawal per unit of GDP from 8.4 m^3/ thousand US$ to 2.3 m^3/thousand US$ by 2050, helping to meet the water withdrawal per unit of industrial value added goal.

At the same time, there is a drop in cumulative CO_2 emissions of 125 Gt (23%), compared to the Baseline without the CO_2 target.

(d) Stricter environmental controls affect energy and water investments

Stricter environmental controls (South Africa Environmental Compliance and China Coal Peak scenarios) reduce investment in coal-based energy supply and associated water supply infrastructure. In the case of South Africa, current emissions regulations requiring FGD for existing coal power plants and new CTL plants prove to be a major water infrastructure investment disincentive. This scenario accelerates the retirement of the existing coal plants starting in 2030, with more solar photovoltaic (PV) coming online to replace it. It also results in a significant reduction in the use of existing CTL and new CTL plant builds, compared with the Reference, where under a carbon constrained world CTL plants are stranded/ abandoned (Fig. 11). This leads to higher imports of petroleum products and a deferment of new water supply schemes in Waterberg that were driven mainly by water demand from CTL plants. Postponement of those schemes could have a wider impact on economic development in the area, such as mines and industries as well as non-energy water consumers that would also depend on the new water supply infrastructure.

For China, a similar scenario driven by air quality concerns looks to limit overall coal use by capping the amount of coal mined. This Coal Peak policy scenario was found to be stricter than the Reference scenario (CO_2 constraint). It forces earlier investments in nuclear and renewable power, retiring coal power plants that are kept operating longer in the Reference scenario, increasing total system cost by 0.6%. This again shows how policies to address environmental concerns/climate change mitigation, if designed properly, will both lower water use by the energy sector, contributing to climate change adaptation, while reducing CO_2 emissions cost-effectively.

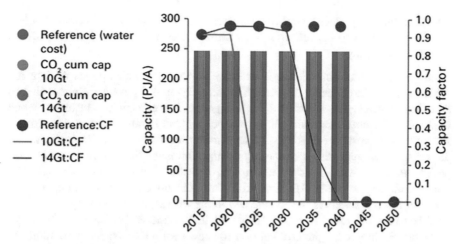

Fig. 11 Impact on CTL (South Africa). *Note* CF = Capacity factors

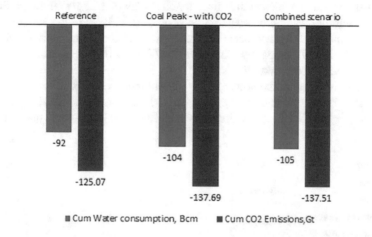

Fig. 12 Summary metrics—cumulative change (2010–2050) from base scenario (China)

Reductions in CO_2 emissions and water consumption for energy are strongly aligned and significant (Fig. 12), with the Coal Peak and Combined scenarios showing greater reductions than the Reference in both water consumption and CO_2 emissions (by 1%, or 12 Gt) beyond current policies, moving China a bit closer to what would be required for 2 °C. However, it must be noted that achieving the NDC reduction of CO_2 emissions has an associated marginal cost of \$75/t CO_2 in 2020 and \$578/t CO_2 in 2050, so deeper cuts may prove to be quite challenging and costly.

(e) **Requiring dry-cooled coal-fired power stations appears economically justified when properly taking into account water, till CO$_2$ mitigation policies come into play**

Even in water-scarce countries such as South Africa, wet-cooled power stations are often considered to be more cost effective than dry-cooled power stations owing to their lower investment costs and higher generation efficiency. However, when regional water costs are taken fully into account, dry-cooled power stations are shown to be a most cost-effective choice. Similar results were seen in China where the assumptions regarding dry-cooling cost/performance there have it more competitive to begin with. Thus in both cases, the current dry-cooling policy of the power company in South Africa (Eskom) and the government in China (for the northern Energy Bases) is in the best economic interest of the country.

When regional water is fully considered in SATIM-W (Reference with water cost), dry-cooling is the preferred option for new coal power plants, particularly in the Waterberg region where the remaining economically viable coal reserves are located (Fig. 13). This dry-cooling policy has a significant impact on the power sector water intensity which could either reach a peak of 1.65 l/kWh by 2050 in the Reference without water cost scenario, or drop to 0.5 l/kWh in the Reference with water cost. In the Reference scenario, the power sector's cumulative water consumption (2010–50) drops by 9338 Mm^3 (77%) with just a modest increase (0.84%) in total energy system cost.

The water intensity of the power sector under other scenarios (Fig. 14) is close to that of the Reference scenario, except in the case of the scenarios requiring greenhouse gas (GHG) reductions, where the model favors use of some

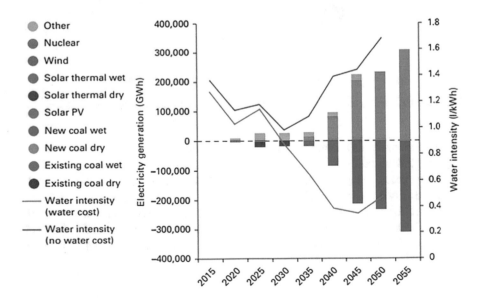

Fig. 13 Difference in electricity generation by type and water intensity (South Africa)

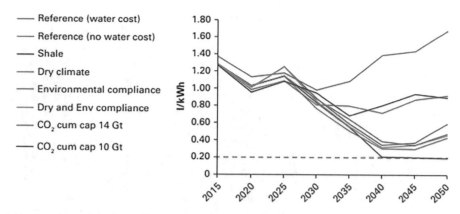

— Reference (water cost)

— Reference (no water cost)

— Shale

— Dry climate

— Environmental compliance

— Dry and Env compliance

— CO$_2$ cum cap 14 Gt

— CO$_2$ cum cap 10 Gt

Fig. 14 Impact on power sector water consumption (South Africa)

concentrating solar plants using wet cooling (for reasons explained in the Case Study). Again, results for China are consistent with those shown here, though since the NDC target is included in all runs, the clustering of the water intensity remains tight across the scenarios.

In South Africa, the Reference scenario produces slightly more CO$_2$ emissions than the scenario without water costs, despite generating 1.3% less electricity with coal and 2% more with renewable energy (RE) technologies. This is due to the higher unit emissions of dry-cooled coal plants, and points out that looking at water policies (to address "climate change adaptation") in isolation is not enough when striving at the same time to properly address climate change and mitigation requirements. In Dry and GHGs mitigation scenarios, the water consumption is decreased, reinforcing the importance of looking at combined policies for climate change adaptation (water consumption reduction) and mitigation (GHG reduction).

Furthermore, as can be seen by the major shift to non-fossil new power plant builds in South Africa (Fig. 15), and the 2050 generation mix in China (Fig. 16) there is virtually no role to be played by coal power plants in a carbon constrained world.

(f) **Regional differences highlight the importance of addressing the nexus at a regional level**

Water and energy are regional commodities that in most cases are not so easy to move around, or costly if doing so. Thus, energy-water trade-offs may take on much different characteristics in one region versus another.

In South Africa's Waterberg region, a large share of water goes to the energy sector. Under the Reference scenario, water consumption by power plants will account for about 40% of all demand in 2050, whereas coal mines consume about 25% and CTL plants and non-energy uses consume equal shares of the remaining 35%. Given that energy sector uses account for more than 80% of the water demand

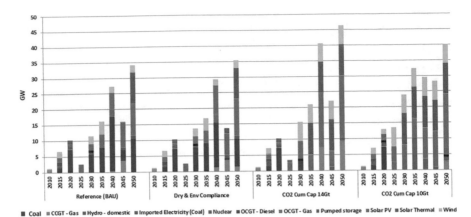

Fig. 15 Power plant builds under dryer climate and to get to 2 °C (South Africa)

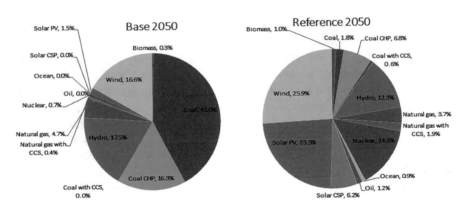

Fig. 16 Change in electricity generation mix 2050—core policies with CO_2 (China)

in the region (Fig. 17), it is no surprise that the region is sensitive to energy policy changes.

This can be illustrated in the results from China by looking at the regional shifts in electricity generation under three different coal policy scenarios (Fig. 18). Reduction in generation takes place in the coal dependent Energy Base, with additional renewable generation coming on in Ordos and hydro plus nuclear in Other.

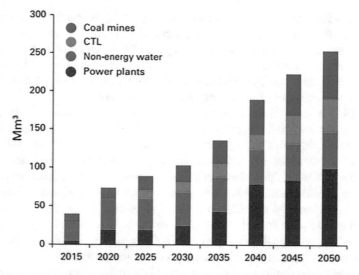

Fig. 17 Water consumption in the waterberg by use in the reference scenario (South Africa)

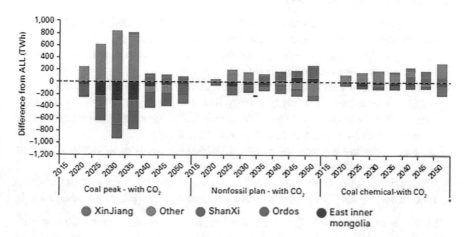

Fig. 18 Change in electricity generation by region—core policies with CO_2 (China)

4 Conclusion

The Case Studies reported here used new tools to examine the water-energy nexus and to explore the possibilities of integrated energy and water planning. The approach applied for South Africa and China can be readily adapted to enable other countries to tackle their water-energy management challenges in a more integrated manner.

These studies demonstrate that important insights can be gained by linking water and energy planning models to foster integrated water-energy strategies aimed at ensuring that these critical long-term aspects of sustainable development are planned in a least-cost manner. This is particularly important as countries gear up to determine how their NDC commitments will be realized in a way that contributes directly to achievement of sustainable development goals.

Perhaps the main finding arising from both case studies is that properly accounting for the cost of water when considering climate change mitigation policies shows that both energy sector CO_2 emissions and water use are substantially reduced, therefore improving overall sustainability of both sectors.

Looking at a water limited and carbon-constrained future, the key technologies comprising a robust sustainable roadmap in both countries are solar (primarily CSP in South Africa, and PV with CSP in China) and nuclear. In China, wind also has an important role to play. In both countries, besides the reductions to coal-fired power plants, the coal chemical industry (CTL/CTG) is all but abandoned, resulting in higher reliance on imports of petroleum products in the future.

The Case Study results have demonstrated key objectives of the Thirsty Energy initiative through the following important findings regarding the broader impacts of modeling the water-energy nexus.

- There is an essential need to account for the cost and availability of water within the framework of an energy planning model, and planning of these key inter-dependent realms must become more integrated with coordinated long-term strategies across both realms.
- The Case Studies shows how a national-level energy systems model can be readily regionalized in terms of energy resource supply and power plant locations, and the regional costs and limitations for water supply infrastructure can be incorporated to create a "water-smart" planning tool.
- An approach that embeds the water infrastructure investment decisions within the "water-smart" energy model is more robust and recommended.
- The process and techniques employed to realize the "water-smart" energy models can be readily adapted to other national energy system planning models.

Furthermore, effects of climate change and climate-related policies, most critically achieving NDC and below 2 °C targets, will have a significant impact on both water and energy planning, so it is more critical than ever that the water-energy nexus be planned in an integrated manner to avoid stranded assets and unused infrastructure in the future and increase sustainability of investments in both sectors. The analyses show what is necessary to first achieve NDC aspirations and then move towards the reductions that countries will need to achieve in order to stabilize global temperature rise below 2 °C, at a cost that could be acceptable with international cooperation as part of a global effort.

The Case Studies also served to identify areas for improving "water-smart" modeling including:

- Incorporating a better representation of non-energy water consumption in order to examine water-reallocation schemes, demand response to cost, and the impact of water-use efficiency and demand side management interventions;
- The aggregation to the basin level in water models, or to the region level in energy models may fail to capture local characteristics and complexities, especially regarding water delivery issues, so further disaggregation of the energy regions may be desirable;
- Incorporating wastewater streams, treatment plants, and related infrastructure is necessary to fully depict the water-energy nexus;
- A regional representation of the demand for energy services, not just supply, is highly desirable;
- Employing more than a single climate model (China) is recommended, and
- Better exploring of the potential impacts and associated risks of future climate change is necessary.

A central goal of the Thirsty Energy initiative is to continue the process of informing decision-makers that such a comprehensive, integrated approach provides for better informed policy formulation and planning processes and needs to become the norm.

References

Chen W, Yin X, Zhang H (2016) Towards a low carbon development in China: a comparison of national and global models. Clim Change 136:95–108

Energy Research Centre (2017) http://www.erc.uct.ac.za/groups/esap/satim

Huang W, Ma D, Chen W (2017) Connecting water and energy: Assessing the impacts of carbon and water constraints on China's power sector. Appl Energy 185:1497–1505

Vuuren Van et al (2011) The representative concentration pathways: an overview. Clim Change 109(1–2):5–31

World Bank (2017a) Modeling the water-energy nexus: how do water constraints affect energy planning in South Africa? World Bank, Washington, DC

World Bank (2017b) Thirsty energy: modeling the water-energy nexus in China. World Bank, Washington, DC

Part IV
The Role of Cities and Local Communities

Challenges Faced When Addressing the Role of Cities Towards a Below Two Degrees World

George Giannakidis, Maurizio Gargiulo, Rocco De Miglio,
Alessandro Chiodi, Julia Seixas, Sofia G. Simoes,
Luis Dias and João P. Gouveia

Key messages

- Not many energy-system related measures can be realistically implemented at city level but many urban-related actions can be taken to cope with city specific issues with significant benefits for the GHG emissions reduction.
- Municipalities are extremely well positioned to steer households towards efficient and low-carbon energy consumption within buildings and transportation.
- Cities have the possibility to allocate land only to projects which prove to be socially desirable and environmentally friendly.

G. Giannakidis (✉)
Centre for Renewable Energy Sources and Saving (CRES), Athens, Greece
e-mail: ggian@cres.gr

M. Gargiulo · R. De Miglio · A. Chiodi
E4SMA Srl, Turin, Italy
e-mail: gargiulo.maurizio@e4sma.com

R. De Miglio
e-mail: rocco.demiglio@e4sma.com

A. Chiodi
e-mail: alessandro.chiodi@e4sma.com

J. Seixas · S. G. Simoes · L. Dias · J. P. Gouveia
CENSE, Center for Environmental and Sustainability Research, NOVA School of Science
and Technology, NOVA University Lisbon, Caparica, Portugal
e-mail: mjs@fct.unl.pt

S. G. Simoes
e-mail: sgcs@fct.unl.pt

L. Dias
e-mail: luisdias@fct.unl.pt

J. P. Gouveia
e-mail: jplg@fct.unl.pt

© Springer International Publishing AG, part of Springer Nature 2018   373
G. Giannakidis et al. (eds.), *Limiting Global Warming to Well Below 2 °C: Energy
System Modelling and Policy Development*, Lecture Notes in Energy 64,
https://doi.org/10.1007/978-3-319-74424-7_22

- The use of the TIMES based City ESM offers an integrated planning tool and can be replicated in other cities for analysing transition pathways.

1 Introduction

The United Nations (2014) estimates that by 2050, the proportion of the European population living in cities will have risen to 80% from 73% today, while two thirds of the global population will live in urban areas. According to IEA (2016) cities today account for two-thirds of the primary energy demand and 70% of total energy-related carbon dioxide emissions. As urbanisation increases the energy and carbon footprint of cities will increase.

Tackling the energy and climate challenge in cities requires answering a set of questions like: How can we deliver a more sustainable energy system at the lowest cost? How can we provide the basis for unlocking the full potential of energy efficiency and large-scale integration of renewable energies in urban areas? How can we improve the cities' planning policy assessment, and the implementation of realistic Sustainable Energy Action Plans?

This challenge has been tackled by many cities and researchers, either using integrated energy system models (Lind and Espegren 2017; Xydis 2012), other energy models (Keirstead et al. 2012) or non-energy modelling approaches (Zanon and Verones 2013; Kostevšek et al. 2013; Yamagataand and Seya 2013). An extensive review of 200 current city mitigation plans concludes that most of the plans are developed by the cities following a relatively simple methodology recommended by organizations as the CoM or Compact of Mayors, which does not consider the integrated energy system (Reckien et al. 2014).

Therefore, the approach developed in the framework on the project "Integrative Smart City Planning"-InSMART, offers an innovative manner to supply answers to these questions by providing a state of the art methodology for energy planning, using effective tools for data gathering, analysis and visualization. InSMART approaches the urban challenges of energy and climate by considering the city energy system as an integrated network of energy flows connecting energy providers with energy consumption in buildings, public spaces, transport and utilities, while taking into account spatial differentiation. Using a detailed characterisation of a local energy system, with simulation tools and the active participation of decision-makers and stakeholders is the essence of the InSMART Integrated City Energy Planning Framework. The end result is the identification of the optimum mix of short, medium and long term measures for a sustainable energy future, addressing the efficiency of energy flows across various city sectors with regards to economic, environmental and social criteria. The InSmart approach is well suited for supporting the cities to address pathways to achieve very strict greenhouse gas (GHG) emissions reductions compatible with a well below 2 °C global warming. Although this was not the goal of the InSmart project, which focused on medium term energy sustainability, all components of the approach here described can be directly used for ambitious urban GHG mitigation objectives.

2 The InSMART Methodology

The InSMART concept brought together cities and scientific organizations and established a methodology for enhancing sustainable planning through an integrative and multidisciplinary planning approach (Fig. 1). The methodology addresses the issue of integration within the cities administration as well as between the decision makers and the different relevant departments in municipalities.

2.1 Understanding the Existing City Energy System

The first step in the methodology (Fig. 1) is an enhanced understanding of the energy system of the city with all information stored in a geographic information system (GIS) based energy database. Buildings and transport are the main energy consuming sectors in a city and therefore they are addressed in detail within the InSMART methodology.

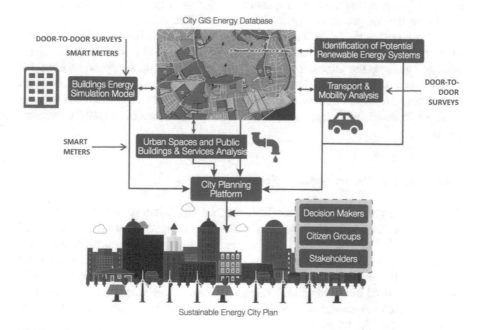

Fig. 1 InSMART methodology overview

Each city selected a base year according to their best possible available data. This was 2012 for Trikala, 2013 for Évora and Cesena and 2014 for Nottingham. The base year is used as a benchmark in the InSMART approach.

2.1.1 Assess Energy Consumption in Buildings

The current status of residential buildings is mapped using door to door surveys which collect data on building shell characteristics, equipment for heating, cooling, lighting and other uses, and behaviour of occupants like hours of operation and set points. The surveys used a representative sample in each city (derived using data from the buildings census). In the following step a number of building typologies, based on similar characteristics, were identified and the results of the surveys were used to define the values of these characteristics (e.g. typical wall construction, typical window frames, etc.). Using building typologies and statistical data is very useful in order to simplify the analysis of energy consumption, alternative scenario analysis and aggregations for the total energy demand over city zones. In the particular case of Évora, where all residential customers are equipped with smart meters, detailed electricity consumption profiles provided an in-depth knowledge of the demand for various uses.

The characteristics of building typologies are used as inputs in a building simulation software (DesignBuilder) to simulate alternative energy savings scenarios and replacement options for heating, cooling, water heating and lighting technologies. The simulation results for energy savings and the corresponding costs are utilised in the Integrated Energy System model (Sect. 2.2).

Energy consumption in municipal buildings is provided directly from the municipalities, while the behaviour of tertiary sector buildings is modelled using average consumption indicators (kWh/m^2 for space heating, for space cooling, for lighting and total consumption in kWh/m^2) from national sources. This is an opportunity for further advancing the methodology, with the application of surveys to these types of buildings, completing the representation of the city's building stock.

2.1.2 Analyse Urban Mobility Needs

A door to door survey for the mobility needs on city inhabitants provides the data for transportation demand between city zones. These inputs are used in a simplified transport model to forecast the future demand of transportation load (defined as a demand of passenger kilometres and ton kilometres) in alternative city development scenarios. The simulation tool is used in order to produce detailed results that will be further utilised in the Integrated Energy System model where the transportation demand drives the energy consumption by vehicles and the need for investing in new vehicles in the future.

2.1.3 Consider Energy Uses in Other Sectors

The InSMART integrative planning approach includes the analysis of all energy consumption sectors in the city. Besides the more commonly studied residential buildings and transport, the following sectors were also studied: municipality managed buildings, tertiary buildings, urban spaces, water and sewage systems, and the waste disposal chain. Current energy consumption profiles for municipality managed buildings, tertiary buildings, public lighting, green areas & public fountains, municipal solid waste (MSW) collection and disposal, water supply and waste water collection and treatment were characterised.

Besides characterizing energy consumption, InSMART mapped and analysed the existing district heating infrastructure (applicable only to Nottingham and Cesena). Finally, the future energy saving possibilities, both technological (e.g. improve efficiency of waste water treatment plants) and behavioural (e.g. impact of increasing MSW recycling rates) were assessed.

2.1.4 Integrate City Renewable Energy Potential

Complementarily, a thorough analysis was also made on the energy supply side for each city assessing the technical potential for urban decentralised energy supply using renewable energy sources (RES). This includes the technical potential for photovoltaic (PV), solar thermal, small-scale hydropower, wind, geothermal and biogas (Table 1). Hydro and wind potential exist in some cities and the values presented in Table 1 come from previous studies.

For the four partner cities, solar was the most important renewable energy resource, and thus InSMART developed and implemented a methodology to assess the potential for installation of solar thermal panels and solar PV systems, both building integrated PV and utility scale PV (i.e. >1 MW). The methodology included a set of limitations to PV installation sites, such as spatial planning constraints for the case of utility-size potential or the available suitable roof areas and angles in urban buildings for the case of rooftop potential. These limitations determine the extent of solar PV capacity to be installed, as well as the efficiency of electricity production. More detailed explanations on the developed methodology can be found in (Dias et al. 2015). The solar thermal potential includes all the buildings where adequate surface area exists for their installation.

2.1.5 Using GIS as a Spatial Energy System Database

The GIS platform plays a central role in the methodology of InSMART in the sense that it is the connecting point between the different sectors that focus into specific parts of the energy demand and supply in each city. The platform integrates and presents the data describing the current energy system on the city's geographical background but it also presents the results of energy scenarios (Sect. 2.2).

Table 1 Maximum RES potentials considered for the four cities

	Cesena	Évora	Trikala	Nottingham
Bioenergy	No new installations of biogas CHPs. Biomass boilers are planned in the time horizon	Not applicable for power production. Available for heating in buildings (up to 47.4 TJ, 46% more than in 2012)	70168 GJ of biomass for energy. 15.1 MW of biogas	3.1 MWe CHP
Hydro	128.7 kWe	Not applicable	Not applicable	Not applicable
Waste heat from power station	Not applicable	Not applicable	Not applicable	15.32 MWth
Wind	Not available	Not applicable	Not applicable	8.55 MWe
PV	82.0 MWe (rooftop and façade)	1.5 GW plant 40.1 MWe rooftop	67.3 MWe	200.7 MWe
Solar thermal	17317 solar water heating systems	65133 solar water heating systems	32405 solar water heating systems	52575 solar water heating systems

2.2 Integrated City Energy System Model

The characterisation of a local energy system, with data analysis and simulation tools, is the cornerstone of a key element of the InSMART Integrated City Energy Planning Framework: the development of city-specific energy system models (City-ESM). All relevant findings from sectoral-specific modelling analysis are used to feed directly the City-ESM and characterize all the energy consuming and producing sectors within the city.

The key goal of such a decision support tool is to analyse and cross-compare multiple future energy scenarios in a medium-term horizon (to 2030); and identify the optimum mix of applicable measures and technologies that will pave the way towards the achievement of the cities' sustainable targets.

The InSMART partners have chosen the TIMES (The Integrated Markal-Efom System) energy system modelling framework to develop the City-ESM. The key reasons for this choice are driven by its intrinsic characteristics, namely the technology explicitness, the technical/economical/environmental dimensions, the geographical scalability (from supra-national to local scale), and the possibility of customization to each city needs and *desiderata*, which will meet the objectives of the project.

2.2.1 Model Structure

The City-ESM has been developed to build a range of medium-term energy and emissions policy scenarios in order to inform local policy and planning decisions. These models are suited to cover all sectors of an urban energy system; however, in this project a particular focus on the buildings and transport sectors was given.

The urban area is represented in zones (nodes in the graph of Fig. 2). Each zone makes explicit the representation of the different building typologies (e.g. detached, semi-detached, blocks, hospitals, schools, etc.) and its energy requirements and refurbishment options; of the stock of appliances (e.g. boilers, lamps, etc.) and technology substitution choices; zone-to-zone mobility needs and its vehicle fleets.

The number and the border of the zones are inherited by the analysis of the building and transport sectors, on the basis of homogenous zones, thus ensuring the full consistency of the spatial representation across the sectors and tools used in the analysis.

Each zonal subsystem is characterised by stacks of individual behaviours (productions, consumptions, willingness to invest, constraints, etc.) of all the agents acting within the zone. As the analysis is mainly oriented to the building and transport sectors, the key agents of the systems are virtually placed in the dwellings meaning that the household, key player in the urban system, is responsible for the investment decisions, key element of the model, and the demand of a certain amount of services, such as space heating, lighting, mobility requirements, etc.

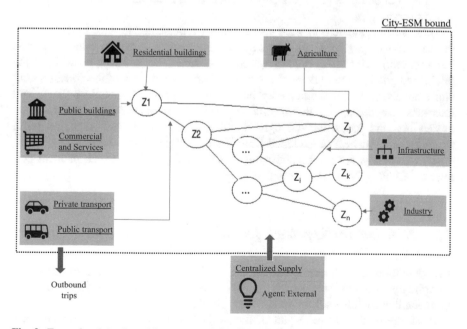

Fig. 2 Example of the "graph" structure of the City-ESM

Investment decisions occur at zonal level, in the zone where the technologies physically operate. Zones of the city may hold different characteristics and affect the investment decisions of agents and the operation of the technologies (e.g. different access to distribution systems, different solar PV potentials, different investments costs, etc.), therefore zone-specific developments/performances may be also analysed.

Dwellings are explicitly represented in the model by typology, zone and period of construction, and so are the available refurbishment options. Savings and the costs of the refurbishment options are calculated making use of a building stocks simulation of the existing building typologies.

Building construction (new demand) and demolishment are defined exogenously. Such assumptions are consistent across all the models used in the complex and integrate analysis, and are captured "twofold" in the city-ESM as they depict different conditions of energy needs (call for different amount of energy, different energy source, in different zones), and at the same time they determine a different matrix of movements (time-dependent) resulting in different levels of private and public transport demands which are fully inherited from the transport specific analysis.

In particular, private mobility demands are allocated to the zones which are at the "origins" of the movements, by assuming that the corresponding investment decisions are taken by the agents located in the zone of origin. Therefore, costs, fuel consumptions and emissions are directly assigned to this zone. The goal of the City-ESM, among the others, is to provide the optimal vehicles mix with respect to that matrix of movements and to any possible sectorial measure/target (scenario) taking into account of the possible integrations of the transport sector with other urban system components. Examples of such integration are presented in the following paragraphs.

Other sectors and activities are also explicitly represented in the City-ESM, with the same zone-specific detail. The key energy services (heating, cooling, public lighting, etc.) of schools, offices, warehouses, and other tertiary, as well as the public and good transport demands, are described to keep track of the consumption/emission levels of the municipality and of the impacts of specific policies and measures. Depending on the specific interest of the municipality and the data availability, the industrial and agricultural activities have been included (with simplified approaches) or excluded from the modelling applications.

The City-ESM covers the emissions of CO_2, CH_4, N_2O and also of local pollutants CO, SO_2, NOx and Particulate Matter (PM).

2.3   Participatory Approach for Scenario Development

The definition of different possible scenarios and the selection of measures accordingly is a key step in the process, along with a participatory workshop approach that includes all the stakeholders in the city.

In all the cities that were part of the project, stakeholders were engaged in a continuous manner throughout the project duration. The key events were the three participatory workshops where stakeholders contributed to designing future energy scenarios, identifying the main issues that should be included in the analysis. Persons from different types of organizations participated in these workshops including regional government authorities; local authorities (municipality and sub-municipal management); market agents (as PV companies, business associations, waste company) and non-governmental organizations (e.g. university, consumer associations and environment protection, association of engineers). The types of scenarios proposed varied from the implementation of specific actions like refurbishment of residential sector buildings to more comprehensive scenarios like focus on urban regeneration or Green Growth scenario. Details on the different scenarios for each city can be found on the InSMART website (InSMART 2016). The stakeholders also defined the criteria and provided the weighting for each criterion that was included in the Multi-Criteria Decision Analysis (MCDA), as described in the next section.

2.4 Multi-criteria Analysis of Alternative Options

Making use of the previously described participatory approach, a MCDA was used to rank the outcomes of the integrated city ESM for each city, which entailed the following steps and criteria (Table 2):

Table 2 Overview of the criteria used in the MCDA for each city

	Cesena	Évora	Trikala	Nottingham
No. assessed scenarios	1 reference + 6 alternative and competitive integrated plans	22	14	1 reference + 7 alternative and competitive integrated plans
No. of criteria used	9	6	9	10
Examples of quantitative criteria	Energy consumption in the building sector, Total particulate emissions	Reduction of energy consumption, GHG emission reduction, Financial effort	Implementation cost efficiency	Energy reduction potential, Overall cost
Examples of qualitative criteria	Complexity of implementation of the strategy (licencing procedures, consistency with laws and regulation, administrative approvals)	Contribution for improving comfort and quality of life of residents, feasibility	City's quality of life improvement	Social acceptability.

(1) Defining criteria together with stakeholders' and allocating weights to the different criteria, according to the previously defined groups of stakeholders (e.g. local authorities, regional authorities, private sector, and civil society);

(2) Definition of the decision-making scheme, regarding the objective function, criteria, sub-criteria and weights' assignment. A Deliberative Multi Criteria Evaluation (DMCE) approach was used and Hinkle's resistance to change method (Hinkle 1965) was applied towards the conclusion of representative preferences that were used as weights in the decision-making scheme afterwards.

(3) Implementing the MCDA by characterizing the results of each tested scenario in the city ESM according to the defined criteria in (1). Some of the model results provided direct inputs in the MCDA (i.e. investment costs), whereas some criteria required qualitative assessment by all stakeholders. The decision-making scheme was solved for each stakeholder group using PROMETHEE method (Brans and Mareschal 1992).

The quantitative criteria were calculated using the results of the City-ESM, while the qualitative criteria where assessed based on the input from the stakeholders. In this way stakeholders participated in the definition of the scenarios which were modeled but also on the ranking of the scenarios based on the quantitative outputs of the ESMs and on their perception of the qualitative characteristics of each scenario.

As a result, a ranking of alternative scenarios was produced which prioritized the actions from the one with the best to the one with the worst compromise among the evaluation criteria. This ranking was discussed with the stakeholders during a second workshop looking in particular for the necessity to change the criteria's weights. Having considered feedbacks from most important stakeholders involved in the decision-making problem, the 1st in ranking options provided the most appropriate and applicable solution for each municipality.

3   Sustainable Energy Plans Using the InSMART Methodology

The InSMART methodology approaches urban planning in an original way. Compared to the approach often used to develop Strategic Energy Action Plans (mainly based on the downscaling of the national/regional planning approaches), the InSMART method allows to explore multiple future planning hypotheses of the "integrated" energy-urban system (explicitly modelled and simulated) and to engage the local stakeholders in all the steps of the decision problem. Table 3 provides a comparison of key elements characterizing the Sustainable Energy Action Plan (SEAP) approach and the InSMART methodology.

Table 3 Qualitative comparison between SEAP and InSMART approach

	SEAP approach	InSMART approach
Approach	Top-down: Downscaling of national targets, policies and measures	Bottom-up: Driven by urban specific needs and integrated with the urban planning
Emissions (location)	Both direct (within the urban area) and indirect (e.g. due to the generation of electricity consumed in the urban area)	Only direct, i.e. all the emissions "directly" generated by the players of the system (e.g. households)
Emissions (type)	Mainly CO_2	CO_2, local pollutants (e.g. particulate)
Measures	Simulation. Cost-benefit analysis of individual stand-alone measures	Optimisation/Simulation (what-if analysis). Integrated system approach
Urban/ Energy	Urban planning and energy planning are carried out separately	Urban planning and energy planning are carried out together

3.1 Cesena

The modelling analysis developed within the InSMART project has identified combinations of measures (planning hypotheses) that are ranked "high" according to the preferences of city stakeholders. Among six alternative planning hypotheses (De Miglio et al. 2016a) the modelling analysis has identified the two hypotheses which perform best against a set of criteria. These alternatives have been composed combining, and simulating, measures in different sectors into comprehensive plans (De Miglio et al. 2016b).

In particular, the most interesting planning option includes the refurbishment of a fraction of the existing building stock, the speed reduction and modal shift from private car transport to cycling, the production of a certain amount of renewable energy, and the promotion of strong information campaigns to support and inform citizens about objectives and opportunities of energy-related rational behaviors. The expected achievements of this plan, as calculated by the models, are reported below:

- Saved energy, with respect to the reference case: 135 TJ in residential buildings.
- Direct CO_2 emissions: 31% below 2013 value, and 17% below 1995 value (reference year of the exiting SEAP).
- CO_2 emissions covered by the analysis: 3.2 t/capita in 2020 and to 2.5 t/capita in 2030.
- Total overnight costs associated with these actions: 120 million Euro approximately, out of which 58% is for the refurbishment of the building stock.

3.2 Évora

By applying the Energy System Model for Évora it became clear from the analysis that not all options contribute equally to energy savings or carbon emissions reduction. For instance, the scenario with higher energy savings is not necessarily the scenario with higher emission reductions. In terms of energy savings, residential buildings have the highest potential and compared with the other modelled measures, these interventions are not the most expensive ones (Simoes et al. 2016a, b). However, the municipality has limited capacity of investing in private residential buildings. Through the application of the MCDA analysis (Simoes et al. 2016a, b) it was found that the best performing options were shifting 15% from private cars mobility to public transportation; fostering use of biofuels in city buses; decrease MSW production per capita; restricting traffic and parking in the city centre, installing insulation and solar thermal water collectors in residential buildings.

Consequently, the municipality developed a plan to implement several actions, such as: review existing municipal programs for private building renovation in the historic centre and provide access to credit schemes for residential owners towards passive energy efficiency measures in the building envelope; increase bike lanes and parking fees, negotiate with the bus company to change fleet to buses running exclusively on biofuels and to increase their frequency, increase the share of MSW collected for recycling via information campaigns and replacement of all street lighting with LED lamps. The adopted measures allow for the following achievements:

- Saved energy in 2030, compared with the reference case in 2030: 5918 GJ/year in residential buildings, 37,285 GJ/year in transport. Note that the total consumption in 2013 in buildings was of 421TJ and in transport of 961TJ.
- CO_2 emissions in 2030: 21–22% below 2013 values.
- CO_2 emissions per capita in 2033: 7% below 2013 values.
- Total overnight costs associated with these actions: 15 million Euro approximately, out of which 48% is for the parking to improve mobility in the city's historic centre.

3.3 Trikala

The analysis of the quantitative results obtained from the Energy System Model for Trikala showed that as far as energy savings are concerned, the building sector has the highest potential. However, these interventions are also associated with a high implementation cost, since the main actions are building shell refurbishment. It is clear that energy efficiency should play a major role in achieving the CO_2 emission reduction targets for 2030 (Giannakidis et al. 2016). On the other hand the estimated RES potential is adequate to cover up to 10% of the local electricity consumption by 2030, with an implementation cost that is quite below that of the building refurbishment actions.

The options that came forward from the MCDA analysis (Stavrakakis et al. 2016) were improved and extended cycling routes and extension of the ring road (with indirect effects on the quality of life in the city centre), implementation of green spaces within the city, refurbishments of all municipal buildings and connection of 80% of the buildings to the natural gas grid. Actions that need to be taken by the municipality include the replacement of all street lighting with LED lamps, the replacement of municipal vehicles and the introduction of incentives for the promotion of hybrid/electric vehicles in the city centre. The calculated achievements of these actions are:

- Saved energy compared to a reference case: 51,098 GJ/year in residential buildings, 7540 GJ/year in transport. The total consumption in 2012 in buildings was of 903 TJ and in transport of 826 TJ.
- CO_2 emissions in 2030: 9% below 2012 values.
- CO_2 emissions per capita in 2030: 15% below 2012 values.
- Total overnight costs associated with these actions: 110 million Euro approximately.

3.4 Nottingham

The MCDA process identified the top ranked scenarios that represent the maximum level of local engagement for the city of Nottingham. Among seven alternative planning hypotheses (Long and Robinson 2016a), the three best-performing scenarios according to local stakeholder's evaluations were shortlisted and further analysed. Each scenario has been properly decomposed in a set of measures and, once the applicability was verified, they form the "ingredients" of a comprehensive action plan (Long and Robinson 2016b).

Specifically, the plan promotes measures to improve efficiency in both the existing and the new residential building stock; to shift from a private car-based mobility towards public transport and cycling; and the highest level of local engagement with 'forced' inclusion of biomass fuelled CHP generation, plant scale PV and low carbon housing.

The expected achievements of this plan (results of the modelling exercise) are reported below:

- Saved energy with respect to the reference case: 650 TJ in residential buildings, 3000 TJ in city-only transport, and 4400 TJ in travel-to-work area.
- Direct CO_2 emissions: 21.17% below 2014 value.
- CO_2 emissions per capita covered by the analysis: reduced in 2030 by 28.51% relative to 2014.
- Total cost associated with the plan over the projected time horizon: 881 million Euro (780 million £ approximately).

4 Challenges Faced When Addressing the Role of Cities Towards a Below 2 °C World

The InSMART integrative participatory method for planning towards sustainable transitions in cities allowed thinking ahead about energy consumption in the cities as an integrated urban energy system and allowed to demonstrate how the energy system can be affected by urban policy making (especially urban planning). This highlighted new priorities, instead of those traditionally taken under municipal management, which is a most fundamental mind set change to ensure an effective city contribution towards a below 2 °C world.

It was found that developing long-term very ambitious mitigation targets for and with medium and small sized cities posed a set of challenges. The most overarching of which is to understand to what extent can cities (and the players acting within the cities) influence roadmaps and policies that effectively get to a below 2 °C world. Many of the most radical energy system transformation, particularly on the energy supply-side, are out of the sphere of influence of any city, especially of medium and small sized cities and of any household agent. Some examples are the implementation of carbon capture and storage options associated to power production, the management of methane venting and flaring and the associated emissions of black carbon, the investments in large scale renewable energy projects. The possibility to design and implement policy instruments such as taxes, subsidies, cap and trade mechanisms, standards, is also quite limited at the level of cities.

Despite increasing worldwide cities commitment for ambitious GHG mitigation, in concrete terms and regarding ongoing daily activities, city planners/decision makers are mostly focused on energy consumption of municipality managed buildings, public spaces and public lighting, for which they are directly responsible and for which they must pay a substantial energy bill. Moreover, the decision criteria used in urban planning do not necessarily include GHG mitigation, or if this is considered, it has relatively low weight. The wellbeing of citizens, including fostering economic development in the city, boosting employment and attracting residents are frequently the most important criteria to be considered. Related to that, despite more or less ambitious long-term GHG mitigation targets, cities tend to develop plans to a period of 10–20 years in the future, which is not always compatible with the long-term thinking of a below 2 °C roadmap and policies.

Similarly, individual preferences are often driven by subjective perceptions, limited knowledge/imperfect information and low willingness to invest (affordability of new efficient technologies), therefore the energy uses of the most energy intensive urban sectors (private buildings and transport) are in many cases hard to be turned into GHG-oriented and sustainable modes.

So what, in fact, can be done at city level to move towards such a global ambitious target?

Although it is very important not to overestimate the contribution and the area of influence of city-agents to the global GHG target, it is undoubted that municipalities are extremely well positioned to steer households towards efficient and low-carbon energy consumption within buildings and transportation (less the industry sector), in itself major CO_2 emitters. Their role in bridging locally the gap between what is perceived/known and what would be economically and technically feasible, is key for the success of a global climate target. With respect to this point, initiatives like awareness campaigns, the promotion of an environmental-friendly city brand (e.g. Indicators for City Services and Quality of Life, ISO37120) or the creation of energy helpdesks to inform people about the benefits of energy savings are generally short-term actions with a long-term perspective/goal: fostering a culture of sustainability through communication and education. In combination with supra national/national-determined technological standards and economic incentives, these actions create the conditions for a more likely success.

But the biggest area of actions for the cities to contribute to the global target, is to find and use the synergies between the incumbent and highly-priority local needs and the global challenge. Not many energy-system related measures can be realistically implemented at city level, while many urban-related actions can be taken to cope with city specific issues with significant benefits for the GHG emissions reduction.

As urban regulators and planners, cities have the possibility to allocate land only to projects which prove to be socially desirable and environmental friendly. For example, the permission of construction for new districts or buildings can be made subject to the strict respect of standards, such as net zero emissions or even negative emissions if trees are planted to absorb CO_2. At the same time, the organisation of work as well as of the school schedule can be adjusted to avoid congestion and inefficient start & stop cycles of private cars in the peak times; some urban streets can be closed to the private mobility thus boosting the public transportation systems (the local fleet can also be turned into a green fleet) and alternative private modes, and/or congestion charges introduced in some area of the city.

Therefore, in order to engage the cities in the global challenge policies and measures should be designed to sustain those changes with the highest synergies between local-specific needs and GHG reduction potentials.

5 Conclusion

InSMART's methodology offers an integrated and participatory process to examine all the energy consumption sectors together with potential local energy generation options and come up with a smart development plan for the city energy sector that is supported by all stakeholders. This approach can be integrated into the process of developing a Sustainable Energy Action Plan or a Sustainable Energy and Climate Action Plan by municipalities participating in the Covenant of Mayors for Climate and Energy and offers the advantage of concrete scientific approaches in local energy planning. The methodology makes use of a well-orchestrated set of tools to identify the optimum mix of short, medium and long-term measures for a sustainable energy future, addressing the efficiency of energy flows across various city sectors with regards to economic, environmental and social criteria and paving the way towards actual implementation of priority actions and a more sustainable city planning.

The tools and experience gained through their implementation may facilitate and support the deployment of local sustainable solutions and decision-making in cities, contributing to the realisation of the Paris Agreement targets. Municipalities can contribute significantly by steering households towards low-carbon energy consumption, but they need to move towards new priorities instead of those traditionally taken until now.

Acknowledgements This project has received funding from the European Union's FP7 programme under grant agreement No 314164.

The authors acknowledge Nottingham City Council, Municipalities of Évora and Cesena, the Municipal Water Supply-Sewage Enterprise of Trikala (DEYAT), the technical partners FCT, Systra Consultancy, E4SMA S.r.l., University of Nottingham and CRES and all the people directly and indirectly involved in the project, as they all contributed to the development of this work.

References

Brans JP, Mareschal B (1992) Promethee V: MCDM problems with additional segmentation constraints. INFOR 30:85–96

De Miglio R, Chiodi A, Gargiulo M (2016a) Report on optimum sustainability pathways—Cesena. http://www.insmartenergy.com/wp-content/uploads/2014/12/D5.4-Optimum-Sustainable-Pathways-Cesena.pdf. Accessed 22 Sep 2017

De Miglio R, Chiodi A, Gargiulo M, Burioli S, Morigi S, Maggioli B, Gentili M (2016b) Mid-Term implementation action plan—Cesena. Available from: http://www.insmartenergy.com/wp-content/uploads/2014/12/D6.5-Mid-Term-Implementation-Action-Plan-Cesena.pdf. Accessed 22 Sep 2017

Dias L, Seixas J, Gouveia JP (2015) Assessment of RES potential at city level. The case of solar technologies. WP4, T4.4. INSMART Integrative Smart City Planning project (ENER/FP7/314164). Available from: http://www.insmartenergy.com/wp-content/uploads/2014/12/I.R.12.WP4-Assessment-of-Solar-Potential.pdf. Accessed 22 Sep 2017

Giannakidis G, Siakkis P, Nychtis C (2016) Report on optimum sustainability pathways—Trikala. http://www.insmartenergy.com/wp-content/uploads/2014/12/D5.2-Optimum-Sustainable-Pathways-Trikala.pdf. Accessed 22 Sep 2017

Hinkle D (1965) The change of personal constructs from the viewpoint of a theory of construct implications. Ph.D. dissertation, Ohio State University

IEA (2016) Energy technology perspectives. http://www.iea.org/etp2016/. Accessed 22 Sep 2017

InSMART (2016) Project website http://www.insmartenergy.com/work-package-5/. Accessed 22 Sep 2017

Keirstead J, Jennings M, Sivakumar A (2012) A review of urban energy system models: Approaches, challenges and opportunities. Renew Sustain Energy Rev 16(6):3847–3866

Kostevšek A, Petek J, Čuček L, Pivec A (2013) Conceptual design of a municipal energy and environmental system as an efficient basis for advanced energy planning. Energy 60:148–158

Lind A, Espegren K (2017) The use of energy system models for analysing the transition to low-carbon cities—the case of Oslo. Energy Strateg Rev 15:44–56

Long G, Robinson D (2016a) Report on optimum sustainability pathways—Nottingham. http://www.insmartenergy.com/wp-content/uploads/2014/12/D5.1-Optimum-Sustainable-Pathways-Nottingham.pdf. Accessed 22 Sep 2017

Long G, Robinson D (2016b) Mid-Term implementation action plan—nottingham. http://www.insmartenergy.com/wp-content/uploads/2014/12/D6.2-Mid-Term-Implementation-Action-Plan-Nottingham.pdf. Accessed 22 Sep 2017

Reckien D, Flacke J, Dawson RJ, Heidrich O, Olazabal M, Foley S, Hamann JJP, Orru H, Salvia M, De Gregorio Hutado S, Geneletti D, Pietrapertosa F (2014) Climate change response in Europe: what's the reality? Analysis of adaptation and mitigation plans from 200 urban areas in 11 countries. Clim Change 122(1):331–340

Simoes S, Dias L, Gouveia JP, Seixas J (2016) Report on optimum sustainability pathways—Évora. http://www.insmartenergy.com/wp-content/uploads/2014/12/D5.3-Optimum-Sustainable-Pathways-Evora.pdf. Accessed 25 Sep 2017

Simoes S, Dias L, Gouveia JP, Seixas J (2016) Report on the multicriteria methodology, the process and the results of the decision making—Évora. http://www.insmartenergy.com/wp-content/uploads/2014/12/D5.7-Report-on-the-Multi-criteria-methodology-Evora.pdf. Accessed 25 Sep 2017

Stavrakakis G, Nychtis C, Giannakidis G (2016) Report on the multicriteria methodology, the process and the results of the decision making—Trikala, Greece. http://www.insmartenergy.com/wp-content/uploads/2014/12/D5.6-Report-on-the-Multi-criteria-methodology-Trikala.pdf. Accessed 22 Sep 2017

United Nations (2014) World Urbanization Prospects: The 2014 revision, highlights (ST/ESA/SER.A/352). https://esa.un.org/unpd/wup/publications/files/wup2014-highlights.Pdf. Accessed 22 Sep 2017

Xydis G (2012) Development of an integrated methodology for the energy needs of a major urban city: The case study of Athens, Greece. Renew Sustain Energy Rev 16(9):6705–6716

Yamagata Y, Seya H (2013) Simulating a future smart city: An integrated land use-energy model. Appl Energy 112:1466–1474

Zanon B, Verones S (2013) Climate change, urban energy and planning practices: Italian experiences of innovation in land management tools. Land use policy 32:343–355

Mitigation of Greenhouse Gas Emissions in Urban Areas: The Case of Oslo

Arne Lind and Kari Espegren

Key messages

- Phase out of fossil fuels in the stationary sector occurs in all climate mitigation scenarios.
- A mixture of different technology choices is utilised for greening the transport sector, where some of them are requiring strong support schemes in order to succeed.
- The use of an urban TIMES model facilitates the dialogue with representative stakeholders at local level.

1 Introduction

Globally, urbanization is taking place rapidly, especially in less economically developed countries. As of 2016, 54% of the world's population lives in urban areas and various projections show shares as high as 75% in 2050 (The World Bank 2017). Sustainable development in urban settlements is, therefore, a challenge with increasing importance. This includes problems such as transport congestion, lack of sufficient housing, and environmental degradation. Based on an energy and climate perspective, the following features are incorporated in a sustainable city:

- Public transport is a feasible substitute to cars;
- Use of renewable resources as an alternative to fossil fuels;
- Waste is recycled if possible, otherwise it is seen as a valuable resource;

A. Lind (✉) · K. Espegren
Institute for Energy Technology (IFE), Kjeller, Norway
e-mail: arne.lind@ife.no

K. Espegren
e-mail: kari.espegren@ife.no

© Springer International Publishing AG, part of Springer Nature 2018
G. Giannakidis et al. (eds.), *Limiting Global Warming to Well Below 2 °C: Energy System Modelling and Policy Development*, Lecture Notes in Energy 64,
https://doi.org/10.1007/978-3-319-74424-7_23

- More energy-efficient buildings, including innovative construction or retrofitting technologies.

Based on the above, an energy systems approach is used in this work to link and study the interactions between the aforementioned features.

Generally, an energy system is more than a technical system (Keirstead et al. 2012), where markets, institutions, consumer behaviors and other factors affect the way infrastructures are constructed and operated. During the last decade, an increased focus on studies of urban energy systems can be seen in literature. In a work by Markovic et al. (2011), different tools to analyse energy, economic and environmental performances of energy production systems, buildings, and community equipment are presented and discussed. The models are divided into the following topics: geography, energy, evaluation and clean energy analysis tools. However, the MARKAL/TIMES modelling framework, as used in the current work, is not described and discussed.

During the last few years, there has been an increased use of TIMES models on the local scale. IEA (2016a) analyzed pathways for a low-carbon society on a local scale, Fakhri et al. (2016) focused on local infrastructure development, and the work by Dias and Simoes (2016) was related to smart city planning. The link between these works is related to development of local TIMES models to analyze a cost optimal transition to more sustainable energy carriers and technologies on a local level.

Other optimization models have also been used for urban analyses. Keirstead and Calderon (2012) compared four European cities by assessing technology pathways to achieve emission reduction targets at minimum cost, including the links between the governance of urban energy systems and the cost of achieving carbon targets. However, the study did not include the transport sector as an integral part of the analysis. The Norwegian electricity system is renewable, and more than 95% of the electricity is based on hydro power, thus there is generally no option for carbon reduction within the Norwegian power sector. Policies and measures to decarbonize the transport sector will therefore have a significant impact on Norwegian cities CO_2 emissions in the future.

As an alternative to overall system optimization, analysis and optimization of individual systems can be combined into an overall energy plan. Lindenberger et al. (2004) analyzed heating systems in various buildings, and combined the results into an overall energy plan for the local community. In general, such an approach can lead to sub-optimal systems because interactions between different systems are ignored.

The current work describes how a technology-rich optimization model is used to analyze how the city of Oslo can be transformed into a low-carbon city. The urban energy system model was used to analyse the following scenarios: a reference scenario (REF) including all current policies, a two-degree scenario (2DS), and a climate target scenario (CLI) including the climate targets for Oslo (50% reduction of greenhouse gas emissions by 2030, and to use no fossil fuels by 2050). The latter is an example of a well below 2 °C scenario.

An advantage of a local area TIMES model is the possibility to integrate it with a TIMES model on a national or regional scale. This allows for comparison of results on a local versus national level, as demonstrated in IEA (2016a, b). The TIMES-Oslo model (Lind and Espegren 2017) used in this work is part of the TIMES-Norway model (Lind et al. 2013); thus, it can be used to enrich the analysis on a national level. As an example, the differences in technology choices between Oslo and the rest of Eastern Norway is more easily identified in two different models than in an aggregated model region. Additionally, the results for Oslo can also be relevant for other Norwegian cities.

2 The Ambitious Climate and Energy Goals of Oslo

Oslo is the capital of Norway and is also the most populous city in the country. Oslo constitutes both a county and a municipality, and it is the economic and governmental center of Norway. In January 2017, the population in the city of Oslo was 666,757 inhabitants (Kommune 2017), whereof almost the entire population lives in urban settlements. In addition, the Metropolitan area of Oslo has a population of just above 1.5 million.

Because of Oslo's northern latitude, the amount of daylight varies considerably, from more than 18 h in the summer, down to six hours in the winter. The city has rather nice summer temperatures, with two out of three days in July with temperatures above 20 °C. Oslo has a varied economy with enterprises covering several sectors. The share of employees is highest for the service sector, reaching just above 90% of total employment (Kommune 2015).

The city of Oslo has ambitions climate goals; to reduce the city's CO_2 emissions by 50% by 2020 and by 95% by 2030 compared to the 1990 level, and to eliminate fossil fuel usage by 2050. Oslo will implement a so-called green transformation, and the focus is how to address climate change challenges by changing the way energy is produced and used in the city. The green transformation represents a transition to a sustainable society based on renewable energy, and the main focus is on innovation, implementation of new technologies, and using existing systems in new and innovative ways. The transport sector has the highest share of CO_2 emissions, and accounts for almost 60% of the CO_2 emissions in Oslo.

The city of Oslo has developed and implemented a climate and energy strategy (Kommune 2016). The strategy focuses especially on the transport sector, wherein pedestrians, cyclists and public transport are prioritized areas. Oslo aims at reducing car traffic by 20% in the period from 2016 to 2020 and by 33% in 2030 compared to 2016 level. Additionally, the strategy focuses on urban development through planning of urban areas and public transport junctions, building of new infrastructure for renewable transport fueling (battery charging, hydrogen and biofuels), and preparing for fossil fuel-free transport in freight, public and private transport.

There is only one small hydropower plant in the city of Oslo. The plant has an installed capacity of 5 MW and a yearly power production of around 16 GWh. Electricity is also produced from waste in a CHP plant, with a maximum annual production of around 160 GWh. Currently, there is also some electricity production from solar PV in the city of Oslo. Additionally, 12 district heating plants are operated by the local energy company within the city limits.

Oslo is a small city in a global context. However, the city wants to take responsibility for the development of sustainable energy systems for the future, and to show how cities can take leadership in the transformation to more sustainable solutions in cities. Roughly half of the world's urban dwellers reside in relatively small settlements of less than 500,000 inhabitants (United Nations 2014). Oslo can be a relevant example of what can be achieved in a medium-sized city, especially in developed countries.

3 Modelling Framework

3.1 The Overall Structure of TIMES-Oslo

The structure of TIMES-Oslo is illustrated in Fig. 1. The base year of the model is 2010, whereas all costs, prices, etc. are given in NOK-2005. Since the Oslo model is based on TIMES-Norway (Lind and Rosenberg 2013; Lind et al. 2013; Rosenberg et al. 2013; Lind and Rosenberg 2014; Rosenberg and Lind 2014), it has the same time resolution as TIMES-Norway (260 time slices per year). Having a time resolution with five time slices per week gives a sufficiently detailed description of the Norwegian hydropower system, while at the same time covering the different demand profiles. The time horizon of TIMES-Oslo is from 2010 to 2050.

The TIMES-Oslo model has been used to analyze how Oslo can be transformed into a low-carbon city by focusing on the focus areas above. Further information regarding the model can be found in Lind and Espegren (2017). Geographically, the TIMES-Oslo model represents the city of Oslo as a single model region. This boundary was needed in order to have a modelling framework representing the city of Oslo's jurisdictional area. Consequently, TIMES-Oslo represents an extraction of one of the geographical regions in TIMES-Norway; the NO1 pricing area in the Nordic spot market (Statnett 2017). Only a small part of the electricity used in Oslo comes from local production facilities. The remaining electricity comes through NO1 and is typically produced in either NO1 or other Norwegian regions. Additionally, it is only the transport within the city limits that is included in the Oslo region of the model.

The model results are based on a linear least cost optimization model assuming perfect competition and perfect foresight. Future technologies not invented are not included in the model. The modeling approach has not taken into account the

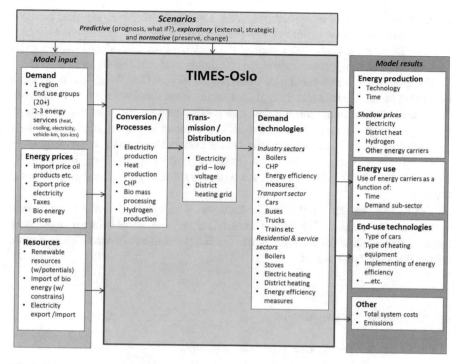

Fig. 1 Schematic drawing TIMES-Oslo model

human behavior aspect, as this is hard to incorporate properly in an optimization model like TIMES. Human behavior is particularly relevant when considering the implementation of energy efficiency measures where the public's willingness to implement such measures is not considered. In addition, infrastructure costs for non-energy items (e.g., tunnels or roads) are not included in the model. This is a weakness as these costs must be added after the analysis. Another shortcoming of the modeling approach is the lack of other emissions than CO_2. In an urban area, the local air quality is clearly of high importance. Modal shifts are also handled exogenously in the model. Despite these limitations, the TIMES-Oslo is still a powerful tool for analyzing alternative pathways to a low-carbon city.

3.2 Increasing End-Use Demands of the City

The demand for various energy services are supplied exogenously to the model. In total, there are 43 different end use demand categories in the TIMES-Oslo model. Load profiles are developed for the relevant demand categories according to the same procedure as presented in Lind et al. (2013). The methodology for energy demand projection is demonstrated in detail in Lind and Espegren (2017).

Assumptions of economic growth, business development, demographics, etc. and development of energy indicators are considered, as well as normative measures (e.g., building regulations). The energy demand is divided into four main sectors with underlying sub-groups: industry, households, service, and transport.

The energy service demand projection for the city of Oslo increases to 18 TWh in 2050 (Fig. 2). While industry and agriculture have a rather constant demand in the future, the transport, households and service sectors have an expected high growth in future demand. In total the energy demand increases with almost 40% compared to 2010. Projections of the future population growth are an important driver of future energy demand. A local population projection, provided by (Statistics Norway 2014), is used as the basis for the energy demand projection for Oslo.

The energy demand projection for transport is calculated based on increase in vehicle-km in transport. The person transport demand in vehicle-km increase with approx. 75% from 2010 to 2015, while the freight demand (v-km) almost double within the same time period.

4 Scenario Assumptions

4.1 Energy Prices and Current Policies

Unless otherwise noted, the analyses in this chapter include all active national measures today. The most important policies are the green certificate market

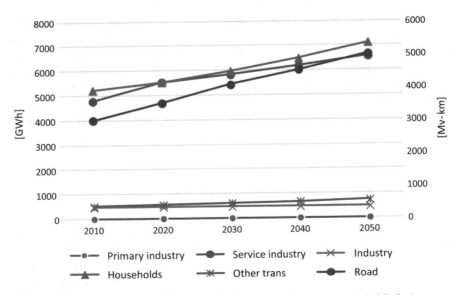

Fig. 2 Energy service demand projection (GWh/year, except road transport in Mv-km)

(GCM) and present policy measures implemented by the public enterprise Enova working towards sustainable energy consumption and generation. The energy taxes are kept constant at the 2014 level until 2050, including the value added tax (VAT), nonrecurring tax for new vehicles, fuel tax for road transport, electricity consumption tax, and various CO_2 taxes.

Development in energy prices for imported energy carriers correspond to the *Current Policy Scenario* of IEA (2013). The prices of electricity imports and exports (to and from Norway) are given exogenously to the model. The electricity trading prices are based on historical prices until 2014, and thereafter kept at the actual 2014 level throughout the analysis. The price assumptions use electricity price variability according to the historical variations observed in 2014. Although it is expected much more volatile prices towards 2050, the annual average electricity price may not necessary increase. We have therefore chosen to keep the future prices at the 2014 level.

The electricity price for Oslo is determined endogenously in the model. The capacity of the power exchange in the existing grid is included in the TIMES-Oslo model. It is possible to invest in new grid capacity to and from Oslo, as well as internally in the region. Associated costs can be found in Lind et al. (2013).

4.2 Climate Scenarios

The city of Oslo identified the following five focus areas for the energy and climate strategy for Oslo: urban development, infrastructure [includes energy stations for renewable fuels in transport (e.g. battery charging, hydrogen, and biofuels)], transport, buildings, energy production and distribution.

Three main scenarios were developed to analyze the transition towards a low-carbon city for Oslo (Table 1). The *reference scenario (REF)* includes all current national policies. This scenario is used to illustrate the effects of the policies analyzed in the other scenarios. The *second scenario is a two-degree scenario (2DS)*. It corresponds closely to the 2DS presented in the Nordic Energy Technology Perspectives 2016 (IEA 2016a). At a global level, it requires an energy system consistent with emission trajectories that would give a high chance of limiting the average global temperature increase to 22 °C. At the Oslo level, a 50% reduction compared to the 1990 level is included in 2030, and in 2050 this

Table 1 Scenario overview

Name	Description
REF (reference scenario)	Current national policies
2DS (two degree scenario)	50% reduction in CO_2 emissions by 2030 from the 1990 level 87% reduction by 2050 from 1990
CLI (climate targets)	50% reduction by 2030 from 1990 no fossil fuels by 2050

reduction constraint is 87%. The *third scenario, the climate target scenario (CLI),* includes halving the emissions of greenhouse gases before 2030 compared to 1990 level, and eliminating fossil fuel usage by 2050, as targeted by Oslo in its climate plans. The targets are added as constraints in the model. In all three scenarios, it is assumed that waste used in the local waste incineration plant has a carbon content of 11 tonnes of CO_2/GWh in 2050. No CO_2 emissions are considered from the use of biomass.

5 Model Results

5.1 Reference Scenario: The Increasing Share of Transport in Energy and Emissions

The reference scenario (REF) is analyzed by keeping the market shares of the different technologies constant within each end-use sector. This means that it is not possible to invest in more efficient technologies or to implement energy efficiency measures in REF, reflecting the behavior of irrational consumers due to lack of knowledge for example. The exceptions are that new buildings can choose any available technology and that new vehicle technologies come with an efficiency improvement. This scenario is not meant to illustrate a realistic baseline, but a development following the same path as the city of Oslo is following today. Relaxing this constraint in climate scenarios means that the appropriate measures would be implemented to overcome the existing barriers.

Total emissions increase from 1.17 Mt in 2010 to 1.46 Mt in 2050 (Fig. 3). The contribution from the transport sector becomes even more dominant in the future. The reason is that the projection of the transportation demand increases more, in relative terms, than the stationary sector. Road freight transport increases the most, driven by the expected economic growth and increased international trade.

Fig. 3 CO_2 emissions for the reference scenario (incl. statistics up to 2010)

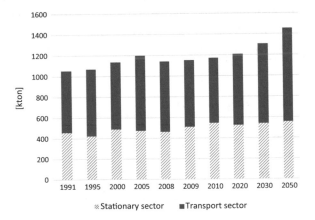

[kton]

Stationary sector Transport sector

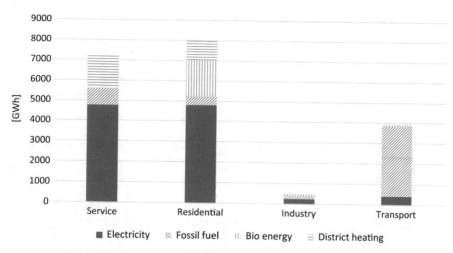

Fig. 4 Energy consumption per sector in 2050 in the reference scenario (GWh)

For Oslo, the total energy consumption in the reference scenario increases from 14.2 TWh in 2010 to 19.7 TWh in 2050 (Fig. 4). Electricity dominates the stationary sector, especially within the service sector. The use of various fossil fuels dominates the transport sector, which clearly emphasizes the significance of this sector in relation to the CO_2 emissions. In 2050, 33% of the energy consumption within the industry sector in the city of Oslo comes from fossil fuels. However, the overall energy use is relatively low for this sector, so the contribution to the total CO_2 emissions is only around 3% in 2050.

5.2 Climate Scenario Results

5.2.1 Drastic Changes in End-Use Sectors

In terms of CO_2 emissions, both 2DS and CLI scenarios comply with Oslo's 2030 target, whereas only the CLI scenario satisfies the 2050 target (Fig. 5). The contribution from the various end-use sectors varies over the analysis period. However, the transport sector and the production of district heat are the main contributors of CO_2 emission for all years for both scenarios.

There is no use of fossil fuels for heating for either of the climate scenarios (Fig. 6). This is stated by definition for the CLI scenario, but interestingly, fossil fuel phase-out also occurs in the 2DS scenario. Compared to REF, there is considerable increase in the use of pellet boilers for both 2DS and CLI, especially in the latter scenario. In the 2DS scenario, the use of electric radiators and oil boilers is reduced compared with REF and compensated by increased use of district heating

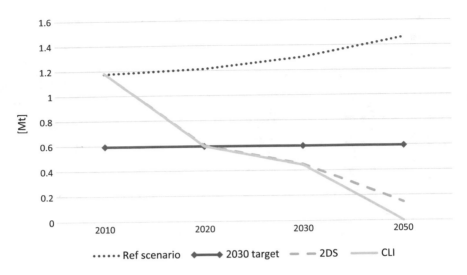

Fig. 5 Summary of CO$_2$ emissions

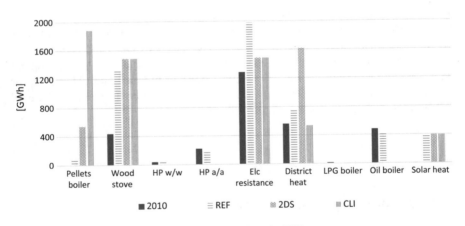

Fig. 6 Household energy consumption per technology in 2050

(1.6 TWh in 2050). District heating is mainly based on waste incineration (which entails greenhouse gas emissions due to the carbon content of waste).

The energy consumption in the transport sector is reduced by more than 40% for the climate scenarios compared to REF in 2050 (Fig. 7). This is due to the use of more efficient technologies, such as electric cars (Table 2). Consequently, the electricity consumption is considerably higher for both 2DS and CLI compared to REF.

As the transport sector is the most important source of CO$_2$ emissions in Oslo, this sector undergoes significant changes in both 2DS and CLI. The results show considerable efficiency improvements through electrification of passenger transport.

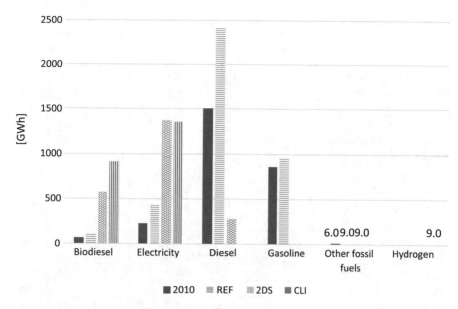

Fig. 7 Use of energy carriers for the transport sector (2050)

This includes both the use of electric vehicles (2DS and CLI) and plug-in hybrid vehicles (only 2DS). For the CLI scenario, long distance travel by car is covered mostly using biodiesel vehicles, whereas hybrid vehicles using electricity and gasoline are used in 2DS. Other differences between 2DS and CLI include the additional use of biodiesel for other transport modes (e.g., construction machines, cranes, etc.) in CLI instead of the use of regular diesel, as well as use of hydrogen in sea transport in the CLI scenario. For freight purposes, a combination of electric (light duty) and biodiesel (heavy duty) vehicles is used in both climate scenarios.

5.2.2 A Relatively Low Abatement Cost

The average abatement cost for the 2DS analysis is just above 256 Euro per ton CO_2 removed. In addition to all the costs related to the energy system (e.g., investment and operational costs), the calculated abatement cost includes costs related to disposal of oil boilers and tanks, as well as costs for retrofitting existing buildings for obtaining better energy performance standards.

The average abatement cost for the CLI analysis is 264 Euro per ton of CO_2 removed. This is slightly higher than for the 2DS analysis, where the increased costs among other are related to the use of advanced construction machineries and hydrogen technologies within sea transport.

One of the key findings from this work is that most of the emissions from the stationary sector can be removed at a low abatement cost (below 56 Euro/ton CO_2),

Table 2 Use of fuel and technologies [TWh] for the transport sector

	2010	2030			2050		
	REF	REF	2DS	CLI	REF	2DS	CLI
Biodiesel (car + blend)	17.9	17.9	0.0	0.0	20.3	0.0	108.1
Electricity (car + hybrid)	0.7	2.6	473.8	473.8	2.9	499.7	475.1
Diesel car	340.7	340.1	0.0	0.0	385.8	0.0	0.0
Gasoline (car + hybrid)	857.8	859.9	18.6	18.6	972.9	19.9	2.8
LD + HD trucks (biodiesel blend)	39.3	49.3	17.3	17.3	82.8	0.0	0.0
LD + HD trucks (diesel)	746.4	937.5	328.9	328.9	1574.0	0.0	0.0
LD trucks (electric)	0.0	10.9	312.8	312.8	0.0	450.3	450.3
HD trucks (biodiesel)	0.0	0.0	0.0	0.0	0.0	397.5	397.5
Other transport (biodiesel)	0.0	0.0	0.0	0.0	0.0	0.0	227.4
Other transport (diesel)	285.7	286.5	286.6	286.6	288.0	288.0	3.7
Public transport (elc)	223.4	271.3	324.1	324.1	432.0	432.0	432.0
Public transport (diesel)	139.7	142.7	151.5	151.5	173.5	0.0	0.0
Public transport (biodiesel)	7.4	7.5	8.0	8.0	9.1	182.6	182.6
Sea transport (fossil fuels)	6.0	6.6	7.5	7.5	9.0	9.0	0.0
Sea transport (hydrogen)	0.0	0.0	0.0	0.0	0.0	0.0	9.0

where most of these actions are relatively easy to implement, such as insulation, phase-out heating system based on fossil fuels, implementation of passive houses, etc.

Transport in the Oslo region must undergo significant changes if the target for 2050 is to be met. The increase in transport demand must, therefore, be covered by either public transport, bikes and/or by walking. Additionally, the public transport in Oslo must be based on renewable energy sources by 2020 and a bicycle strategy must be implemented by 2025 to see 15% of the daily travels by bike. Additionally, renewable fuels must be used for all transport modes by 2050 in order to reach the overall target. Another finding from the analysis is that strong support schemes for hydrogen infrastructure are needed for decarbonizing the entire transport sector. This action comes with a medium feasibility, and with an abatement cost approaching 167 Euro/ton CO_2.

5.2.3 Need to Support Innovation in Fuels and Technologies

For the two climate scenarios (2DS and CLI), we observed a high utilization of new fuels and technologies. In general, urban areas are more energy-efficient and have more technology options available to mitigate climate change. This is due to smaller apartments, shorter travel distances which allows for biking, walking or public transport etc. The results showed that biofuels are required in the medium to long-term to decarbonize heavy duty vehicles for freight transport. In urban areas, driving distances are typically shorter and charging infrastructure can be more easily constructed due to urban compactness.

To reduce or eliminate the dependence on fossil fuels in road freight, deployment of alternative technologies, such as hybrid vehicles, battery electric vehicles (BEV) or hydrogen fuel cell vehicles (FCEV), use of ethanol or compressed natural gas are needed. The use of electric vehicles (BEV and FCEV) is an option that has the potential to eliminate fossil fuel dependency. Another alternative is biofuels produced in a sustainable matter, not just for freight transport, but for the entire transport sector. As discussed for the 2DS scenario, a combination of using hybrid and biofuel vehicles contributed to a significant reduction in the CO_2 emissions for freight transport. Furthermore, by introducing traffic management schemes, emissions from the transport sector can be reduced even more. This could include park and ride schemes, cycle lanes, congestion charging schemes, carpooling, and low emissions zones.

6 Conclusion

The city of Oslo has ambitious climate targets for next 35 years, including a 50% reduction of greenhouse gas emissions before 2030 and to eliminate fossil fuel use by 2050. To meet these targets, there is a need for new energy infrastructure, new transport solutions and more energy-efficient buildings. Along with other Norwegian cities, Oslo has a shared responsibility to lead by example, and it is important that the results from Oslo are transferable to other cities, both nationally and internationally.

We have developed and used a local TIMES model to be able to analyse how the city of Oslo cost-effectively can be developed into a renewable and low-carbon society

One of the key findings from the work is that most of the emissions from the stationary sector can be removed at a low abatement cost (below 56 Euro/tonne CO_2), where most of these actions are relatively easy to implement. Transport in the Oslo region must undergo significant changes if the target for 2050 is to be met. The increase in transport demand must, therefore, be covered by either public transport or by bikes. Additionally, the public transport in Oslo must be based on renewable energy sources by 2020 and a bicycle strategy must be implemented by 2025 to see 15% of the daily travels by bike. Additionally, renewable fuels must be used for all transport modes by 2050 in order to reach the overall target. Another finding from the analysis is that strong support schemes for hydrogen infrastructure are needed for decarbonizing the entire transport sector. This action comes with a medium feasibility, and with an abatement cost approaching 167 Euro/ton CO_2.

As demonstrated in this study, bottom-up optimization models like TIMES are well-suited for analyzing how urban areas can develop sustainable energy systems in the future. Indeed, the use of a long-term investment model helps avoiding suboptimal systems also at local level. A few shortcomings must be kept in mind and deserve more work. First, future technologies not invented yet are not included in the model, while they may play a key role in ambitious future scenarios.

Second, human behavior aspects are approximately represented. This is particularly relevant when considering the implementation of energy efficiency measures where the public's willingness to implement such measures is not considered. In addition, infrastructure costs for non-energy items (e.g., tunnels or roads) are not included in the model. Emissions besides CO_2, more particularly local air pollutants, would also need to be added, given the high importance of local air quality in urban areas. Despite these limitations, the TIMES-Oslo is still a powerful tool for analyzing alternative pathways to a low-carbon city.

References

Dias L, Simoes S (2016) Integrative smart city planning—energy system modelling for the city of Evora. ETSAP workshop May 2016. Cork, Ireland

Fakhri AS, Ahlgren EO, Ekvall T (2016) Modelling economically optimal heat supply to low energy building areas—the importance of scales. ETSAP Workshop, Cork, Ireland

IEA (2013) World energy outlook 2013. International Energy Agency, Paris

IEA (2016a) Nordic energy technology perspectives 2016—cities, flexibility and pathways to sustainability. International Energy Agency, Paris

IEA (2016b) Energy technology perspectives 2016—towards sustainable urban energy systems. International Energy Agency, Paris

Keirstead J, Calderon C (2012) Capturing spatial effects, technology interactions, and uncertainty in urban energy and carbon models: retrofitting newcastle as a case-study. Energy Policy 46:14

Keirstead J, Jennings M, Sivakumar A (2012) A review of urban energy system models: approaches, challenges and opportunities. Renew Sustain Energy Rev 16(6):19

Kommune O (2015) Oslo kommune statistikkbanken. From www.oslo.kommune.no

Kommune O (2016) Climate and energy strategy for Oslo. From www.oslo.kommune.no/politikk-og-administrasjon/miljo-og-klima/miljo-og-klimapolitikk/klima-og-energistrategi/

Kommune O (2017) Folkemengde og endringer. From www.oslo.kommune.no/politikk-og-administrasjon/statistikk/befolkning

Lind A, Espegren KA (2017) The use of energy system models for analysing the transition to low-carbon cities—the case of Oslo. Energy Strategy Reviews 15:44–56

Lind A, Rosenberg E (2013) TIMES-Norway model documentation. Insitute for Energy Technology

Lind A, Rosenberg E (2014) How do various risk factors influence the green certificate market of Norway and Sweden? Energy Procedia 58:6

Lind A, Rosenberg E, Seljom P, Fidje A, Lindberg K (2013) Analysis of the EU renewable energy directive by a techno-economic optimisation model. Energy Policy 60:13

Lindenberger D, Bruckner T, Morrison R, Groscurth H-M, Kümmel R (2004) Modernization of local energy systems. Energy 29:11

Markovic D, Cvetkovic D, Masic B (2011) Survey of software tools for energy efficiency in a community. Renew Sustain Energy Rev 15(9):6

Nett H (2012) Lokal Energiutredning 2011 for Oslo Kommune. From www.hafslund.no

Rosenberg E, Lind A (2014) TIMES-Norway—oppdateringer 2013/2014. Institute for Energy Technology

Rosenberg E, Lind A, Espegren KA (2013) The impact of future energy demand on renewable energy production—case of Norway. Energy 61:12

Statistics Norway (2012) Energiregnskap og energibalanse—8: Beregning av fornybar energian-delen for Norge, totalt 2004–2010. Statistics Norway

Statistics Norway (2014) Population projection 2012.06.20

Statnett (2017) Elspot areas. From http://www.statnett.no/en/Market-and-operations/the-power-market/Elspot-areas–historical/Elspot-areas/

The World Bank (2017) Urban population. From https://data.worldbank.org/indicator/SP.URB.TOTL.IN.ZS?view=map

United Nations (2014) World urbanization prospects: 2014 revision

Achieving CO$_2$ Emission Reductions Through Local-Scale Energy Systems Planning: Methods and Pathways for Switzerland

Mashael Yazdanie

Key messages

- An approach combining clustering techniques and cost optimization modeling is applied to assess local energy systems within Switzerland under a climate policy.
- Key local energy system archetypes are analyzed and demonstrate a significant collective decarbonization potential in the long-term.
- CO$_2$ taxes, building renovations, and decentralized generation technologies (e.g., solar photovoltaics) are instrumental in reducing local-scale carbon emissions.

1 Introduction

Today, urban areas accommodate the majority of the world's population and account for more than 70% of global energy-related CO$_2$ emissions (Edenhofer et al. 2015). As urban populations are expected to grow even larger in the long-term, local energy systems planning has the potential to play a key role in carbon mitigation efforts. That is, in order to achieve a mean global temperature increase of less than 2 °C by the turn of the century, carbon mitigation strategies should involve the active participation of policymakers not only on an international or national level, but also on a local scale.

Given the broad focus of climate and energy strategies, energy system models tend to focus on analysis at the national or international level. However, methods are needed to characterize and evaluate the range of local energy systems within a

M. Yazdanie (✉)
Laboratory for Energy Systems Analysis, Energy Economics Group,
Paul Scherrer Institute, Villigen PSI, Switzerland
e-mail: mashael.yazdanie@psi.ch

© Springer International Publishing AG, part of Springer Nature 2018   407
G. Giannakidis et al. (eds.), *Limiting Global Warming to Well Below 2 °C: Energy
System Modelling and Policy Development*, Lecture Notes in Energy 64,
https://doi.org/10.1007/978-3-319-74424-7_24

nation, enabling policymakers to define effective local climate policies within international climate agreements.

This study presents an approach which applies clustering techniques to identify characteristic urban, rural, and suburban energy systems in a case study for Switzerland. These characteristic communities (or archetypes) are then analyzed using a flexible, community cost optimization energy systems model built using the TIMES framework. The performance of each characteristic community is analyzed under climate-stringent national energy policy conditions, enabling policymakers to identify key community players in Switzerland's climate policy arena.

Switzerland ratified the Paris Agreement on Climate Change in 2017, committing itself to reducing national emissions by 50% in 2030 compared to 1990 levels (Swiss Federal Council 2017). Switzerland has further outlined intentions to reduce emissions by 70–85% by 2050. These reductions correspond to CO_2 emission rates of approximately 3 tonnes of CO_2 per capita in 2030, and 1–2 tonnes of CO_2 per capita in 2050 (Swiss Federal Council 2015). Although these targets do not aim for carbon neutrality by 2050, they reflect Switzerland's targets within the Paris Agreement and it is of interest to investigate the local-scale pathways by which these emission reductions can be achieved.

The New Energy Policy (NEP) outlined by the Swiss Federal Office of Energy in (Kirchner et al. 2012) follows the aforementioned reduction targets. The NEP strategy assumes strong efficiency improvements over time with respect to end-use technologies and building space heat demand via renovation measures. High CO_2 taxes are also assumed, reaching approximately 140 Swiss Francs (CHF) per tonne of CO_2 by 2050.

The energy system model in this study focuses on the heat and electricity demands of communities, across residential, commercial, industrial, and agriculture sectors. The cost optimal role of decentralized generation and storage technologies (DGSTs), and local energy resources are evaluated in the long-term (until 2050) in archetype communities under the NEP strategy.

Overall, the presented clustering and cost optimization community modeling approach can be utilized by policymakers to identify which types of communities have the most potential to contribute to national emission reductions, and what the pathways are to achieving these reductions cost optimally. Although demonstrated for Switzerland, the approach is general and can be adapted and applied to different scopes and scales.

2 Methodology

The approach consists of two main parts. Clustering techniques are first applied to municipal data sets in order to identify characteristic community energy systems (archetypes). Key archetypes are then modeled individually using a least-cost optimization, community energy systems model. This enables an evaluation of the cost optimal role of DGSTs and local energy resources in the long-term across key archetypes under NEP scenario conditions.

The clustering and modeling steps are described in the following sections. Further details on the approach and data inputs are provided in Yazdanie (2017).

2.1 Community Energy System Characterization Using Clustering

Clustering algorithms aim to separate data sets into unique groups of similar objects based on given criteria. A wide range of clustering algorithms and applications exist. With respect to the energy sector, clustering techniques are often applied to characterize building performance (Santamouris et al. 2007; Xiao et al. 2012; Nikolaou et al. 2012; Gao and Malkawi 2014). They have also been applied to characterize municipal heating systems in a small Swiss canton (Trutnevyte et al. 2012). However, clustering techniques have not been applied to characterize local energy systems on a national scale, with respect to the municipal characteristics considered in this study, to the best of the author's knowledge.

Swiss municipal energy systems are characterized with respect to three criteria: the developed environment classification (DEC) (urban, rural or suburban), local energy resource potentials (ERPs), and current energy usage shares (EUS). These criteria have been selected based on available data. Decentralized generation ERPs include rooftop solar irradiance, municipal waste, local wood production, small hydro, and manure. EUS refer to the contribution of six energy carriers (oil, natural gas, wood, district heating, other heating, and electricity) to the current total heat and electricity supply mix across the four evaluated sectors. Input data sets for Switzerland's approximately 2300 municipalities are determined using several sources (Hertach 2012; Swiss Federal Office of Energy 2013; Eymann et al. 2014; Buffat 2016; National Forest Inventory 2016; Panos and Ramachandran 2016; Swiss Federal Office of Statistics 2016a, b; Vögelin et al. 2016).

K-means clustering is applied to data sets using Lloyd's algorithm (Lloyd 1982). This algorithm partitions data sets into k unique clusters based on the Euclidean distance, minimizing the within-cluster sum of squared errors (WCSS):

$$\text{WCSS} = \sum_{i=1}^{k} \sum_{x \in C_i} \|x - \mu_i\|^2,$$

where C_i is the ith of k clusters, x is the set of data points belonging to cluster C_i, and μ_i is the average (or centroid) of cluster C_i. Cluster centroids are first initialized; a range of methods can be applied for this purpose (e.g., cluster centroids may be randomly initialized). Data points are then assigned to the nearest cluster centroid. Cluster centroids are recalculated and data points are reassigned to the nearest cluster centroid. This process is repeated until convergence, whereby the final clusters and centroids are established.

In this study, each data point represents one municipality, which is defined by its characteristic data. The value for k is predetermined using the silhouette criterion (Rousseeuw 1987). K-means is repeated several times and cluster centroids are initialized using the heuristics-based k-means++ algorithm (Arthur and Vassilvitskii 2007). Averaged centroids over these repeated runs serve as initial centroids for a final run. Further details on the clustering approach are provided in (Yazdanie 2017).

The clustering approach yields several characteristic archetypes. Each archetype consists of several municipalities and is distinguished by a unique combination of DEC, ERP, and EUS cluster characteristics. An archetype denotes an average representation of the municipalities belonging to the cluster. The key archetypes selected for modeling are defined as those archetypes which represent the largest shares of national energy demand in the base year.

2.2 Community Energy System Modeling Using TIMES

The TIMES framework (Loulou et al. 2016) is utilized to develop a parameterized community energy systems model. This model is an abstract representation of a community, defined in terms of characteristic parameters (which are specified for each archetype via the clustering approach). Thus, the modeling framework can be readily adapted to represent different communities, enabling the evaluation of a wide range and large number of archetypes. This customizable formulation and application of a TIMES model is atypical, as TIMES models are traditionally designed to represent specific energy systems. TIMES energy models also traditionally represent large-scale energy systems (i.e., on a regional, national, or international level) (Goldstein and Tosato 2008; Gago da Camara Simoes et al. 2013; Amorim et al. 2014; Ramachandran and Turton 2016; Pattupara and Ramachandran 2016). The application of TIMES to develop community-scale models is less common.

The community model in this study minimizes the total system cost and provides details on capacity planning and dispatch over the modeling time horizon, from 2015 to 2050, for the set of technologies considered. Five-year time steps are utilized and end-use demands are modeled on an hourly basis for an average weekday and weekend across four seasons.

Model input parameters describe a single archetype community. Exogenous inputs to the community model reflect national NEP conditions until 2050. Inputs include a carbon tax, national grid electricity costs, fuel costs, efficiency improvements for end-use devices, and renovation potentials. Exogenous values are based on the NEP scenario described in (Kirchner et al. 2012) and results from the national Swiss TIMES model in (Ramachandran and Turton 2013). Overall, it is of interest to assess overarching trends and the collective CO_2 emissions reduction potential of the evaluated archetypes under NEP conditions; additional CO$_2$ constraints are not imposed.

End-use electricity and heat demands are determined according to national projections. End-use heat includes space heat, domestic hot water, and process heat across sectors. Hourly electricity demand is modeled by sector, while hourly heat demand is modeled by building category. Building categories include residential single and multi-family homes, commercial buildings, and industrial/agricultural buildings.

Technology investment options in the model relate to heat and electricity generation technologies, local storage, energy infrastructure, fuel conversion, renovation measures, and end-use devices. Local generation technologies in the electricity sector include rooftop photovoltaic (PV) panels, combined heat and power plants (small and micro CHP) for different fuel types (e.g., natural gas, wood, oil, and waste), and small hydro. Dedicated heat generation technologies include boilers (natural gas, wood, oil, electric, and waste-fueled), solar thermal panels, and air-source heat pumps, while storage includes building-level batteries and heat storage. Energy infrastructure includes electricity transmission and distribution grids, natural gas networks, and district heating networks. Gasification and processing technologies are utilized to convert biomass (wood, manure and waste) into biomethane or wood into pellets. Different renovation measures are also available, ranging from low to high cost options (e.g., window replacement to full building renovation). Electric end-use devices include five categories: lighting, cooking, refrigeration, cooling, and other appliances. Technology input data is primarily based on data applied in Swiss national energy system models, documented in (Ramachandran and Turton 2011, 2013, 2014; Panos and Ramachandran 2016). Cost and technology assumptions are further detailed in (Yazdanie 2017; Yazdanie et al. 2017).

3 Results

The archetypes defined using the clustering approach are described in the following text. The sections thereafter discuss archetype modeling results and technology trends with respect to heat and electricity demand under the NEP scenario.

3.1 Archetype Definition

The clustering approach yields more than 120 archetype groups (Fig. 1). The archetypes are ranked in order of their contribution to the total national energy demand; the 20 largest contributors are selected for subsequent cost optimization modeling.

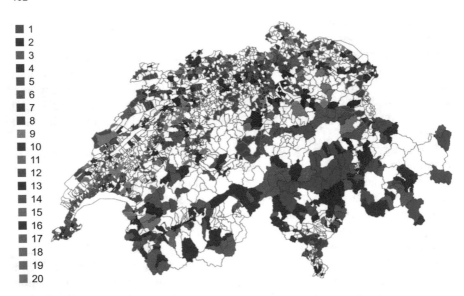

Fig. 1 Geographic representation of archetype clusters selected for modeling

Each of the 20 archetypes is modeled using the TIMES community model. The modeling results for these archetypes represent approximately 80% of Swiss municipalities and (heat and electricity) energy demands. The selected archetypes and their contributions to the national energy demand are described in Table 1. Archetype modeling results are presented in the following sections.

3.2   Heat Generation: Oil Substitution in All Archetypes

The cost optimal heat generation mix over the modeling horizon across a range of archetypes is presented under the NEP scenario (Fig. 2). These archetypes have been selected in order to illustrate key trends and the diversity of observed results. The heat generation share across all 20 modeled archetypes in 2050 is provided in the Appendix.

The long-term generation mix differs notably between archetypes and is driven by differences in local energy infrastructure access and energy resource potentials. Natural gas forms a significant share of the heating mix in archetypes with access to the national gas network (for example, in Fig. 2a). Gas[1] technologies also provide a significant share of the heating mix in archetypes with access to relatively large biomass resources; in Fig. 2c, for instance, biomethane provides approximately

---

[1]Gas refers to methane from the natural gas network or to locally generated and delivered bio-methane (or a combination of the two).

Table 1 Selected archetype details and modeling results

Arche-type ID	Archetype cluster characteristics			National energy share[b]	2050 DGT generation share using local resources		CO$_2$ emissions reduction in 2050 relative to 2015		
	DEC	ERP	EUS[a]		Heat (%)	Elec. (%)	Oil (%)	Gas (%)	Total (%)
1	Urban	High solar/waste	Oil, com	6.2	25	12	97	0	97
2	Urban	High solar/waste	Gas, com/ind	29.1	29	19	53	21	74
3	Urban	High wood, med. solar	Gas, com/ind	3.2	24	6	55	20	76
4	Rural	Low potentials, some hydro/wood	Oil, res	2.0	95	59	100	0	100
5	Rural	Low potentials, some hydro/wood	Oil, ind	2.3	71	22	100	0	100
6	Rural	Low potentials, some hydro/wood	Oil, com	1.4	73	32	100	0	100
7	Rural	High solar/waste	Oil, ind	3.6	49	14	96	0	96
8	Rural	High manure	Oil, ind	2.8	84	15	100	0	100
9	Rural	High wood, med. solar	Gas, com/ind	1.5	59	15	55	27	83
10	Suburban	High solar/waste	Oil, res	2.6	64	30	95	0	95
11	Suburban	High solar/waste	Oil, com	2.2	38	17	97	0	97
12	Suburban	High manure/waste	Oil, res	1.4	87	39	100	0	100
13	Suburban	High manure/waste	Oil, ind	2.7	74	16	100	0	100
14	Suburban	High manure/waste	Gas, com/ind	4.0	55	17	55	33	88

(continued)

Table 1 (continued)

Arche-type ID	Archetype cluster characteristics			National energy share[b]	2050 DGT generation share using local resources		CO$_2$ emissions reduction in 2050 relative to 2015		
	DEC	ERP	EUS[a]		Heat (%)	Elec. (%)	Oil (%)	Gas (%)	Total (%)
15	Suburban	High waste	Oil, res	1.1	80	35	100	0	100
16	Suburban	High waste	Oil, ind	2.4	60	14	96	0	96
17	Suburban	High waste	Oil, com	1.2	65	19	100	0	100
18	Suburban	High waste	Gas, com/ind	4.1	48	16	55	28	84
19	Suburban	High manure	Oil, res	1.2	86	38	100	0	100
20	Suburban	High manure	Gas, com/ind	1.5	54	15	55	32	88

[a]Significant EUS fuel and sector; sector abbreviations: res (residential), com (commercial), ind (industry)
[b]National heat and electricity energy share represented by archetype results in the base year

50% of heat through gas technologies by 2050 (however, communities with relatively large biomass resources collectively represent a small share of the national energy demand). On average, across the evaluated archetypes, gas heating technologies generate approximately 20% of municipal heat in 2050.[2] Approximately 75% of the gas across these archetypes is sourced from the national gas network, while 15% is sourced from local manure and 10% is from municipal waste.

Archetypes which have limited access to gas (i.e., archetypes without national gas network access or insufficient local biomass resources to generate biomethane) rely more heavily on wood for heating. Wood replaces heating oil under the relatively high NEP scenario carbon tax. Only 2% of the heat generation mix is supplied by oil in 2050 on average across the considered archetypes under the NEP scenario.

Heat pumps generate a significant share of heat in 2050 across the archetypes, supplying up to 50% of municipal heating in 2050 (as observed in Fig. 2b). On average, across the modeled archetypes, heat pumps generate more than 35% of municipal heat in 2050.

By comparison, solar thermal technologies generate 13% of municipal heat in 2050 on average across the archetypes. Solar thermal heating plays a larger role in communities which have lower total ERPs relative to total demand and/or limited energy infrastructure access (for example, in Fig. 2b, d).

---

[2]Presented average figures are based on a weighted average considering archetype contributions to total national energy demand.

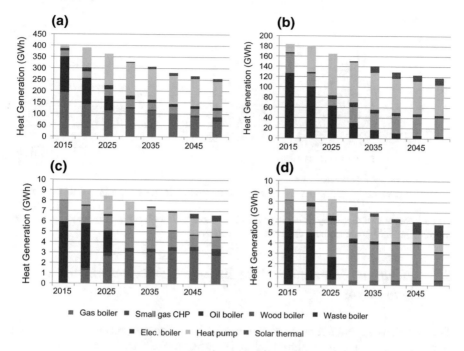

Fig. 2 Heat supply breakdown by technology over the modeling horizon under the NEP scenario for archetypes 3 (**a**), 1 (**b**), 12 (**c**), and 4 (**d**) (see Table 1 for archetype descriptions)

Building-level heat storage investments occur across all sectors. The operation of heat generation and storage technologies on a winter weekday in 2050 is illustrated exemplarily in Fig. 3 for archetype 1. In general, heat storage co-operates with solar thermal panels to store heat during the day and with heat pumps to store heat during low electricity price hours overnight. Heat is then discharged primarily during peak heat demand hours in the morning and evening, enabling peak shaving. The heat storage capacity represents between 5 and 14% of the heat demand during a winter weekday (i.e., the day with maximum heat demand over the year) in 2050 across the archetypes.

Renovation measures also play a vital role in reducing space heating demand over time. Renovations are largely deployed and enable the generation reduction observed over time in Fig. 2.

Local energy resources contribute significantly to local heat generation. On average, approximately 45% of heat generation is supplied by local ERPs in 2050 across the modeled archetypes (Fig. 4). Table 1 also provides the contribution of local resources to heat generation by modeled archetype. Solar, wood, and waste resources form the largest share of locally-sourced energy carriers for heating.

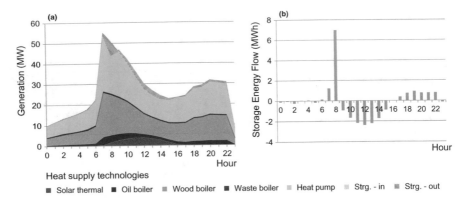

Fig. 3 Hourly heat supply by technology **a** and net storage flows **b** across all sectors on a winter weekday in 2050 in archetype 1 under the NEP scenario

3.3 Electricity Generation: The Emergence of Decentralized Generation

The electricity generation mix over the modeling horizon is presented for the same archetypes as in the preceding section (Fig. 5). The electricity generation share across all 20 modeled archetypes in 2050 is provided in the Appendix.

A range of supply mixes is observed across the archetypes. Communities rely on the national transmission network to meet the majority of electricity demand in most cases; however, decentralized generation technologies (DGTs) also provide a significant share of local electricity generation. National grid imports are reduced by 45% in 2050 compared to 2015, on average across the modeled archetypes. This result is driven by technology switching to DGTs and efficiency improvements in end-use technologies.

Solar PV panels generate approximately 15% of local electricity in 2050 across the modeled archetypes, on average. Small gas CHPs also contribute to local generation in communities where fuel is available (e.g., in Fig. 5a, c); these investments occur primarily in the industrial sector.

Small hydro potentials are generally available in small rural communities, such as the archetype in Fig. 5d. Small hydro plants are able to provide a significant share of local electricity generation in these archetypes; however, communities with similar small hydro potentials represent only 6% of national energy demand in the base year.

Battery storage investments occur across all sectors. The operation of electricity generation and storage technologies on a winter weekday in 2050 is illustrated exemplarily in Fig. 6 for archetype 1. In general, batteries store solar PV electricity during the day and grid electricity during low electricity price hours overnight. Batteries discharge largely during peak electricity demand hours in the evening.

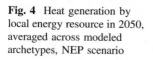

Fig. 4 Heat generation by local energy resource in 2050, averaged across modeled archetypes, NEP scenario

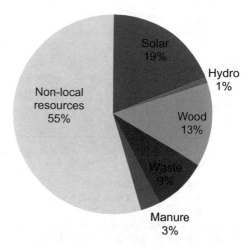

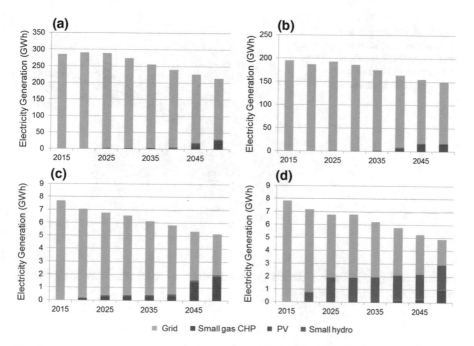

Fig. 5 Electricity supply breakdown by technology over the modeling horizon under the NEP scenario for archetypes 3 (**a**), 1 (**b**), 12 (**c**), and 4 (**d**) (see Table 1 for archetype descriptions)

The total battery capacity is equivalent to 7–11% of the total electricity demand during a winter weekday in 2050 across the modeled archetypes.

Local resources contribute significantly to municipal electricity generation in 2050. Approximately 20% of electricity is generated using local ERPs on average across the modeled archetypes in 2050, with solar PV providing the bulk of local

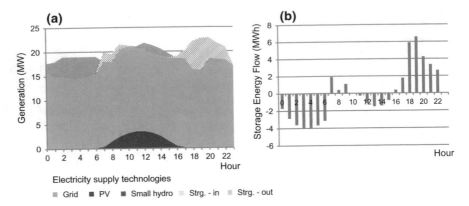

Electricity supply technologies

▧ Grid ■ PV ■ Small hydro ▨ Strg. - in ▨ Strg. - out

Fig. 6 Hourly electricity supply by technology **a** and net storage flows **b** across all sectors on a winter weekday in 2050 in archetype 1 under the NEP scenario

Fig. 7 Electricity generation by local and non-local energy resource in 2050, averaged across modeled archetypes under the NEP scenario

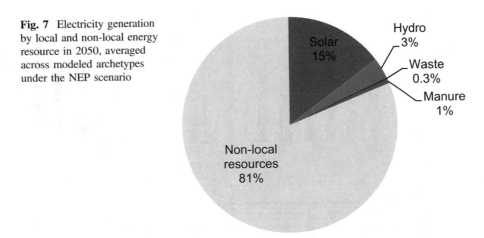

generation (Fig. 7). Table 1 details the contribution of local resources to electricity generation by modeled archetype.

3.4 CO_2 Emission Reductions: Over 95% for Most Archetypes

CO_2 emissions are evaluated for the combustion of fossil fuels within a community only. The implementation of high CO_2 taxes in the NEP scenario enables a drastic reduction in fossil fuel (i.e., oil and natural gas) usage across archetypes (Table 1). The majority of emission reductions are attributed to the reduced use of heating oil. The largest total reductions are observed for smaller rural and suburban archetypes, many of which demonstrate complete decarbonization of local heat and electricity generation in 2050 in the NEP scenario. Larger (e.g., urban) archetypes also demonstrate significant CO_2 emissions reductions, but continue to rely partially on fossil fuels (primarily natural gas) in order to meet large local demands that are not fully satisfied by local ERPs. The national gas network could be further decarbonized using biomethane in the future in order for these archetypes to achieve higher relative CO_2 emission reductions.

The CO_2 emissions reduction is 85% on average across the modeled archetypes in 2050 relative to 2015, with most archetypes demonstrating emission reductions of over 95%. This figure is even larger compared to 1990 levels; however, municipal-level data was not available for this year at the time of the study. Indicatively, national CO_2 emissions per capita were reduced by approximately 30% in 2015 compared to 1990 (The World Bank Group 2017). Overall, the average reduction meets Switzerland's targets within the Paris Agreement with respect to the evaluated sectors and archetypes.

4 Conclusion

Policymakers must employ a range of measures in order to achieve drastic global CO_2 emission reductions and limit the average global temperature increase to well within 2 °C this century. This set of measures must include local energy planning initiatives which focus on the deployment of decentralized generation and storage technologies, local renewable energy resources, and efficiency measures in the built environment.

This study presents an approach to evaluate a range of local energy systems within a larger region, enabling the identification of characteristic local energy system architectures and providing an adaptable cost optimization modeling framework to analyze key communities. The application of TIMES to develop this small-scale, parameterized community model differs from the conventional scope of TIMES models. The overall method is demonstrated for the Swiss case, but it can be adapted for different scales and evaluation criteria.

The proposed approach can be utilized by decision-makers ranging from the local to the international scale. The information gained using the method can inform the development and implementation of national policies that support local-scale decarbonization. The approach can also be employed by local policymakers to identify and encourage the uptake of suitable efficiency measures, DGSTs, and local ERPs.

Drastic reductions in local CO_2 emissions and energy demands are observed under the NEP scenario. The combustion of fossil fuels for local heat and electricity generation is reduced by 85% in 2050 relative to 2015, on average across the modeled key archetypes. This reduction exceeds Switzerland's CO_2 emissions reduction strategy in the context of the Paris Agreement, with respect to the evaluated archetypes and sectors. Several factors influence these results and should be considered by Swiss decision-makers at both the national and local level:

- Local-scale CO_2 emission reductions in the heating and electricity sectors in the long-term are driven largely by the implementation of relatively high CO_2 taxes.
- Space heating demand and CO_2 emission reductions are also driven by renovations in the building sector. As Switzerland's building sector currently accounts for over 40% of energy consumption and CO_2 emissions (Swiss Federal Office of Energy 2017), renovations must be encouraged by municipal governments in order to achieve ambitious climate targets.
- Several generation technologies and local energy resources facilitate the transition to low-carbon local energy systems as well, which should be supported by decision-makers. Heat pumps provide efficiency gains in the heating sector and form a significant part of the cost optimal heating mix by 2050. Local wood and waste resources, as well as solar thermal heating play an important role in decarbonizing the heating sector as well. Solar PV technologies form a key part of the decentralized electricity generation mix by 2050 across the modeled archetypes. PV, together with the adoption of energy-efficient end use devices, enables significant reductions in grid electricity demand by communities.

The aforementioned points should be considered as part of local climate policies. However, supplemental case-specific studies should also be developed in order to formulate individual municipal policies as the presented approach aims to identify key local archetypes and overarching trends. Still, the approach in this study provides policymakers with a powerful tool to assess local energy systems within a large region. The results of this work stress the value and importance of including local energy systems planning as part of national and international dialogues to mitigate climate change.

Appendix

The heat and electricity generation mix in 2050 across the modeled archetypes is illustrated in Figs. 8 and 9.

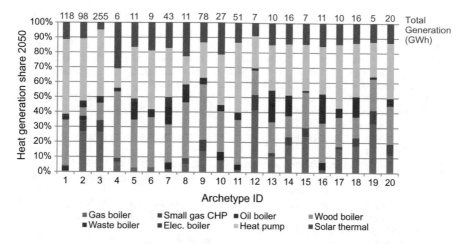

Fig. 8 Heat supply mix in 2050 across archetypes

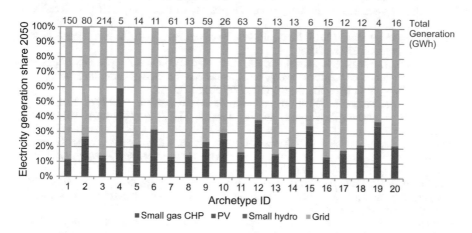

Fig. 9 Electricity supply mix in 2050 across archetypes

References

Amorim F, Pina A, Gerbelová H et al (2014) Electricity decarbonisation pathways for 2050 in Portugal: A TIMES (The Integrated MARKAL-EFOM System) based approach in closed versus open systems modelling. Energy 69:104–112. https://doi.org/10.1016/j.energy.2014.01.052

Arthur D, Vassilvitskii S (2007) K-means ++: the advantages of careful seeding. In: SODA '07: proceedings of the eighteenth annual ACM-SIAM symposium on discrete algorithms, pp 1027–1035

Buffat R (2016) Spatial allocation of Swiss municipal waste and manure resources (personal communication)

Edenhofer O, Pichs-Madruga R, Sokona Y et al (2015) Climate change 2014 mitigation of climate change summary for policymakers and technical summary

Eymann L, Rohrer J, Stucki M (2014) Energieverbrauch der Schweizer Kantone. Waedenswil

Gago da Camara Simoes S, Wouter N, Ruiz Castello P et al (2013) The JRC-EU-TIMES model—assessing the long-term role of the SET plan energy technologies. http://publications.jrc.ec.europa.eu/repository/handle/JRC85804. Accessed 22 Apr 2016

Gao X, Malkawi A (2014) A new methodology for building energy performance benchmarking: an approach based on intelligent clustering algorithm. Energy Build 84:607–616. https://doi.org/10.1016/j.enbuild.2014.08.030

Goldstein G, Tosato G (2008) Global energy systems and common analyses—final report of annex X (2005–2008)

Hertach M (2012) Dokumentation Geodatenmodell Kleinwasserkraftpotentiale der Schweizer Gewässer. http://www.bfe.admin.ch/geoinformation/05403/05570/index.html?lang=de&dossier_id=05569. Accessed 25 Oct 2016

Kirchner A, Bredow D, Dining F et al (2012) Die Energieperspektiven für die Schweiz bis 2050. www.bfe.admin.ch/php/modules/publikationen/stream.php?extlang=de&name=de_564869151.pdf. Accessed 22 Apr 2016

Lloyd S (1982) Least squares quantization in PCM. IEEE Trans Inf Theory 28:129–137. https://doi.org/10.1109/TIT.1982.1056489

Loulou R, Remme U, Kanudia A et al (2016) Documentation for the TIMES model. http://www.iea-etsap.org/docs/Documentation_for_the_TIMES_Model-Part-I_July-2016.pdf. Accessed 16 Dec 2016

National Forest Inventory (2016) National forest inventory. http://www.lfi.ch/resultate/anleitung-en.php. Accessed 25 Oct 2016

Nikolaou TG, Kolokotsa DS, Stavrakakis GS, Skias ID (2012) On the application of clustering techniques for office buildings' energy and thermal comfort classification. IEEE Trans Smart Grid 3:2196–2210. https://doi.org/10.1109/TSG.2012.2215059

Panos E, Ramachandran K (2016) The role of domestic biomass in electricity, heat and grid balancing markets in Switzerland. Energy 112:1120–1138. https://doi.org/10.1016/j.energy.2016.06.107

Pattupara R, Ramachandran K (2016) Alternative low-carbon electricity pathways in Switzerland and its neighbouring countries under a nuclear phase-out scenario. Appl Energy 172:152–168. https://doi.org/10.1016/j.apenergy.2016.03.084

Ramachandran K, Turton H (2011) Documentation on the development of the Swiss TIMES electricity model (STEM-E). http://www.psi.ch/eem/PublicationsTabelle/2011_Kannan_STEME.pdf. Accessed 22 Apr 2016

Ramachandran K, Turton H (2013) A long-term electricity dispatch model with the TIMES framework. Environ Model Assess 18:325–343. https://doi.org/10.1007/s10666-012-9346-y

Ramachandran K, Turton H (2014) Switzerland energy transition scenarios—development and application of the Swiss TIMES energy system model (STEM). https://www.psi.ch/eem/PublicationsTabelle/2014-STEM-PSI-Bericht-14-06.pdf. Accessed 22 Apr 2016

Ramachandran K, Turton H (2016) Long term climate change mitigation goals under the nuclear phase out policy: the Swiss energy system transition. Energy Econ 55:211–222. https://doi.org/10.1016/j.eneco.2016.02.003

Rousseeuw PJ (1987) Silhouettes: a graphical aid to the interpretation and validation of cluster analysis. J Comput Appl Math 20:53–65. https://doi.org/10.1016/0377-0427(87)90125-7

Santamouris M, Mihalakakou G, Patargias P et al (2007) Using intelligent clustering techniques to classify the energy performance of school buildings. Energy Build 39:45–51. https://doi.org/10.1016/j.enbuild.2006.04.018

Swiss Federal Council (2015) Switzerland targets 50% reduction in greenhouse gas emissions by 2030. https://www.admin.ch/gov/en/start/dokumentation/medienmitteilungen.msg-id-56394.html. Accessed 22 Nov 2017

Swiss Federal Council (2017) Instrument of ratification deposited: Paris climate agreement to enter into force for Switzerland on 5 November 2017. https://www.admin.ch/gov/en/start/documentation/media-releases.msg-id-68345.html. Accessed 22 Nov 2017

Swiss Federal Office of Energy (2013) Small-scale hydropower. http://www.bfe.admin.ch/themen/00490/00491/00493/?lang=en. Accessed 25 Oct 2016

Swiss Federal Office of Energy (2017) Measures for increasing energy efficiency. http://www.bfe.admin.ch/energiestrategie2050/06447/06457/index.html?lang=en. Accessed 28 Nov 2017

Swiss Federal Office of Statistics (2016a) Geostat. https://www.bfs.admin.ch/bfs/de/home/dienstleistungen/geostat.html. Accessed 25 Oct 2016

Swiss Federal Office of Statistics (2016b) Gebäude- und Wohnungsregister. In: Gebäude- und Wohnungsregister. https://www.bfs.admin.ch/bfs/de/home/register/gebaeude-wohnungsregister.html. Accessed 25 Oct 2016

The World Bank Group (2017) CO$_2$ emissions. https://data.worldbank.org/indicator/EN.ATM.CO2E.PC?locations=CH. Accessed 11 Dec 2017

Trutnevyte E, Stauffacher M, Schlegel M, Scholz RW (2012) Context-specific energy strategies: coupling energy system visions with feasible implementation scenarios. Environ Sci Technol 46:9240–9248. https://doi.org/10.1021/es301249p

Vögelin P, Georges G, Noembrini F et al (2016) System modelling for assessing the potential of decentralised biomass-CHP plants to stabilise the Swiss electricity network with increased fluctuating renewable generation. Zurich

Xiao H, Wei Q, Jiang Y (2012) The reality and statistical distribution of energy consumption in office buildings in China. Energy Build 50:259–265. https://doi.org/10.1016/j.enbuild.2012.03.048

Yazdanie M (2017) The role of decentralized generation. Storage and Local Energy Resources in Future Communities, ETH Zurich

Yazdanie M, Densing M, Wokaun A (2017) Cost optimal urban energy systems planning in the context of national energy policies: a case study for the city of Basel. Energy Policy 110:176–190. https://doi.org/10.1016/j.enpol.2017.08.009

Printed in the United States
By Bookmasters